SECURITY FOR COMPUTER NETWORKS

WILEY SERIES IN COMMUNICATION AND DISTRIBUTED SYSTEMS

Satellite Communications Systems
G. Maral and M. Bousquet

Integrated Digital Communications Networks (Volumes 1 and 2)
G. Pujolle, D. Seret, D. Dromard and E. Horlait

Security for Computer Networks, Second Edition
D.W. Davies and W.L. Price

SECURITY FOR COMPUTER NETWORKS

An Introduction to Data Security in Teleprocessing and Electronic Funds Transfer

Second Edition

D.W. Davies
Data Security Consultant

and

W.L. Price
National Physical Laboratory, Teddington, Middlesex

JOHN WILEY & SONS
Chichester · New York · Brisbane · Toronto · Singapore

Wiley Editorial Offices

John Wiley & Sons Ltd, Baffins Lane, Chichester,
West Sussex PO19 1UD, England

John Wiley & Sons, Inc., 605 Third Avenue,
New York, NY 10158-0012, USA

Jacaranda Wiley Ltd, G.P.O. Box 859, Brisbane,
Queensland 4001, Australia

John Wiley & Sons (Canada) Ltd, 22 Worcester Road,
Rexdale, Ontario M9W 1L1, Canada

John Wiley & Sons (SEA) Pte Ltd, 37 Jalan Pemimpin #05-04,
Block B, Union Industrial Building, Singapore 2057

Reprinted August 1992
Reprinted September 1993
Reprinted February 1994

Library of Congress Cataloging-in-Publication Data:

Davies, Donald Watts.
Security for computer networks : an introduction to data security in teleprocessing and electronic funds transfer / D.W. Davies and W.L. Price. — 2nd ed.
p. cm. — (Wiley series in communication and distributed systems)
Includes bibliographies and index.
ISBN 0 471 92137 8
1. Data transmission systems—Security measures. I. Price, W. L. II. Title. III. Series.
TK5105.D43 1989
005.8—dc20

89-14760
CIP

British Library Cataloguing in Publication Data:

Davies, D. W. (Donald Watts, *1924–*)
Security for computer networks : an introduction to data security in teleprocessing and electronic funds transfer. — 2nd ed — (Wiley series in communications and distributed systems)
1. Computer systems. Security measures. Cryptograms
I. Title II. Price, W. L.
005.8′2

ISBN 0 471 92137 8

Printed and bound in Great Britain by
Bookcraft (Bath) Ltd

To Diane and Marjorie

CONTENTS

ACRONYM LIST

ABA	American Bankers Association
ANSI	American National Standards Institute
AP	authentication parameter
ASCII	American Standard Code for Information Interchange
ATM	automatic teller machine
BACS	Bankers' Automated Clearing Services
BSC	binary synchronous communication
CBC	cipher block chaining
CCITT	International Telegraph and Telephone Consultative Committee
CFB	cipher feedback
CHAPS	Clearing Houses Automated Payments System
CHIPS	Clearing Houses Inter-bank Payments System
CLCB	Committee of London Clearing Bankers
CRC	cyclic redundancy check
DCE	data circuit terminating equipment
DEA	Data Encryption Algorithm
DEE	data encryption equipment
DES	Data Encryption Standard
DK	data key
DMA	direct memory access
DSA	decimal shift and add
DTE	data terminal equipment
EAROM	electrically alterable read-only memory
ECB	electronic code book
EDC	error detection code
EFT	electronic funds transfer
FAR	false alarm rate
FIPS	Federal Information Processing Standard
HDLC	high level data link control
IPR	impostor pass rate
ISO	International Organization for Standardization
ITAR	International Traffic in Arms Regulations
IV	initializing variable
KGM	key generation module
KK	key encrypting key
KKM	master (key encrypting) key
KLM	key loader module
KTM	key transport module

lcm	least common multiple
LSI	large scale integration
MAC	message authentication code
MAR	MAC residue
MDC	manipulation/modification detection code
NBS	National Bureau of Standards
NIST	National Institute of Standards and Technology
NPL	National Physical Laboratory
NSA	National Security Agency
NSM	network security module
OFB	output feedback
OSI	Open Systems Interconnection
OWF	one-way function
PCM	pulse code modulation
PIN	personal identification number
PKC	public key cryptosystem
p-MOS	p-channel metal oxide semiconductor
PROM	programmable read-only memory
PSS	Packet Switchstream
PVV	PIN verification value
RAM	random access memory
ROM	read-only memory
RSA	Rivest, Shamir and Adleman (PKC)
SDLC	synchronous data link control
SID	S.W.I.F.T. interface device
SNA	systems network architecture
S.W.I.F.T.	Society for World-wide Inter-bank Financial Telecommunications
TRM	tamper resistant module
TTL	transistor-transistor logic
XOR	exclusive-OR

PREFACE TO THE FIRST EDITION

Blackmail, fraud and the stealing of commercial secrets are examples of crimes using information as the medium. There was a time, not many years ago, when it was necessary to alter the account books in order to cover up a fraud; now the same effect can sometimes be produced at a computer terminal. The widespread introduction of information technology into business inevitably leads to its misuse for crime, which data security aims to prevent.

Essentially, security means controlling access to data, depriving the blackmailer of his information, protecting commercial secrets and preventing the falsifying of records. The need to control access in computers first became serious when 'time-sharing' began to operate. Undergraduates regarded the security features of these systems as a challenge and soon found that the existing methods of control were defective. This taught designers a useful lesson: that data security needs care and attention.

In 1983 the film *War Games*, and the publicity it created, introduced people to a new and surprising cult, the 'computer hackers' who spend their time obsessively trying to break into computer systems. Their devotion to a basically tedious pastime is extraordinary, and it shows up one advantage of an amateur attacker over the professional defenders. The attacker's time is apparently unlimited and uncosted; the defender's time is expensive. The systems that hackers have broken into were protected only by weak password schemes, but the hackers' success and the excitement it generated prepares the way for more advanced attacks when simple hacking fails to satisfy them. A special feature of illegal attacks on computer systems is that there is no tradition of associated guilt-feelings. The law, in most countries, has not begun to define the new kinds of crime. With no likelihood of punishment, the probing of the defences of computer systems is considered to be a game. It could become the tool of organized crime.

Cryptography is a long-established way to keep information secret. Now that the majority of information systems transport data from place to place (as well as processing and storing it) the place of cryptography in data security is assured. It is beginning to be accepted that cryptography is the basic tool for achieving data security. This book is about the use of cryptography to protect data in teleprocessing systems, not only keeping data secret but also *authenticating* it, preventing alteration and proving its origin. These go beyond the classical uses of cryptography. Some of the main concerns of data security in banking and other commercial information systms are those of authentication.

After the introduction in Chapter 1, ciphers are described in Chapter 2 and then the Data Encryption Standard in Chapter 3. This standard is used as one of the principal tools of data security in what follows. The first step is to define a number of standard ways of using the cipher — extending its properties and making it more useful. These are the 'modes of operation' described in Chapter 4. The application of ciphers to the authentication of messages is the subject of Chapter 5. Each application needs a scheme to manage encipherment keys and two of these key management schemes are described in Chapter 6.

Teleprocessing systems are used by people at widely spread terminals. Only if the identity of these people can be determined at a distance will it be possible to control access to data. Verification of personal identity is the subject of Chapter 7.

A breakthrough in encipherment technique occurred in 1976 with the discovery of the 'public key ciphers' by Diffie and Hellman. The properties of these ciphers make them specially useful for access control and they have unique virtues for key distribution and for a strong form of authentication called a 'digital signature'. Chapter 8 describes public key ciphers and Chapter 9 describes digital signatures.

The largest commercial use of cryptography and other data security techniques is in banking and the payment systems operated by banks. Chapter 10 describes how data security measures can be applied to 'electronic funds transfer' in its various forms, leading us to some new concepts for future payment systems.

Finally, the wide use of these methods will only be possible if systems can interwork and a range of compatible equipment can be manufactured. Chapter 11 describes the progress made towards standards for data security.

We have tried to make the approach pragmatic, illustrating the principles with examples. At present the subject is essentially pragmatic in that only the insecurity of systems can be demonstrated, by breaking them, but their security is demonstrated only by the absence of demonstrated flaws.

The information in this book represents the best practice known to us but it cannot carry any guarantee of security — nothing can. When a cipher, or another part of a security scheme, resists attacks for a long time, this gives the users confidence, but there is no absolute or provable basis for that confidence. The possibility of provable security (in some sense yet to be defined) still exists but, with a few exceptions, proofs elude us. In this respect there has been no big change as a result of new technology — the security of ciphers has always been tenuous and this is still true of nearly all aspects of data security.

Mathematical notation is often the clearest way to describe unambiguously how a cipher operates, or how it is used. Some ciphers, notably public key ciphers, cannot be described in any other way. Whenever possible in this book, intuitive arguments have been given as well. We aim in this book to help those people who will actually put security schemes into practice. More formal and complete treatments will sometimes be found in the original papers listed in references at the end of each chapter.

PREFACE TO THE SECOND EDITION

Concern for computer security has much increased since this book was first published. Reports of successful hacking into computer networks appear regularly in the press. Among those more closely concerned there is perhaps a better understanding that 'computer crime' does not usually employ technological means and that hackers are not the most serious threat. Nevertheless the successes of hackers and the considerable publicity they obtain makes them significant. At one time their activities seemed entirely innocent but it now appears that the bulletin boards they use to exchange information are also used for credit card abuse and drug traffic, particularly in USA where the police have penetrated these illegal networks. In some universities, large numbers of people were found to be involved.

The extent of white-collar crime remains controversial but there is no doubt that financial institutions which have not taken adequate precautions do suffer from major frauds.

Attacks by modifying software have increased — the 'Trojan horse' attack. Spectacular forms of computer virus have been much publicized, making this the most feared type of software threat. As with hackers, the virus has acquired a mystique that goes beyond reality. Protecting software against corruption, by quite conventional means, must be a concern of all those who need to make their computer systems secure. This is a special problem for users of personal computers who tend to exchange programs freely and pay little attention to software security.

Most of this book in both editions is concerned with the basic principles of security and in this respect little updating has been needed for the second edition. Cryptography and its applications are by now a relatively stable technology. There are three main areas in which the second edition has been changed to include new material. The first is Open Systems Interconnection where the seven-layer architectural model now has a security architecture which specifies how security services will be applied in the model and what mechanisms they will employ. The work of producing the specifications that will introduce these services is well under way but only a few standards have been completed. In the area of financial applications there is a little more progress.

The second area of development is that of key management. An international standard has been produced for key management in wholesale banking and this is likely to be a model for future standards, therefore it is worth studying even for those outside the sphere of banking. We have added a full description of the principles of ISO 8732, the new standard.

The third area of development is electronic funds transfer at the point of sale or EFT-POS. Though we are still some way from international standards (and these may continue to elude us) there are now a few schemes with well-developed security that make them worthwhile examples of how this difficult environment can be dealt with. We have added descriptions of three schemes, the transaction key, derived unique key per transaction and the use of public key cryptography.

We have taken the opportunity to bring some of the terminology in line with international standards, but without losing touch with accepted terminology. Also, a set of protocols for authentication, associated with the names of Fiat and Shamir, are described in this new edition.

Because this book is mainly concerned with the application of new technology to information security and deals with basic methods rather than specific implementations, the ideas it describes are remarkably stable.

The second edition has allowed us to make worthwhile changes and additions to the text while the main shape of the book and the concepts it introduces have remained unchanged.

Finally, we have received no solutions for the cryptogram given at the end of the preface to the first edition and reproduced here.

```
ZSYTT  IFXVJ  NVWRS  OHTTR  ESUIJ  VGEJB  DDO.A  TZMKV
QUYVS  QTMVX  NGDVX  .AWIP  TKVTZ  KMMSG  DVITF  NKXRF
BDYDL  HHRDK  DNTQE  AKJDK  HGFST  DU.MO  CCNXY
```

Chapter 1 DATA SECURITY

1.1. THE NEED FOR DATA SECURITY

Each major advance in information technology changes our ideas about data security. Consider the use of written mesages to replace those carried in the memory of a messenger. Written messages are less prone to error, and since the courier need not know the message, but can destroy it in an emergency, there could be better security. Yet written messages can be concrete evidence of conspiracy or spying, whereas a carrier of a spoken message might escape unsuspected. The need to hide the content of a written message must therefore have been realized very soon, and there is evidence that codes and ciphers appeared almost with the beginning of writing.

When telegraphy began, Wheatstone and others thought that ciphers would be needed by the senders of telegrams and there was a flurry of new ideas for cryptography. But the need did not show itself. Cryptography has traditionally been used by kings, popes, armies, lovers and diarists and has moved into business and commerce only recently. Now that information is processed, stored and transmitted in large quantity, the need for data security is greater and more varied. We shall be dealing with the secrecy of transmitted information, using encipherment, and also with authentication of information, verifying the identity of people, preventing the stealing of stored information and controlling access to both data and software. Such varied data security problems arise from the more recent advances in information technology — large stores and microprocessors which give us processing wherever we need it. The microprocessor and the floppy disc create a new and urgent problem of safeguarding software. The requirements summed up by the phrase 'data security' do not stay the same; they change as the technology changes.

Nowadays, almost everything we do with information is assisted by some new invention. The fundamental operations of storage, processing and transmission seem to be undergoing a relentless and rapid improvement, so fast that the application of the technology cannot keep up with the rate of advance. The introduction of teleprocessing into all aspects of business, commerce and industry brings new data security problems to the fore. Undoubtedly, electronic banking has experienced the first impact, with spectacular frauds in the shape of falsified money transfers. Banks have been the first to apply modern methods to make their systems secure. The rest of commerce and industry is well behind banking in its attention to data security.

In order to avoid repeating the same phrases we will use conventional names for the actors in the security drama. The bad guys who are trying to do something with the system which the designers would like to avoid will be called 'the enemy' and their activities will be called 'attacking' the system.

These words have only a conventional meaning. The enemy may not, for example, be a remorseless or professional adversary. For a long time, students were the principal finders of weaknesses in operating systems, because they enjoyed the intellectual challenge. Almost certainly, a lot of the effort that went into 'phone freaking' was motivated in this way. None the less, it is potentially dangerous because the techniques developed in innocent attacks can be exploited by criminals.

New methods of payment and transfer of funds using communication networks and intelligent tokens are developing fast and are usually called 'electronic funds transfer'. These are an obvious target for fraudulent manipulation of data. As electronic message systems of all kinds come into wider use, carrying data which has already been captured in the 'electronic office', there must inevitably be a proportion of sensitive information carried as 'electronic mail'. This is another area which needs attention to security, both for secrecy of information and to preserve its authenticity.

The advance of information technology has sparked off a public debate on the subject of individual privacy. Like many public debates it expresses both rational and irrational fears. Everyone is now alerted to the real dangers of inaccurate personal information and uncontrolled access to files. Most advanced countries have introduced laws to enforce a reasonable degree of individual data privacy and others have such laws in preparation. If we think of *privacy* as the legal concept, then *security* of data is one of the means by which the privacy can be obtained. The implementation of these new laws is likely to produce applications for data security techniques.

Of the three main operations carried out in information systems, storage, processing and transmission, it is undoubtedly data transmission that carries the greatest of security risks. A communication network consists of numbers of cables, radio links, switches and multiplexers in a variety of locations, all of these parts of the system being potential targets for 'line taps' or 'bugs'. It is impossible to make a widespread network physically secure, therefore security measures depend on information processing techniques such as cryptography.

Storage of data is next on the list of vulnerabilities because data spends much more time in storage than in processing. The protection of stored data can use the same techniques of cryptography but with different twists. For most purposes, processing is the least vulnerable part but it may be worth attacking. Perhaps an enemy can gain more by discovering which parts of a file are active than by stealing a whole file. For example a police file of intelligence about crime activities contains a lot of data about dormant criminals and perhaps even more about dubious suspects. A team contemplating a big crime could learn more from the data being accessed than by searching the whole data base.

No data system can be made secure without physical protection of some sort of the equipment. The effect of good design is to concentrate the need for physical security, not to circumvent it entirely. In particular, processing of data usually (perhaps always) requires those data to be in clear form, not enciphered, therefore processors themselves must be protected from intrusion, such as the attachment of radio bugs. In many systems the data and operations needing the greater degree of security can be contained in one box of modest size, physically strong and designed to destroy its stored secrets when it is opened. This is called a *tamper-resistant module* and it is a central element of many systems which this book will describe.

A tamper-resistant module typically protects a process which is essential to the security of the system such as the handling of cryptographic keys and the cryptographic operations which they enable. Usually there is a procedure to be performed inside the module and this requires a small processor. The interface between the module and the rest of the system must be designed carefully to avoid any compromise of secret data and this interface is also controlled by the processor within. The software employed within the module must be trusted — software integrity is discussed in the next section. The module is protected by being enclosed in a strong box but usually the strength of this box cannot ensure that it will never be broken into. Consequently, the module is provided with a number of sensing devices which will detect various forms of tampering. These are sensitive to vibration, penetration of the protective box, tilting or temperature, for example. When the sensors detect an attack, at a certain level of severity all the important data contained within the enclosure are destroyed by over-writing. By this technique of 'tamper-responding' the secret data can be protected without an exceptionally strong outer case. Details of the techniques for making tamper-responding modules are best kept confidential because knowledge of the sensing devices helps an attacker to overcome them and this increases the cost for future generations of tamper-resisting modules.

1.2. ASSESSMENT OF SECURITY

Security is a complex property and difficult to design or optimize. Designing a system to be efficient, convenient or cheap is an optimization problem which, though complex in practice, has a mathematical structure which is easy to understand. The problem is to choose the design parameters in order to maximize a figure of merit. Designing a system for *security* means analysing an adversary problem where the designer and the opponents are each independently thinking out their strategies. The outcome of the contest is a result of their combined choices. The mathematical theory for such problems is the 'theory of games'. The kind of game which fits the situation is a 'two-person, non-zero-sum game with imperfect information'. The 'non-zero-sum' condition appears because the system attacked may lose more than the attacker gains. 'Two-person' means that we consider one attacker at a time, to simplify the problems as much as possible. Games theory, beyond the most trivial cases, is extremely difficult and there seems to be no way in which the theory of games can actually be useful in analysing security — it merely serves to illustrate the underlying complexity of the problem and the inadequacy of a naive 'risk analysis' approach.

Every kind of threat to a system should be assessed, yet it is very difficult to enumerate the complete range of attacks. First of all, the motives and intentions of the attackers have to be guessed. Stolen information can be used for fraud, espionage, commercial advantage, blackmail, and probably in many other ways. Enciphering information on a magnetic tape can make the tape useless to an enemy as information, but, if there are no other copies of this tape and the information is essential to its owner, the stolen tape can be used as a hostage, to extort payment for its return. This was a threat used in a crime in Europe, where an ex-employee stole the tapes from a large company's computer centre. A great deal of imagination is required to think out all the ways in which a system could be attacked. In a bank in the USA, each cheque book was supplied with magnetically encoded forms for paying money into its account. If the customer had forgotten to bring his forms, he could pick up similar forms in the banking hall, which were not encoded. At the first stage of sorting according to the magnetic characters, uncoded forms were rejected, then operators inserted the account number of the payee along with the other details that had been written on the form. An ingenious fraud was devised by distributing coded paying-in forms among the forms in the banking hall. Customers did not notice the presence of the coding and unwittingly paid their money into the fraudster's account.

The existence of so many methods of attack makes the protection of an information system very difficult indeed. It is doubtful that a systematic method to analyse the problem will ever be found, though checklists can be a help. It is unusual to have an information system described in sufficient detail that all the potential weaknesses can be identified. Therefore, a good investigator works from observation and by discussion with those who know the system well, not from paper designs and specifications. The ability to find security flaws depends on an attitude of mind which takes nothing on trust.

Security is essentially a negative attribute. We judge a system to be secure if we have not been able to devise a method of misusing it which gives some advantage to the attacker. But we might just have failed to identify the nature of the enemy or missed out a method of attack.

A security investigation should never be based on the designer's concept of the system. He has thought about the security and convinced himself that every aspect has been covered. Probably, he has a good theoretical reason for believing that the security goals have been achieved. What is then required is 'lateral thinking' which questions his assumptions and finds different approaches. The designers may have concentrated their attention on some parts of the system and forgotten others. On the other hand, systems can be so complex that only those who implemented them can fully understand their details. With a complete change of attitude, the implementers can be valuable allies in a security study. They must first be convinced that there are flaws and persuaded to search for them.

The security measures applied to a system always cost something and in an extensive network the total expense may be very large. Before embarking on a programme of installing security features, managements will want to know that the expense is justified. For this purpose elaborate procedures have been developed, called *risk assessment*. A number of different threats are tabulated, together with measures of the probability that each will happen and the likely losses that would arise from a successful attack. From these data, in an obvious way, the 'annual loss expectancy' associated with each type of threat can be evaluated. The idea behind this assessment is that a comparison of the loss expectancy with the cost of providing protection will be a guide for the managers in deciding what security features to install.

The data which enter into a risk assessment are known only with very low accuracy, rarely better than an order of magnitude and sometimes even less precision. Consequently, risk analysis can hardly be considered a scientifically based procedure. But managements need some guidance and the many risk-assessment methodologies on the market offer them a degree of help.

There cannot be any prescription for obtaining data security. The system designer could miss one vital point, perhaps one which had nothing to do with information technology. For example, an automatic teller machine (no longer in production) delivered bank notes through a slot which was contained in a neat recess. Someone built an attractive looking plate which covered the recess and presented to the user a dummy slot from which the money would never emerge. He fitted this plate to an automatic teller outside a bank, when the bank was open, stood back and watched what happened. The user went through the correct procedure but received no money, then went into the bank to complain. The crook took away his plate and the money which was concealed behind it. This is a variant of an old trick which has been used with returned-coin chutes since vending machines and public telephones began; it pays off better with automatic teller machines. Some of the earliest attacks on automatic teller machines were to pull them out of the bank wall, illustrating that the security of a system may have nothing to do with its data handling.

When all the precautions have been taken against illegal access to data in communications and storage, there remains an indefinite number of other threats. There are two areas of sufficient importance for special mention, system software and the people operating the system.

Software integrity

The complexity of an information system is built mainly into its software, in order that the hardware can be simpler and composed mainly of standard units such as microprocessors and stores. The virtue of software is that it can be changed both during development and subsequently to give the system new properties. This flexibility is a severe threat to system security.

The first difficulty with software is understanding it sufficiently to be sure that it functions correctly in the first place. Where equipment is built with several processors, each handling a restricted range of functions and interacting with others according to carefully devised rules or protocols, the verification of software may be a tractable problem. At the other extreme, mainframes with their complex operating systems are never fully understood. If the security requirements are strict, it must be assumed that any normal operating system has weaknesses. In particular these flaws could be exploited by the software designers, who know the details well enough to find the flaws. In this respect, large operating systems have improved a little and the provision of special hardware to assist access control has made them proof against attacks which were common in the early days but no complex system can be regarded as entirely secure, even if the designers and builders were all trustworthy and had only that aim in mind.

A dishonest designer could provide 'hooks' in the software ready for the attachment of modifications to undermine its security. The auditing of software by independent software designers should look for unnecessary features which might be used in this way.

During software design, special tools are provided to help the implementers modify programs at various levels, from binary patches up to the top level of specification. The very convenience of these tools makes them dangerous, particularly if a software writer is working for an enemy. Retaining these tools when the software is in operation constitutes a major security weakness because it allows post-design modifications which could introduce deliberate trapdoors into the security-handling parts of the system. The severity of this threat is greater if it is planned early in the system design process, hence the importance of trustworthy system designers. There may be a need to keep one or two highly instrumented versions of the system for software testing and updating but those systems which go outside a secure environment should be made as difficult to modify as possible.

If the software of a system is entirely reliable and trustworthy, the next problem for the designer is how to keep it that way. Some software can be physically protected, for example by storing it in a read only memory (ROM) or in a store which, though it could be overwritten is protected by a physical 'write protect' device. Preventing the over-writing of software by a programmed feature (such as an access control mechanism in the operating system) is no real protection because that can be altered. The more ingenious attacks on software integrity employ the operating system. This is not only the most powerful part of the software but also the least understood and therefore the most difficult place for the discovery and prevention of attacks.

Dedicated processors, carrying out a single task with a fixed program, can be reasonably secure if they are protected against physical attack.

The least secure program environment today is the microcomputer because the philosophy of designing these systems is to make them highly adaptable, with the greatest ease for changing their programs. Users of microcomputers can obtain free software from a number of sources. If they accept and run it they can endanger their entire system, for example doing irreparable damage to the contents of their fixed disc. This attack by a program which appears innocent and does something useful but contains a hidden enemy is called the *Trojan horse* attack.

Even more insidious is the software *virus*. This is designed to infect other programs so that the treacherous program is replicated and, in some cases, moves from one computer to another.

Just before Christmas 1987, in West Germany, a Christmas greeting message was sent into an electronic mail system on a European Academic Research Network. The message produced a pretty display and invited the user to run the program. When this was done, the program accessed the standard files containing the names of the user's correspondents in the mail system and sent out a copy to all of these people. Though basically a harmless type of virus, it spread through at least three networks and the quantity of electronic mail it produced swamped the networks' capacity.

More insidious viruses can spread in the same way and remain dormant until some criterion is met (such as time and date) when they begin a destructive phase, such as over-writing storage. In order to remain dormant and yet spread from program to program, they can be hidden in the operating system or in parts of files which are employed by the operating system but invisible to the user.

Maintaining the integrity of software is not technically difficult but it can conflict with the way in which computer systems are used and with the freedom to make changes or adopt software from any source. It is not only the software that needs to be protected. The security properties of a system can also be incorporated in tables giving the rights of access to files for each user. These tables must be protected as carefully as the software itself. The tool for protecting the contents of any file against unauthorized access is authentication, which is the subject of chapter 5. The authentication mechanism itself cannot be in software; therefore an unmodified general purpose system such as the typical personal computer is never a secure environment.

Security and people

The owners of a system depend for its security in the first place on the integrity of the supplier, which itself depends on the people who design, build and maintain the system. When it is in operation, keys and passwords are introduced which protect it from enemies outside, including enemies in the ranks of the suppliers, unless software changes have subverted the protection. With good design, the security then depends on the people who operate the system and carry out its security-related procedures.

Some operators will be in positions of very great trust, such as those who load the system with master keys and those who transport keys from one part of the system to another. Improved design can reduce the need for some of this trust, up to a point. For example, the transport of cryptographic keys can use hardware modules which make the keys inaccessible to the couriers. With these means, it requires a much more elaborate attack to obtain the keys illegally, though it is never impossible. Some kinds of privileged operation cannot be avoided. Someone has to write into the system the rules of access control at the top level, giving special privilege to certain people, such as database managers or security controllers. Careful design can reduce the residual threat from these directions. For example, a security controller may be responsible for enforcing the correct procedure in using the system but he need not know the values of any cryptographic keys. In this way, though some trust is placed in a number of people, there are few opportunities for individuals to subvert the whole system. Where there is a particularly sensitive requirement, such as a top-level master key, the responsibility can be split between several individuals, with the effect that it would require a conspiracy of all of them to undo the security of the system as a whole. In a well-designed system it should be clear who is being trusted and to what extent, but there is no way to make the system proof against unlimited deceit.

1.3. THE EFFECTS OF TECHNOLOGY

As information technology becomes more complex, the opportunities for its misuse are greater. Complexity of design is not a protection against interference with data systems. At first sight it might seem that a complex protocol makes it more difficult to extract information from a communication link or change data in transit. In practice, increased complexity is made workable by greater standardization and then standard software and hardware becomes available to unravel the complexities. For example, it is now possible to buy instrumentation which will interpret the binary pattern on a communication line, enabling a specialist to determine the link layer protocol and extract the contents of frames, also to interpret the packets of the X.25 protocol and the transport, session and document layers of the Teletex procedure. With this compact equipment, all the complexities of many levels of protocols can be stripped away for diagnostic purposes. Using the same equipment, an enemy can interpret what is passing on the line and, with some additional software, can capture information, alter it and re-launch it with the correct format and procedure. This is not easy but it is not made prohibitively difficult just by protocol complexity.

In fact, the variety of new technologies increases the possibilities of attack. If a local ring or Ethernet is used, all of the data passing between terminals and computing resources in this local area can be made accessible from a single line tap. The Ethernet broadcasts all messages to all terminals. In most designs of ring, all the data pass completely round the loop and return to the sender. A line tap attached to an Ethernet or ring network gives access to all the information it carries. For example, some large organizations employ a set of interconnected Ethernets carrying information of many different levels of sensitivity. The station logic attached to the Ethernet tap filters out messages addressed to each destination, but a small change to that logic can make it extract information for other destinations, giving access up to the highest level of sensitivity for the information in that network. Without additional precautions such as encipherment, today's local area networks make a very insecure environment.

Microwave radio and satellite systems similarly expose large quantities of data to easy access. Together with the increasing use of electronic mail in the future, this provides good opportunities for large-scale tapping of traffic.

In the past, the mere quantity of data would protect most of the secrets because the target of the enemy would be buried in uninteresting information. But since the source and destination of packets and messages travels with them in their headers, extracting information travelling to or from a target is not difficult. This target may be a terminal, a person or a computer process or port, depending on the level of addressing. Furthermore, unprotected messages can be selected according to their content, extracting messages which contain certain words or bit patterns for review by the enemy. It can be seen that modern technology increases rather than decreases the opportunities.

Improved technology of information processing has aided the protection of information by making good ciphers available in the form of single-chip devices and by allowing complex systems to be made more compact and thus given adequate physical protection. But the same technology is available to the enemy who can buy low-cost processing and storage in order to investigate systems and exploit their weaknesses.

1.4. THE NOTATION FOR ENCIPHERMENT

Codes and ciphers are part of the subject of cryptography which is the art of hiding information in a secret way so as to preserve its confidentiality. We shall only be concerned with ciphers so the operations that are carried out are called encipherment and decipherment.

The original purpose of ciphers is to conceal information while it is being transmitted. Stored information can be concealed in a similar way. But the security of information systems is much wider in scope than merely concealment and includes the integrity of messages and the authentication of one communicating party by another. Wherever a high level of security is needed a technique based on cryptography can be found, so cryptography proves to be the fundamental tool for building secure systems. The ways in which ciphers can be used for these purposes form a main theme of this book. For the present, we need to know how to describe these basic operations in cryptography.

We need a notation for the two operations of encipherment and decipherment. The pictorial convention we shall use is shown in Figure 1.1. The two operations can be described as functions of two variables, or they can be described by an algorithm — a systematic procedure for calculating the result when the values of the variables are given.

The *plaintext* mentioned in the figure is the set of data before encipherment. The result of encipherment is the *ciphertext*. The result of applying decipherment to the ciphertext is to restore the plaintext. These words 'plaintext' and 'ciphertext' are relative to a particular encipherment — it could easily happen that the plaintext shown here is the result of a previous encipherment but for our purpose, since it enters into the encipherment function, it is called the plaintext. The capital letters D and E are reserved for the decipherment and encipherment functions respectively. In the mathematical notation we shall use, if x is the plaintext then

$$y = \mathrm{E}k(x)$$

the result of encipherment, and the inverse function

$$x = \mathrm{D}k(y)$$

expresses decipherment of the ciphertext to produce the plaintext. In both functions, the other variable k is the *key*.

The key is an essential feature of a cipher. In principle, if the function $y = \mathrm{E}(x)$, without a key, were kept secret, this might serve to conceal the value of x but, if the secret of the function E is lost, nothing can be done to restore the usefulness of the cipher. An entirely new cipher is needed. Since the design and testing of a cipher is a very big task, its loss in this way would be expensive. Furthermore, all those who took part in the design and testing or who wrote programs for the encipherment and decipherment procedures would have to be trusted indefinitely. These secrecy problems are made much easier by employing a *key*, a parameter which, in effect, allows a large class of ciphers $y = \mathrm{E}k(x)$ to be defined. If a key is compromised (discovered by an enemy, or by someone untrustworthy) it can be changed and the cipher function can remain in use.

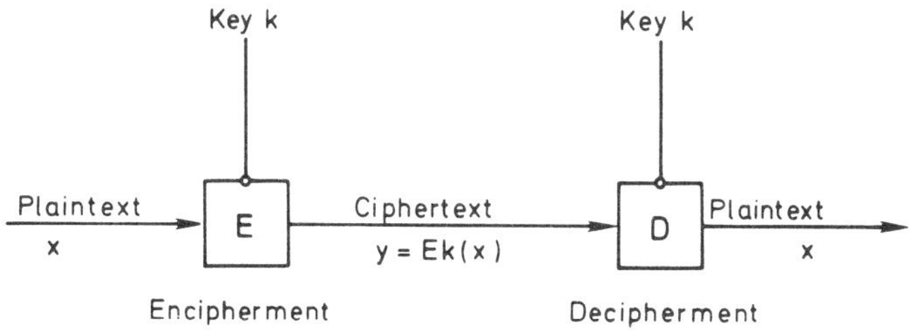

Fig. 1.1 Notation for encipherment and decipherment

Secrecy of the function or algorithm $Ek(x)$ as well as the key is a useful security measure, but, considering how many people must know its form, a secret function cannot be reliable by itself. It is the secrecy of the *key* that matters. This cipher must remain effective even if the enemy knows the function $Ek(x)$. When $Ek(x)$ is a *standard* cipher, like the Data Encryption Standard, everyone who is interested can find out the function.

The notation $Ek(x)$ is derived from one that has now been widely used and it probably started with k as a suffix $E_k(x)$ to denote that k is a parameter, which is kept constant for a number of encipherments. In the pictorial form of the notation it is necessary to distinguish between the two inputs to the boxes which denote the functions E and D. Therefore we use the small circle shown in Figure 1.1 to show which is the *key* input.

Other notations for encipherment have their uses. If complex expressions or lists are enciphered and when the procedure of decipherment is not shown explicitly, the notation $\{\ldots\}^k$ is often used, where the data or text in the curly brackets is enciphered with key k. This is the notation that has been used widely in discussions of authentication. We have preferred to keep a uniform notation throughout this work and $y = Ek(x)$ is our choice. If there are several keys in use these could be shown with suffices as k_{ab}, k_s etc, but in this book we put these symbols on the same line. Thus $Ekab(x)$ means that x is enciphered, using as key the value *kab*. This *kab* is similar to the 'identifier' in a programming language — it does not, in this context, mean that *k*, *a* and *b* are separate variables. All our key identifiers will begin with the letter k.

The domain of values from which the key k is taken is called the *key space*. In a typical cipher, the key might be 20 decimal digits, for example, which gives a key space of size 10^{20}. Alternatively the key might be expressed as a binary number of 50 digits with a size of key space 2^{50}. In other ciphers, the key is a permutation such as the letters A, B ... Z in some sequence, consequently the key space has size 26!

The notation $y = Ek(x)$ treats x and y also as single variables, which have values in a finite domain. Since x represents, in general, a block of text, a cipher of this kind is called a *block cipher*. For example, eight characters in ISO 7-bit code with parity make up a block of 64 bits. The cipher could be designed for a 64-bit block, but the redundancy of the ISO code makes the information content much less. If the characters are taken from text in a natural language such as English, the true information content of x is smaller still.

We also need a notation for the encipherment of messages of indefinite length. The single variable x is replaced by a string of variables. For this situation we write the cipher text as:

$$Ek(x_0,\ x_1,\ \ldots\ x_i,\ \ldots)$$

where x_0, x_1 ... is a string of variables whose size is given by the context. They could be parts of the message or its header, such as the source identifier or a sequence number. The simplest case is the one in which each variable x_i is a character with its value taken from a finite set of values known as the *alphabet*. For example the alphabet might be, literally, A, B, ... Z or it might contain the 256 different values of the octet (i.e. an 8-bit character). Ciphers which are designed to encipher a string of such characters with no theoretical limit to the length of the string are sometimes called *stream ciphers*. Unfortunately, the terminology has developed haphazardly and there is no clearly understood and agreed definition of this term. It is usually employed when the alphabet is fairly small — the extreme case is the bit stream with an alphabet of 0 and 1. Obviously, the ciphertext for small alphabets must be calculated from more than just the current character or it becomes trivially simple.

The need for key distribution and management

When encryption is used to make data secure in communications, there must be prior agreement between the communicating parties about all aspects of the procedure. A cipher algorithm, and the method of using it and initializing it, must be agreed. The most difficult requirement is that a key must be chosen and made available at both ends of the communication path. Before the enciphered data flow over the line, the value of the key must make a similar journey. Keys can be enciphered using other keys but in the end at least one key has to be distributed by some other means. The cipher can therefore be regarded not as a creator of secrecy but as a means of extending the secrecy applied to the distribution of the key into the transport of a much larger volume of data enciphered with that key.

When encipherment is used for protecting stored data, key handling is easier because the encipherment and decipherment are carried out at the same location, so the key need not travel, but secrecy of the stored key is still important.

Choosing the key in the first place and making it available only to authorized users are aspects of *key management*. After encipherment methods have been decided, key management becomes a major task for system designers because the security of the whole system has then been concentrated on the keys. Chapter 6 is devoted to key management.

1.5. SOME USES FOR ENCIPHERMENT

Encipherment conceals the content of a message, except for those intended to receive it who, knowing the key, can decipher it. Thus it controls access to the message content, making it readable to those who know the key. If the ciphertext is made public and the cipher algorithm is well known, knowledge of the key is the equivalent of access to the message. Knowing the key also allows a person to construct another message so, under the right circumstances, a message can both be read and written. By using a 'public key cipher' it is possible to have separate control of reading and writing, as we shall see in chapter 8.

Without a knowledge of the key an enemy cannot usefully alter a message. He can probably alter the ciphertext but when that is deciphered it will read as a random pattern, because lack of the key prevents the enemy having control of its deciphered form. Assuming that the message has some redundancy, the authentic message produced with the aid of the key can easily be distinguished from the random pattern. In this way its authenticity can be proved to whoever knows the secret key. We discuss authentication in more detail in chapter 5. It is one of the principal uses of encipherment methods. These operations of *access control* and *authentication* can be applied to transmitted messages or stored files.

Together they sum up the various uses of encipherment but, like any generalizations they conceal the great variety of ways that ciphers can be used. To illustrate this, consider the protection of a claim to priority in an invention.

Before the institution of patents for inventions, if an inventor published his methods he lost the ability to exploit it before others could. Some inventors wanted to put their brainchild on record and establish their priority without revealing it to competitors. To do this they wrote a description and enciphered it, publishing the cipher and keeping the key secret. Later, when the invention became well known, others might claim to be its originator. Then the true inventor revealed the key so that anyone could verify that he published it at the earlier date.

In chapter 8 we describe a novel kind of cipher in which both sender and receiver have secret keys but neither sends his key to the other (page 224). This was first described in the context of 'Mental Poker' — a game of poker played between people who are all in different places and joined only by telecommunication channels. This is a vivid illustration of a use of encipherment in which secrecy and authentication are not the whole story. We can illustrate this by a simpler case, the tossing of a coin by two people at a distance, using only messages to link them.

The two people are Ann and Bill. Deliberate bias in their coin-tossing can be avoided by a convention in which Ann announces her call C_a with its value 'heads' or 'tails' and Bill similarly announces C_b then the outcome of the procedure is a head if the two calls coincide and a tail if they differ. The simple analysis of this 'game' shows that both Ann and Bill will choose between head and tail at random. If either shows a bias the other can improve their chance of winning.

If Ann announces her call first then Bill can arrange to win, and vice versa. If no trusted third person is available to receive the calls and adjudicate, the two parties can use encipherment in the following way.

Ann chooses one of two highly redundant statements such as 'My call for this game is heads' or 'My call for this game is tails', enciphers whichever phrase she chooses and keeps the key secret. Because of the redundancy she cannot find two keys that will generate either message from the same ciphertext. Of course there must be no other clues such as the length of the mesage. Ann and Bill exchange their enciphered calls. When Ann has received Bill's call she reveals her key and vice versa. If each finds the deciphered message to be one of the two phrases, the calls stand and the outcome is as defined. The use of encipherment authenticates the calls after they have been exchanged and ensures that they cannot be modified after learning the other's call.

A large part of the practice of data security consists of procedures (or protocols) using encipherment of which the coin-tossing procedure is just a simple example. The procedures range from the very simple to rather complex schemes such as those of key management described in chapter 6.

1.6. GENERAL PROPERTIES OF CIPHER FUNCTIONS

The most useful ciphers are those which do not expand the text. A block cipher for an n-bit block maps the 2^n different plaintext values onto the 2^n different ciphertext values. In effect, it is a permutation of these 2^n values, one which varies with the value of key. The possible number of permutations is $(2^n)!$ which is usually larger than the key space, so many of the possible permutations do not happen. For all practical purposes the permutations seem to be chosen at random in a good cipher because, if there were obvious regularities, the cryptanalyst could exploit them.

For a given value of key k, if x ranges over all its possible values, $y = \mathrm{E}k(x)$ also ranges over all its possible values. On the other hand if x is held constant and k ranges over all its key space, the y values generated can be regarded as random choices. Some y values may not occur at all, others may occur more than once.

When there is no data expansion the identity $x = \mathrm{D}k[\mathrm{E}k(x)]$ implies another identity $x = \mathrm{E}k[\mathrm{D}k(x)]$. In other words, the encipherment and decipherment can be interchanged. This is illustrated in Figure 1.2 and it is not a trivial result. If there is data expansion then $\mathrm{D}k(x)$ *reduces* the size of the text and the second identity cannot be true. Bear in mind that, when encipherment and decipherment are exchanged, though they have all the formal properties of a decipherment/encipherment pair, some of their useful properties may disappear. Figure 1.2 is a trivial example based on a 3-bit character.

The type of function shown in Figure 1.2 is called a *bijection*. This kind of function maps a set of values into a set of equal size. Each value maps into just one corresponding value, both for the function and its inverse. This is a one-to-one mapping.

There is a special kind of cipher which is an *involution*, that is a function $f(x)$ which has the property that $f[f(x)] = x$. In the case of a cipher function which is an involution,

$$\mathrm{E}k[\mathrm{E}k(x)] = x$$

implies that $\mathrm{E}k(x) = \mathrm{D}k(x)$, encipherment and decipherment are identical. In the days when ciphers were produced by mechanisms a 'self-inverse' cipher was an advantage because the mechanism needed no adjustment between sending and receiving. Many examples exist of involutions as cipher functions.

The product of two involutions is not necessarily an involution, as Figure 1.3

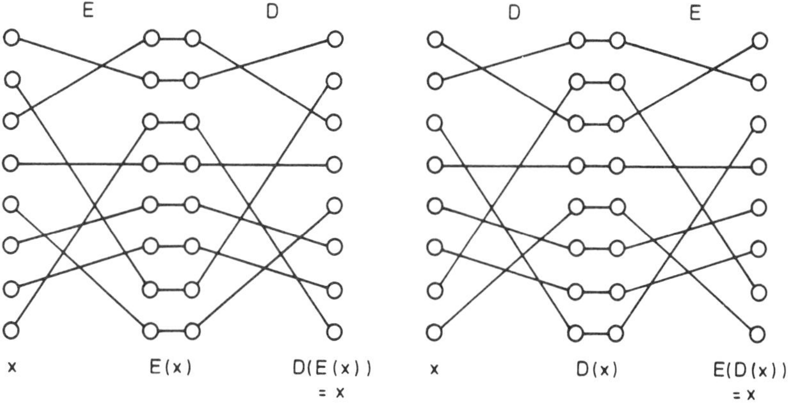

Fig. 1.2 Exchange of cipher and decipher functions

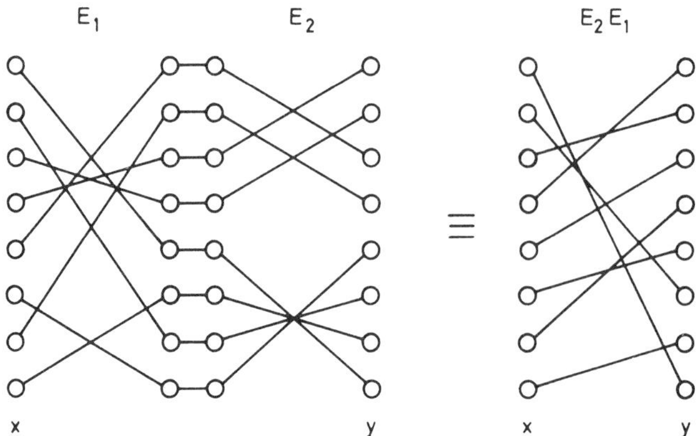

Fig. 1.3 The product of two involutions

illustrates. The inverse of such a product can easily be found if its components are known. If encipherment is

$$y = \mathrm{E}_2[\mathrm{E}_1(x)]$$

and both E_1 and E_2 are involutions, the decipherment is

$$x = \mathrm{E}_1[\mathrm{E}_2(y)]$$

A series of involution transformations can produce a secure cipher which is easily inverted, and this principle underlies the construction of the Data Encryption Standard described in chapter 3.

Chapter 2 CIPHERS AND THEIR PROPERTIES

2.1. INTRODUCTION

The known history of codes and ciphers is long and fascinating, dating back almost 4 000 years to the time when the Egyptians, possibly out of respect for the dead, used a hieroglyphic code for inscriptions on tombs. It is likely that the art of concealment of meaning of written communication goes back even further, though the Egyptian funerary inscriptions are amongst the earliest known examples. Many and various have been the techniques employed over the centuries. By today most of the 'classical' techniques are thought to be too weak for serious application; however, many of their basic principles can still be identified as forming part of modern techniques and therefore it is worth examining how some of them worked. This chapter aims to give a flavour of classical techniques without attempting to be exhaustive. For a comprehensive treatment readers should consult David Kahn's book *The Codebreakers*.[1] Another useful reference on classical techniques is the book *Cryptanalysis* by Helen Fouché Gaines.[2]

Concealment may take two forms, either the message is hidden in such a way that the very fact of its existence is obscured or the message is transformed so as to be unintelligible even though its existence is apparent. The first of these techniques has been called 'steganography' (literally: covered writing), whilst the second is known as 'cryptography' (literally: hidden writing); the literal distinction between these terms seems very fine, but their technical significance makes a rather useful distinction.

A steganographic method said to have been popular in classical times involved shaving the head of a slave, writing the message on the shaven head, waiting till the hair grew again and then despatching the slave to have his head shaved by the intended message recipient. This technique has little to commend it, least of all the slow rate of communication (unless early Greek hair grew much faster than does ours). Another method used pictures to carry hidden messages; Figure 2.1 illustrates an example. At first glance this seems to be just another drawing of a shrub; however, a concealed message is contained within the picture. Reading clockwise around the outline of the shrub (beginning at the trunk) and recording a '1' for each twig bearing a fruit and a '0' for each empty twig, we obtain a binary message. Enclosed regions are ignored — they only add to the pictorial effect — but in a more elaborate 'tree-structured' reading of the figure they could be used, adding 16 bits to its message.

Lord Bacon's method of steganography used two type founts, differing slightly in appearance. Effectively one fount stood for binary 0 and the other for binary 1, permitting encoding of letters of the secret message within a totally unrelated text. Much effort has gone into attempts to prove that Bacon was the true author

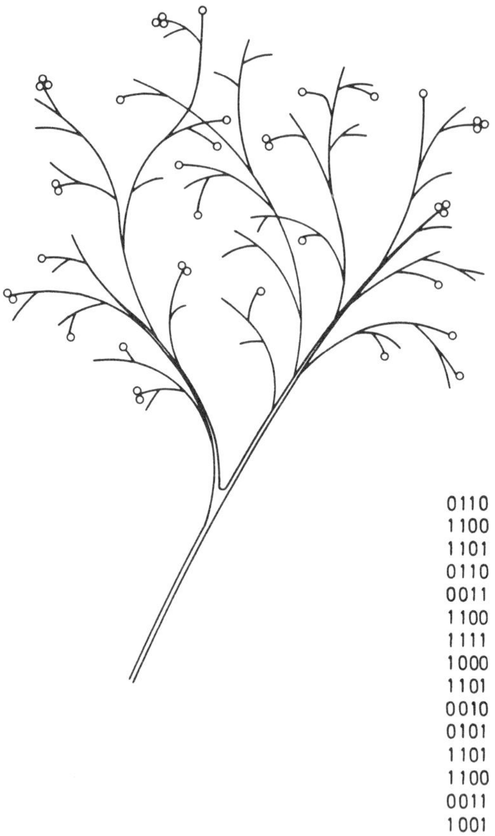

Fig. 2.1 A picture containing a hidden message — steganography

of the plays attributed to Shakespeare; one approach was to scan the first folios of these plays for messages hidden by Bacon's method. The famous American cryptologist, William Friedman, began his cryptographic career in a fruitless search for these messages at the Riverbank Laboratories, Geneva, Illinois. Under Friedman's leadership the Riverbank Laboratories became highly respected in cryptology but their owner, George Fabyan, never gave up his belief that the Baconian ciphers could be found in Shakespeare.

In yet again another steganographic method, messages have been expressed by patterns of goods in shop windows. Writing a message in invisible ink could be said to use steganography. The variations on the steganographic theme are endless. However, the nearest thing to a steganographic technique that is useful for modern data protection is that of filling the spaces between encrypted messages with meaningless garble when these messages are transmitted over communication links. The intention is to present the appearance of a link busy at all times and thus conceal the passage of messages; unlike real steganography, the existence of a communication link is evident, only the traffic level is concealed.

We shall find cryptography of far greater significance than steganography as we consider methods of achieving security of data. Here the important thing is so to change the message that its meaning is completely concealed from anyone not entitled to read it. We therefore need some kind of message transformation and for this purpose two basic methods have been devised — codes and ciphers.

To encode a message its words or groups of words are looked up in a code book containing code equivalents which may be alphabetic or numeric; the book is arranged in sequential order of words or word groups. Replacement of the component parts of a message by their code equivalents constitutes the entire process of encoding. To decode an encoded message efficiently a second, matching code book is required, this time indexed according to the code elements. A brief example, based on the international radio 'Q' code, is given in Figure 2.2. Codes are often applied today in ordinary commercial correspondence in order to achieve message compression and economize on transmission costs, by using single-word equivalents to represent plaintext word groups. Many such codes have been devised and offered for sale; company letter headings often carry a list of codes used; such codes are public knowledge and can therefore not be used for message concealment.

Secret codes have held an honourable place in the techniques of general-purpose cryptography. They can be made more effective for this purpose by ensuring that multiple equivalents are listed in the code books, so that the code element used for a particular word or word group is not always the same, but is chosen at random among the alternatives. In this way an important word which frequently occurs does not leave a tell-tale frequency in the coded text. For the protection of computer data, codes are far from ideal, because the equivalent of a code book would make bulky storage demands and the process of look-up might be time consuming. The generation of an adequate code book is itself a formidable task. Protection of stored code tables, vital for security, could itself be difficult to achieve. The secure and rapid transport of bulky code tables to the locations where they are required is potentially a difficult problem, especially if frequent change of code is required in order to maintain security.

Unlike cryptographic codes, ciphers are not concerned with whole words or groups of words. In the past they have operated, generally speaking, at the level of the individual letter; modern cipher techniques applied in a computer context operate sometimes with big units of many characters, sometimes at the character or letter level, and sometimes at the level of the individual binary digit. Some of the classical techniques which have operated at the letter level are considered in the next section.

Before leaving the subject of codes, the important class of secrecy systems known as *nomenclators*[1] must be mentioned. These had their origin round about the year 1380 and at first consisted simply of code substitutions for important elements

Are you busy?	QRL	QRH	Does my frequency vary?
Does my frequency vary?	QRH	QRL	Are you busy?
Is my keying correct?	QSD	QRO	Shall I increase power?
Shall I increase power?	QRO	QRZ	Who is calling me?
What is the exact time?	QTR	QSD	Is my keying correct?
Who is calling me?	QRZ	QTR	What is the exact time?

Fig. 2.2 Brief extract from a two-part code book

such as names of places or people, the rest of the text being left in clear. After about 20 years the system was extended to include code substitutions for important word or letter groups. As an example, the phrase 'of the' or the fragment 'ing' might have been included. At the same time a letter-by-letter substitution cipher was introduced to apply to the portions of text not appearing in the code lists. It is noteworthy that the designers of this combination of code and cipher were sufficiently aware of the possibility of cryptanalysis by frequency counts that they included multiple equivalents (known as 'homophones') in their substitution cipher.

Nomenclators survived in popular use until the middle of the nineteenth century, when commercial codes were developed in order to economize on message lengths in telegraph transmission and improved understanding of cipher cryptanalysis led to the development of ciphers that were stronger than simple substitution.

The two basic components of classical cipher techniques are *substitution* and *transposition*. In substitution, letters are replaced by other letters, whilst in transposition letters are arranged in a different order.

2.2. SUBSTITUTION CIPHERS

The Caesar cipher

If we wish to replace letters of a message by other letters in a systematic fashion, then one of the simplest ways is to express the alphabet in cyclic form (Figure 2.3) and then select letters shifted by a specified number of places in a particular direction from each plaintext letter. An example of an enciphered message is given in the figure. Here the letter displacement is by three places to the right and the cipher corresponds to one credited to Julius Caesar. Another classical cipher of this kind is said to have been due to Caesar Augustus, Julius' nephew, who chose a letter displacement of one position. Such a cipher is extremely easy to operate;

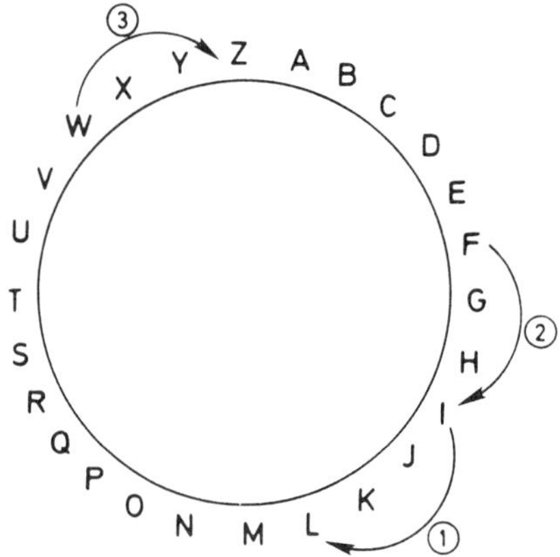

Plaintext I F W E W I S H T O R E P L A C E L E T T E R S O F A _ _ _ _
Ciphertext L I Z H Z L V K W R U H S O D F H O H W W H U V R I D

Fig. 2.3 Caesar's cipher, $c = p + 3$ *modulo 26*

```
           GDUCUGQFRMPCNJYACJCRRCPQ
           HEVDVHRGSNQDOKZBDKDSSDQR
Plain -    IFWEWISHTOREPLACELETTERS
           JGXFXJTIUPSFQMBDFMFUUFST
           KHYGYKUJVQTGRNCEGNGVVGTU
Cipher -   LIZHZLVKWRUHSODFHOHWWHUV
           MJAIAMWLXSVITPEGIPIXXIVW
           NKBJBNXMYTWJUQFHJQJYYJWX
```

Fig. 2.4 Breaking a cipher of the Caesar type by searching twenty-six keys

the alphabet written on two concentric discs displaced by the required amount makes a useful tool for encipherment and decipherment. The cryptologist may even attempt to make things more difficult for the cryptanalyst by changing the amount of the displacement from time to time. If, in Julius Caesar's cipher, the letters A, B C . . . are given numbers 0, 1, 2 . . . the process of encipherment can be expressed as $c = p + 3$ (modulo 26), where the addition requires 26 to be subtracted if the result exceeds 25.

Unfortunately such a cipher, which must surely have been re-invented by countless schoolboys, is ridiculously weak. If the amount of the displacement is known, then there is no secret and the plaintext is readily retrievable; even if the displacement is not known, then it can be discovered very easily. All one need do is to write out the ciphered message followed by all its possible displacements by from one to twenty-five letters. One of the transformed messages will correspond to the plaintext and will be obvious to the reader from the meaning conveyed, assuming that there is adequate redundancy in the message (a plaintext message consisting entirely of arbitrary characters would present a problem). In Figure 2.4 we give an example of what is involved; the displacement here is of three letters, as in the Caesar cipher. In Figure 2.4, and indeed in all subsequent figures in this chapter, spaces between words are omitted; they are ignored in the cipher operation. This convention is customary when simple ciphers are used. They could have been treated as a twenty-seventh letter and enciphered like all the others, but their high frequency weakens any simple cipher.

The technique of trying all possible displacements to solve a Caesar-type cipher is effectively a key search; the displacement is the key and possible values are tested until the right one is found by inspecting the 'plaintext' produced in each case.

Monoalphabetic substitution

Somewhat greater security may be obtained if the letter substitution is carried out with a 'jumbled alphabet'. In Figure 2.5 we show the message of Figure 2.3

```
     Plain   ABCDEFGHIJKLMNOPQRSTUVWXYZ
    Cipher   DKVQFIBJWPESCXHTMYAUOLRGZN

 Plaintext   IFWEWISHTOREPLACELETTERS
Ciphertext   WIRFRWAJUHYFTSDVFSFUUFYA
```

Fig. 2.5 A monoalphabetic substitution

encrypted using such a method. Effectively we are making the displacement of the cipher letter depend on the particular plaintext letter. This technique is known as a monoalphabetic substitution because one jumbled alphabet is used. Whereas the number of possible displacements and therefore the number of possible encipherments under a Caesar-type cipher is 25, the number of different jumbled English alphabets is 26! or about 4×10^{26}. With such a large number of possibilities monoalphabetic substitution is seemingly a very strong cipher system. Sadly its apparent strength is a delusion. Two reasons can be found for the weakness: the statistical clues given to the cryptanalyst are very strong and the alphabet can be discovered piece by piece. Even if the alphabet used has many wrong letters, the cipher can be read, like a badly transmitted telegram, and the wrong letters can be corrected. In a good cipher, getting the key nearly right reveals nothing of the plaintext.

The monoalphabetic substitution illustrates the fallacy of relying on double encipherment for increased strength. Double encipherment with different alphabets is equivalent to single encipherment with an alphabet formed by applying the second substitution to the first alphabet.

Cryptanalysis of a message enciphered with a monoalphabetic substitution depends on the fact that each plaintext letter is always transformed into the same ciphertext equivalent. In natural language, letter frequency tables can be set up which show the relative order of use of letters of the alphabet. For English text the order is generally that shown in Figure 2.6 (after Gaines). Depending on the text from which the frequency tables are created we may find small variations

	%		%		%
E	13.1	D	4.1	G	1.4
T	9.0	L	3.6	B	1.3
O	8.2	C	2.9	V	1.0
A	7.8	F	2.9	K	0.4
N	7.3	U	2.8	X	0.3
I	6.8	M	2.6	J	0.2
R	6.6	P	2.2	Q	0.1
S	6.5	Y	1.5	Z	0.1
H	5.9	W	1.5		

Letter frequencies in English text

Digrams	Trigrams
TH	THE
HE	AND
AN	THA
IN	ENT
ER	ION
RE	TIO
ES	FOR
ON	NDE
EA	HAS
TI	NCE

The ten most frequent digrams and trigrams in English text

Fig. 2.6 Frequencies of letters, digrams and trigrams in English text

in the ranked order. For cryptanalysis, the letter frequency count of the enciphered message is calculated; from this it should be apparent which are the most frequent letters and which are the least frequent; the middle-frequency letters convey the least information in the process of cryptanalysis. By trial and error the cryptanalyst tries out possibilities for the plaintext letter equivalents, relying on redundancy in the plaintext to indicate when he is hot on the scent. Aid may be obtained from tables of digraph and trigraph frequencies in English, which will show what letters are most likely to be adjacent in text. Figure 2.6 also shows a few of the most frequent English digraphs and trigraphs. If the cryptanalyst is not even sure that he is dealing with an English enciphered message, then the frequency count, if the ciphertext is long enough, may give him a clue. For example the letter 'Q' ranks very low in English and German, but rather higher in French and Italian; the vowel 'O' ranks relatively high in English, Italian, Spanish and Portuguese, but rather lower in French and particularly low in German. Figure 2.7 shows letter frequency profiles for the more frequent letters of English, German and Italian (the spot values are joined together to assist in reading the profiles). For the four most frequent letters in each language the profile is very distinctive; beyond this the profiles rapidly become indistinguishable. From such

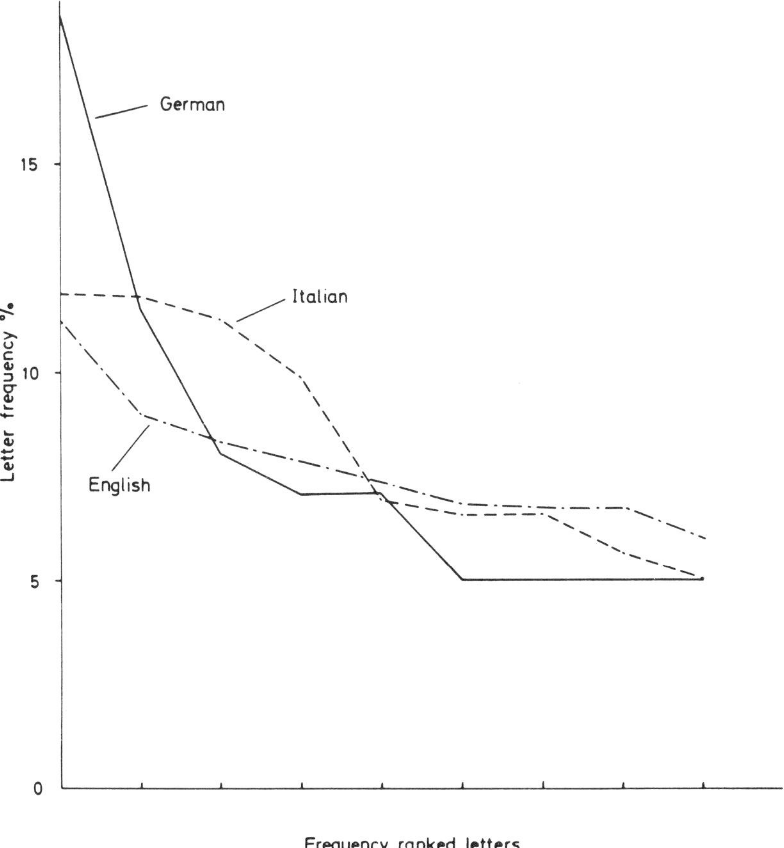

Fig. 2.7 Letter frequency profiles for English, German and Italian

indications of the letter frequency distribution it may be possible to make an informed guess at the language of the plaintext even before the message has been cryptanalysed.

Polyalphabetic substitution

Realizing this weakness of monoalphabetic ciphers the cryptologists went one better — they created polyalphabetic ciphers. Here there is not one jumbled alphabet for use as a substitution table, but a number of them which are used in succession on the successive plaintext letters. When the last alphabet has been used, we start again with the first. Because each letter of the plaintext is no longer always represented by the same letter in the ciphertext, it is not possible to take a simple frequency count and deduce the shape of the substitution — more subtlety is required. It is of great help to the cryptanalyst to find out how many alphabets are in use; for this there are various techniques available. One of them depends on the frequent repetition of letter groups in text; the letters *the* are very common in English. In a sufficiently long text there is a very good chance that corresponding repeated letter groups may be found in the ciphertext, and they will sometimes be located at intervals of some multiple of the number of alphabets in use. These can be found by scanning the ciphertext and, from their spacing, intelligent guesses made as to the number of alphabets. A short text enciphered with five different alphabets is shown in Figure 2.8; note the repeated letter groups in the ciphertext occurring at a multiple of five letters. This method of determining the number of alphabets in use was made public in 1863 and is due to Kasiski.[1] Once the number of different alphabets is known, the cryptanalysis is simple if the text is long enough. All one does is to take frequency counts of the ciphered letters enciphered under each individual alphabet. The process then proceeds by trial and error as in the case of monoalphabetic substitution.

In both monoalphabetic and polyalphabetic substitution the common secret information that must be shared by sender and receiver of an enciphered message is the mixed alphabet or alphabets. This information forms the cryptographic key, which has to be distributed to both sender and receiver. A compact form of key is more convenient. For this purpose the alphabet can be generated from a key word as demonstrated in Figure 2.9. A key word of ten or twelve letters, in which each letter occurs once, might be suitable. In creating the mixed alphabet the key

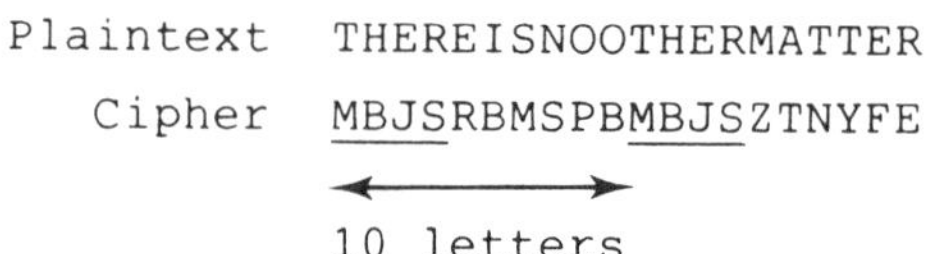

Fig. 2.8 A polyalphabetic substitution with five alphabets

Plain	ABCDEFGHIJKLMNOPQRSTUVWXYZ
Cipher	EXHAUSTINGBCDFJKLMOPQRVWYZ

Fig. 2.9 Method of generating a substitution alphabet using a key word

word is first written, then all the remaining letters in their normal order. To communicate such a mixed alphabet to a correspondent all that is necessary is to send the key word. Evidently there is economy of data transmission, which might be important if several mixed alphabets are to be communicated. Another advantage is that the key word may be remembered easily, useful if it is being carried in someone's memory. Note that at the end of the sequence some of the letters of the cipher alphabet are unchanged from the plain alphabet. This is sometimes avoided by an additional cyclic shift of the alphabet, as in the Caesar cipher.

The Vigenère cipher

Key words play an important part in a particular form of polyalphabetic cipher known as the Vigenère after its inventor and first published in 1586.[1] Here the multiple alphabets are not jumbled but displaced from the normal with a cyclic shift (compare the construction of the Caesar cipher). Application of the Vigenère cipher is facilitated by using a tableau, in which all the possible displaced alphabets are tabulated. Figure 2.10 shows such a tableau.

In a Vigenère encipherment the key word determines which displaced alphabet is used to encipher each successive letter of the plaintext message. This process can also be expressed as a modulo 26 addition of the letters of the key word (repeated as many times as necessary) into the plaintext letters. In Figure 2.11 we show a short section of plaintext enciphered with the key word 'cipher'. The process of modulo 26 addition can be carried out conveniently by means of a device

```
     ABCDEFGHIJKLMNOPQRSTUVWXYZ

A    ABCDEFGHIJKLMNOPQRSTUVWXYZ
B    BCDEFGHIJKLMNOPQRSTUVWXYZA
C    CDEFGHIJKLMNOPQRSTUVWXYZAB
D    DEFGHIJKLMNOPQRSTUVWXYZABC
E    EFGHIJKLMNOPQRSTUVWXYZABCD
F    FGHIJKLMNOPQRSTUVWXYZABCDE
G    GHIJKLMNOPQRSTUVWXYZABCDEF
H    HIJKLMNOPQRSTUVWXYZABCDEFG
I    IJKLMNOPQRSTUVWXYZABCDEFGH
J    JKLMNOPQRSTUVWXYZABCDEFGHI
K    KLMNOPQRSTUVWXYZABCDEFGHIJ
L    LMNOPQRSTUVWXYZABCDEFGHIJK
M    MNOPQRSTUVWXYZABCDEFGHIJKL
N    NOPQRSTUVWXYZABCDEFGHIJKLM
O    OPQRSTUVWXYZABCDEFGHIJKLMN
P    PQRSTUVWXYZABCDEFGHIJKLMNO
Q    QRSTUVWXYZABCDEFGHIJKLMNOP
R    RSTUVWXYZABCDEFGHIJKLMNOPQ
S    STUVWXYZABCDEFGHIJKLMNOPQR
T    TUVWXYZABCDEFGHIJKLMNOPQRS
U    UVWXYZABCDEFGHIJKLMNOPQRST
V    VWXYZABCDEFGHIJKLMNOPQRSTU
W    WXYZABCDEFGHIJKLMNOPQRSTUV
X    XYZABCDEFGHIJKLMNOPQRSTUVW
Y    YZABCDEFGHIJKLMNOPQRSTUVWX
Z    ZABCDEFGHIJKLMNOPQRSTUVWXY
```

Fig. 2.10 A Vigenère tableau

Plaintext	THISPROCESSCANALSOBEEXPRESSED
Keyword	CIPHERCIPHERCIPHERCIPHERCIPHE
Ciphertext	VPXZTIQKTZWTCVPSWFDMTETIGAHLH

Fig. 2.11 A Vigenère encipherment

called a 'Saint-Cyr' slide (invented by Kerckhoffs and named after the French military academy), rather like a slide rule (Figure 2.12). The same device carries out the modulo 26 subtraction needed for decipherment.

An attack on a Vigenère-enciphered message can be carried out by many of the methods applicable to polyalphabetic ciphers in general. One method of attack is particularly powerful against the Vigenère cipher — the 'probable word' attack. The cryptanalyst chooses a word which he considers is likely to be in the plaintext message and then carries out modulo 26 subtraction of that word from the ciphertext in all possible locations, i.e. a scan is made of the entire ciphertext. If at any position the result of the subtraction yields a word of natural language or a fragment of such a word, then it is very likely that the key word or part of it has been discovered. The whole message can then be deciphered easily. In Figure 2.13 we take the ciphertext of Figure 2.11 and show how the key word 'cipher' is revealed by guessing that the word 'process' is in the plaintext. Only when 'process' is subtracted from the ciphertext in the right position do we see fragments making up a plausible key word. A procedure like this employs the strong redundancy of natural language texts.

A particular form of Vigenère cipher uses a 'running key'. For this purpose a text is chosen by the sender and notified secretly to the receiver; the text must be something that is readily available, for example, part of a standard work of reference or a well-known novel. To encipher the message it is combined letter by letter by modulo 26 addition with the chosen text; decipherment is carried

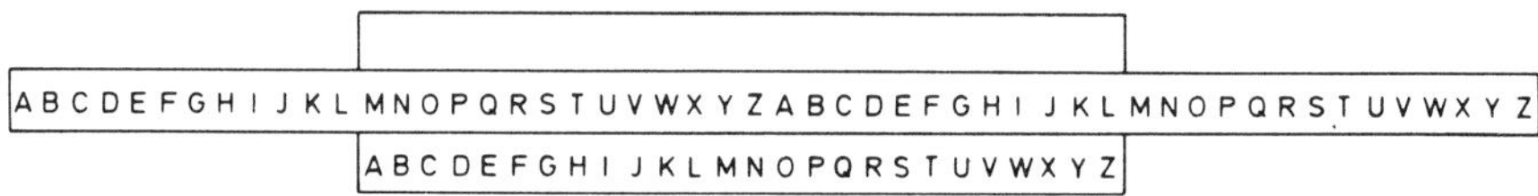

Fig. 2.12 The Saint-Cyr slide

Ciphertext	VPXZTIQKTZWTCVPSWFDMTETIGAHLH
—	PROCESS
Result	GYJXPQY
—	PROCESS
Result	AGLREYS
—	PROCESS
Result	IOFGMSB
—	PROCESS
Result	KCUOGBH
—	PROCESS
Result	ERCIPHE ← Key word revealed

Fig. 2.13 A probable word attack on a Vigenère ciphertext

out by modulo 26 subtraction of the same text. This method is a favourite with new inventors.

With a natural language text as the encipherment key the cipher may not be secure against attack, but if the key text were a truly random series of characters, absolute security against cryptanalysis would be assured. The problem for the communicators is to establish the same random series at both ends of the communication channel. The channel which carries the key must carry as much data as the enciphered channel. The use of a pseudo-random generator at the two ends may be an acceptable alternative, providing that the pseudo-random cycle is very long indeed. This point is considered again in the context of the Vernam cipher.

It is generally recommended that a running key be used only once for encipherment, because certain cryptanalytic attacks can take advantage of its repeated use. Where two cipher texts are enciphered with the same key and are in phase, they are said to be enciphered 'in depth'. If the start points are not known it is possible to make use of a property known as the 'index of coincidence' in order, when shifting one ciphertext against the other, to determine when the two ciphertexts are in depth. The index of coincidence is a concept due to William Friedman, who defined it in connection with solutions[3] of the Vogel and Schneider ciphers. The concept makes use of letter frequency distribution in natural language. If two streams of letters are made up of totally random sequences (with equal probability for all letters), then the probability of any two corresponding letters being the same is $\frac{1}{26}$ or about 0.0385. If, on the other hand, the letters in the two streams obey the natural language letter frequency distribution (e.g. if they are meaningful plaintext streams), then the probability of coincidence in any letter pair from the two streams is about 0.0667 (for English text). If two ciphertexts are examined for coincident letter pairs, then the probability of coincidence will be about 0.0385, as for the random streams, unless they are enciphered with the same running key and are in phase, when the probability of coincident letters will be about 0.0667. Given amounts of ciphertext sufficient to establish the index of coincidence reasonably accurately, the difference in the two measures is sufficiently pronounced for the condition of encipherment in depth to be detected.

2.3. TRANSPOSITION CIPHERS

Transposition ciphers aim to hide the contents of a message by taking the individual characters or bits and rearranging their order. For decipherment an inverse process of unscrambling takes place. The word 'transposition' is commonly used for this kind of operation, though 'permutation' might be better.

A very early cipher based on transposition was the Greek 'scytale' (literally: staff). To operate this a narrow strip of writing material, such as parchment, was wrapped spirally around the staff. The message was then written in rows of characters longitudinally along the staff. Unwrapped, the strip appeared to carry meaningless sequences of characters. Only by carefully wrapping the strip around a staff of equal diameter would the plaintext message be apparent. The method is neat but not secure because the relation between strip width and staff diameter allows only a small key space.

Simple transposition

The simplest possible transpositions take successive letter groups and rearrange each in a regular fashion. For example the letters of a message might be taken in groups of five and rearranged in the order 41532; this sequence constitutes the encipherment key. Figure 2.14 shows a message enciphered in this way. A frequency count taken on such a message should show the letters occurring in their normal places in the ranking, suggesting that a transposition and not a substitution has taken place. Solution of a transposition cipher of this kind is on a par with the kind of anagramming which is carried out by crossword enthusiasts.

A convenient way to express the permutation in easily memorable form is by a key word. This operation is illustrated in Figure 2.15 where the key word is 'computer'. The letters of the key word are numbered according to their alphabetical order — 14358726. The plaintext letters are then written in rows of eight letters beneath the keyword, filling in the final row, if that is incomplete, with null characters. As a general rule, the user of a transposition cipher based on a letter array would be well advised to use garble rather than unusual letters standing as nulls. The use of a string of unusual letters helps cryptanalysis, because the cryptanalyst can find column and row arrangements bringing together these unusual letters. To create the ciphertext, each row is read out in succession in the order of the letters of the key word. Given that a message has been trans-

```
Plaintext     THESIMPLESTPOSSIBLETRANSPOSITIONSXX

Key                     4 1 5 3 2

                        S T I E H
                        E M S L P
                        S T S O P
                        E I T L B
                        S R P N A
                        T O I I S
                        X O X S N

Ciphertext    STIEHEMSLPSTSOPEITLBSRPNATOIISXOXSN
```

Fig. 2.14 A simple transposition cipher

```
Plaintext     ACONVENIENTWAYTOEXPRESSTHEPERMUTATION

Key word              C O M P U T E R

                      1 4 3 5 8 7 2 6

                      A N O V I N C E
                      E W T A O T N Y
                      E R P E T S X S
                      H E P R T U E M
                      A O I N Z Z T Z

Ciphertext    ANOVINCEEWTAOTNYERPETSXSHEPRTUEMAOINZZTZ
```

Fig. 2.15 Transposition based on a key word

posed in this way, the cryptanalyst can count the number of letters in the message and find the factors of this number. He can guess which of these factors was a likely length for a key word. Writing the ciphertext in rows of the width of this word, he would then attempt to re-order the columns in such a way as to make the text meaningful in all rows. A probable word helps the cryptanalyst because its letters do not get distributed very far and can be found and used to determine the transposition or part of it. If the word 'permutation' is thought to be present, the underlined letters are found. The same transposition in PER and ION establishes which E to use and the size of the permuted arrays.

The Nihilist cipher

Slightly more complex is the operation of the 'Nihilist' cipher which is a combination of columnar and row transposition. The plaintext is again written in rows under a key word, but this time the key word is also written vertically at the side of the columns (Figure 2.16). Ciphertext readout takes place row by row as in the previous example, but with the order of row selection being determined by the key word written vertically alongside; the effect of this process is to spread the mixture of plaintext letters over a domain equivalent in size to the square of the size of the key word. The key word is repeated as many times as necessary in the vertical dimension, the transposition process being taken separately for each key word repetition.

A cryptanalytic attack on a Nihilist cipher depends on column and row rearrangement, with much trial and error in seeking to build up words, aided by intuition. To program this kind of attack, use can be made of digraph and trigraph letter frequencies when testing trial rearrangements.

An alternative method of operating this kind of cipher is to write in the plaintext according to the sequence controlled by the key word and to read out the ciphertext according to some complex order, such as diagonally and in opposite directions for alternative diagonals, as also shown in Figure 2.16.

```
          Plaintext   NOWISTHETIMEFORALLGOODMEN

                                 L E M O N
                                 2 1 3 5 4

                      L 2        O N W S I
                      E 1        H T E I T
                      M 3        E M F R O
                      O 5        L A L O G
                      N 4        D O M N E

 Nihilist Ciphertext  HTEITONWSIEMFRODOMNELALOG
 Diagonal Ciphertext  ONHETWSEMLDAFIITRLOMOOGNE
```

Fig. 2.16 A Nihilist transposition and variant read-out

2.4. PRODUCT CIPHERS

We have now seen how some of the simpler forms of substitution and transposition ciphers operate and it is evident that they are not secure. Therefore, ciphers of this kind cannot be considered seriously for protecting data, though they may once have served a useful purpose.

Cryptanalysis of simple ciphers depends on the use of letter, digraph and trigraph frequency counts and on the ability to rearrange letter groups easily, looking for patterns of plaintext. How much more complex would the task of cryptanalysis be if substitution and transposition techniques were used in combination? Because of the substitution, it would no longer be possible to rearrange the rows and columns of a matrix by inspection. Because of the transposition, digraph and trigraph counts would be broken up.

Attractive though a product cipher might be from the point of view of cryptographic strength, it is too complex for human effort, unaided by a machine. The risk of making a mistake is too high and the encipherment and decipherment processes are slow. To use ciphers of this complexity requires cipher machines.

2.5. CIPHER MACHINES

Cryptographers have used mechanical aids of one kind or another for centuries; these began with simple devices like the Saint-Cyr slide and developed into very complex electromechanical systems by the time of the Second World War, before electronics took over.

The Jefferson cylinder

An interesting cipher device is credited to the American president, Thomas Jefferson, who in the 1790s developed a mechanism for applying polyalphabetic substitution. This consisted of thirty-six discs, each of which carried a jumbled alphabet on its periphery, which could be mounted on a common axis (Figure 2.17). The discs were numbered so that their order on the axis could be specified and agreed between communicators. The disc order constitutes the key; the number of different ways in which thirty-six discs can be arranged is of the order 10^{41}, therefore the key space is very large. In operation a message would be set up on the device by rotating the discs until the message letters stood on the same row against an index bar. To read out the ciphertext the bar could be moved to any of the other twenty-five positions around the cylinder; the row indicated would then be read out as the ciphertext. It is immaterial which row is chosen,

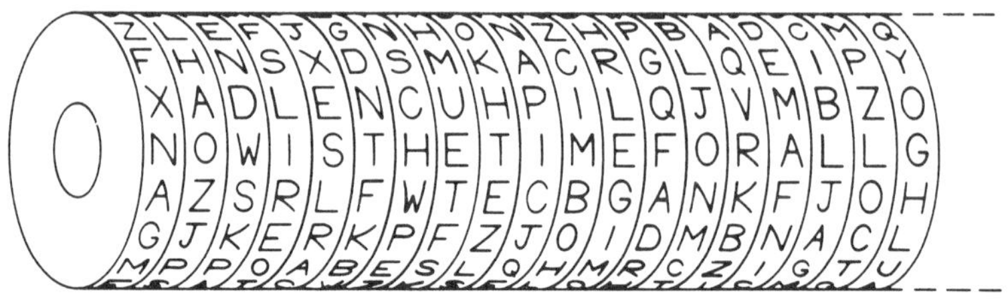

Fig. 2.17 The Jefferson wheel cipher device

because the intended recipient of the enciphered message would simply set up the row of ciphertext and then scan round the cylinder until a meaningful row of letters was found. It appears that this cipher was not used seriously until the 1920s, when it seems to have been re-invented and used by the US Navy for several decades.

The Wheatstone disc

Of slightly later date was the concentric disc type of cryptograph, originally invented by Wadsworth in 1817, but developed by Wheatstone in the 1860s. This type of device is effectively a polyalphabetic substitution cipher, but the way in which the various alphabets are used depends on the plaintext letter sequence. Two concentric discs carry the letters of the alphabet around their peripheries. Wheatstone's outer disc had twenty-six letters in alphabetical order plus space, whilst the inner disc had just twenty-six letters in random order (Figure 2.18). Two pointers were provided, geared together in such a way that when the outer had traversed twenty-seven letters so had the inner on its scale of twenty-six; if the pointers were coincident at the outset, after one complete revolution of the outer pointers the inner had moved by one revolution plus one letter, thus separating the pointers by one letter position. For encipherment the outer pointer was moved successively to the letters of the plaintext (moving always clockwise and including space as a character), whilst the ciphertext letters were indicated

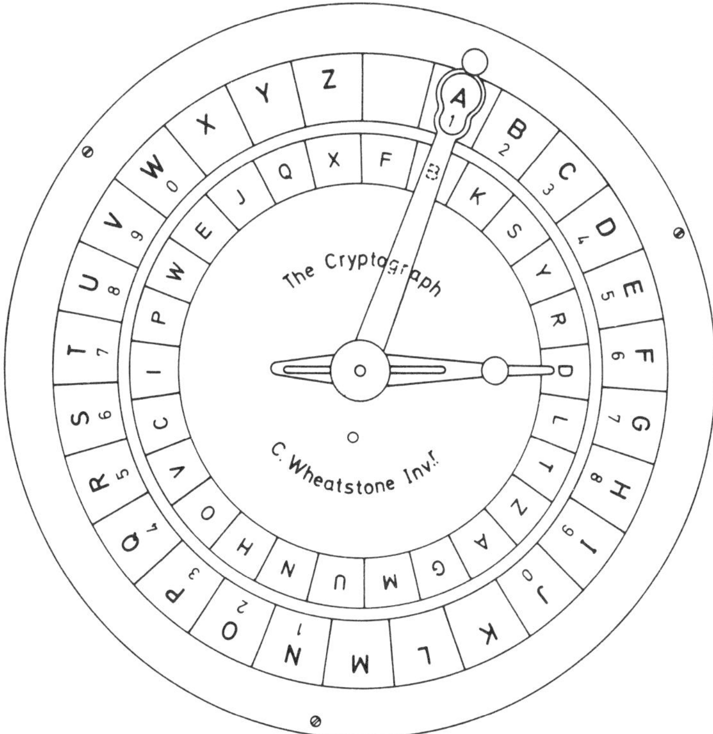

Fig. 2.18 The Wheatstone disc cipher device

by the inner pointer. The cipher is of the polyalphabetic substitution type with a change of alphabet at least at each word boundary (because of the space) and also within a word wherever successive characters of the plaintext are in reverse alphabetical order (or repeated). The cipher has the interesting property that the ciphertext representing a particular word is influenced by the preceding plaintext. This feature has properties similar to those of 'chaining', to be defined in chapter 4; chaining is of great importance in present-day applications.

Rotor machines, the Enigma

A very important class of cipher machines is based on various rotor designs in which rotors with internal cross-connections are used to achieve substitution with a continually changing alphabet. The internal links have been expressed both by pneumatic and electrical connections. Fibre optics came too late for the rotor machines.

One of the most famous polyalphabetic substitution machines was the Enigma (wiring shown schematically in Figure 2.19), a development from a Dutch invention dating from about 1920 and used by the German forces in the Second World War. This was a machine in which three, or in later machines four, rotors turned on a common shaft. Each rotor had twenty-six positions. The mode of rotation was similar to the action of an odometer with the rightmost rotor, R1, moving by one position for each letter enciphered and each rotor stepping the one to its left as it passed a certain position. Unlike a standard odometer the rotor, R2, in the middle stepped on itself by one additional position as it communicated a stepping action to the next rotor. Rotor R2 rested for only one character on its 'stepping' position and the effect was to give a three-rotor machine a cycle of $26 \times 25 \times 26$ characters. The four-rotor machine had the same cycle because its left-hand rotor did not move — no messages were long enough to make this matter.

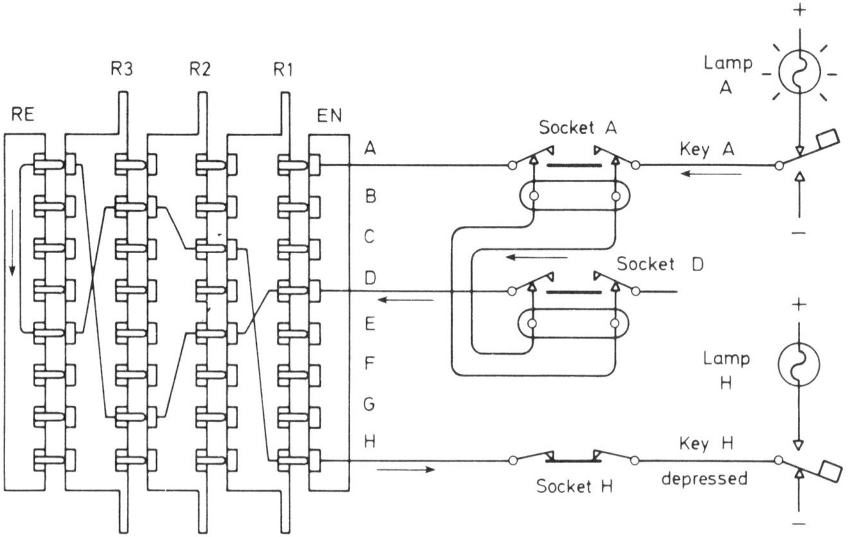

Fig. 2.19 Schematic of the Enigma cipher machine

Each Enigma rotor had twenty-six electrical contacts on each face, joined by internal wiring and the wiring pattern was different for each rotor. The fixed commutator, RE, to the left of the slowest-moving rotor produced a reflection of the electrical circuit back through the moving rotors by a different path. To make good contacts, one face of each rotor had sprung pins, as the figure illustrates.

When the original Enigma machine was adopted by the German forces, a plugboard was added to it. This had twenty-six sockets labelled A to Z and a number of double-ended cables were provided to join them in pairs. For most of the time ten cables were used, leaving six sockets unplugged. The plugboard was wired between the set of rotors and the set of keys and lamps. As shown in the figure for the two sockets A and F, the effect of a cable and its two plugs was to transpose one pair of connections that otherwise went straight through.

The figure shows only eight connections to represent the twenty-six connections of the actual machine. Depression of key H is shown, which disconnects that key from the lamp H and connects it, via the plugboard, to pin H on the fixed-entry commutator EN, since socket H is not used in this case.

Lamp A is connected through its key (undepressed) to socket A which is plugged to socket D. The effect of the cable is to transpose connections A and D. In the position of the rotors shown it happens that D and H are joined together so the connection to the negative supply at key H results in lighting lamp A. The plugboard permutes the connections from lamps/keys to EN and the permutation happens to be an involution (A to D implies D to A, as the figure shows). There is no logical reason why it should be an involution and late in the war a twenty-six position switch, the Enigma Uhr, was introduced which removed this restriction and allowed the plugboard setting to be changed each hour.

The figure shows the eight connections laid in a row instead of a ring of twenty-six. Rotation of any rotor completely changed the permutation of the twenty-six wires. The machine provided a sequence of $26 \times 25 \times 26$ substitutions before it repeated. In the fixed commutator, RE, on the left are thirteen connections. Since the key H operates lamp A it follows that, at the same rotor position, key A would operate lamp H, therefore each substitution is an involution — and this does not depend on how the plugboard is wired. It is a consequence of the thirteen wires in RE which themselves define an involution that is transformed by the rest of the equipment. It is also clear that no letter can be enciphered into itself because the electrical connections prevent that happening.

It was possible to replace Enigma rotors on the shaft from a collection or 'basket' of several different rotors. The initial setting of the machine required the selection and installation of rotors with their ring settings and then setting up the plug board. These operations constituted the current encipherment key. For each message, the initial rotor positions were chosen by the operator and transmitted in the message header. This 'indicator' was the weak point exploited in cryptanalysis. The German forces changed the method twice.

Operationally the Enigma machine was inconvenient. It needed at least two people, one to type the text on the keyboard and the other to read the lamps and write down the enciphered message. A printer attachment was developed but not widely used. There was also an extension attachment for the lamps so that only one person need see the plaintext.

It is now well known that, during the Second World War, messages enciphered with the Enigma machine were regularly cryptanalysed, first in Poland, then in France and finally, on a large scale and with great effect, at the Government Code and Cipher School at Bletchley Park.[4]

Printing cipher machines

With the invention of the typewriter in the latter part of the nineteenth century it seems natural that efforts should be made to develop cipher machines in which the plaintext could be typed on a keyboard and the ciphertext produced directly by a printing mechanism in the machine. The first known machine of this type was due to De Viaris in the 1880s. Many other machines with this property followed, some of which were also able to transmit ciphertext directly onto a communication link. A notable example of a printing cipher machine was the celebrated M-209 machine of Boris Hagelin, used in large numbers during the Second World War. A full description of the M-209 is given by Beker and Piper.[5] The M-209 was only one of the very many different cipher machines invented by Hagelin.

On-line cipher machines take the automatic handling of text one stage further than the printer. In these, the plaintext is typed on a machine which enciphers it and transmits the ciphertext (via a telegraph circuit or a radio-telegraph channel) to the receiver. Here, the same type of machine deciphers the received text and prints it in clear form. No human handling of the ciphertext is needed.

One of the most highly developed of the on-line telegraph cipher machines was produced by Siemens and Halske from the early 1930s to 1945. The machine was called type T52 and was built around a standard electric teleprinter. It was probably also called the Geheimfernschreiber or Geheimschreiber. It was used very effectively, mainly on telegraph circuits, in the Second World War. In its simplest form this machine used 10 wheels, each wheel stepped on by one place for each letter sent. The sizes of wheels were 47, 53, 59, 61, 64, 65, 69, 71 and 73, relatively prime, potentially giving a repeating cycle of length about 9×10^{17}.

The wheels produced letter substitution on the 5-bit Baudot code by a combination of running-key addition from five wheels and permutation of the five code elements by five other wheels. The key setting of the machine was at first a plug and jack field for permuting the ten channels from the cams on their way to the encipher/decipher circuits. Later versions of the T52 produced an irregular movement of the ten wheels by stepping the wheels according to logical functions (derived by magnet circuits) of the outputs of the cams, taken from a different place on the cam. A further complication was that the ten binary encipherment channels, after permutation according to the key, were subjected to linear logical operations (XOR) so that each channel to the encipher/decipher circuits depended on four cam outputs. A detailed study of the development of the T52 machine has been made by Davies.[6]

Modern cipher machines

A wide range of cipher machines is currently available from many manufacturers. These machines are mostly communications oriented, provided with appropriate interfaces for connection to modems and communication lines. Their design details are usually secret, even to customers who purchase them, the design being a commercial property of their manufacturers. Whether secrecy of design is desirable in cipher machines is a moot point. Should the security strength of the system depend on the secrecy of the encryption algorithm? If this were the only element on which security depended, then clearly the whole integrity of the protected communication would depend on preventing an intruder from discovering the details of the algorithm. It is realistic to assume that an intruder may be able to steal a machine and study it. Any machine worthy of consideration can be expected to possess a large key domain from which a secret key can be chosen; this is the other component in attaining adequate strength. As we saw in connection with monoalphabetic substitution ciphers, the size of the potential key domain is not necessarily a good indication of cryptographic strength.

Substitution in modern ciphers

A substitution can replace a 'word' of n bits by another word either as a mathematical (logical or arithmetical) function or using an arbitrary function expressed in a table of values.

We shall later meet examples of mathematical functions, such as those based on modular arithmetic. Addition, modulo 26, provides the Vigenère substitution. In chapters 8 and 9 exponential functions with very large moduli are used. However, the simplest mechanism for a cipher is an arbitrary table of values.

As an example consider an 8-bit substitution. There are 256 different values of the independent variable and for each value the 8-bit output must be specified. This can be done in 256 bytes of memory, which is a small expenditure in a micro-based encipherment device. In a programmed form, it usually requires two or three instructions and of the order of 1 μs of time in a cheap 8-bit microprocessor. This is an excellent component for a simple cipher, though 8-bit substitutions alone are not very secure.

If we are willing to employ more of the address space of a microprocessor in a substitution table, then large substitution tables are possible. In an 8-bit machine with 16-bit addressing it is unlikely that more than 16 Kbytes could be given over to a substitution table. This puts the practical limit at 14 bits. If the table has to be loaded using the key, the time to do this is also important.

Building a 64-bit cipher (for example) out of 8-bit substitutions requires some way of mixing together the outputs of several substitutions. This illustrates the importance of permuting bits or characters, the so-called 'transposition cipher'.

Keyed substitution

A fixed substitution, for example one built into a read-only memory (ROM), can be a powerful component of a cipher, but a substitution that is unknown to the enemy because it is derived from the key can be even better. Fortunately it can be shown[7] that almost all 8-to-8 bit substitutions are acceptable ones, free of the flaws of linearity or insensitivity to certain input bits.

Often the requirement is for an invertible substitution or bijection, which means that the set of 256 byte values (for example) are a permutation of the set of numbers 0, 1, ... 255. Efficient methods of generating such permutations exist[8], but economy of key bits is not a consideration if each key bit is being used for multiple purposes. Simpler methods are available.

For example, let $y = S(x)$ be the substitution to be generated, in which x and y take the values 0, 1, ... 255. Initially make $S(x) = x$. Then use the key, or part of the key, as a 'seed' to generate a pseudo-random integer sequence $R(i)$ for $i = 0, 1, \ldots 255$ where $0 \leq = R(i) \leq 256$. These random integers are used in turn to permute $S(i)$ with $S[R(i)]$. In this way each element of $S(x)$ is permuted with a randomly chosen other element of the array. If desired, the permutation process can be continued, to ensure a random result.

Transposition in modern ciphers

We have seen how substitution can be provided by table look-up, which is a fast operation either in a programmed microprocessor or purpose-built hardware. Transposition of bits in purpose-built hardware need be no more than a crossing over of connecting wires on a chip or board. In a serially operated device, it requires a number of delays and feedback paths and becomes complex.

Frequently the effect of transposition is provided by shift registers and by loading a set of 8-bit 'registers' in one sequence and reading them in another. The carry propagation in arithmetic operations can exploit the functions available in a processor to produce the effect of transposition, but the average length of a carry sequence is small, about log n bits for an n-bit number.

Variable or key-dependent transpositions of the classical form (bit or letter transposition) can be difficult to arrange in hardware. If all the permutations are to be available, a device like a miniature-scale telephone switch is required. In programmed form it is a little easier to do, but bit transpositions always tend to be inefficient and slow, hence the value of hybrid methods based on shifting, adding and XORing whole words.[9]

2.6. ATTACKS AGAINST ENCIPHERED DATA

Until the advent of digital computers the task of decipherment of cryptograms depended entirely on human skill; it was an esoteric art requiring considerable intuition on the part of the practitioner and a great deal of tedious work. Digital computers have changed this. For all the classical methods, analysis of enciphered text to identify the type of cipher can be carried out rapidly; once the cipher mechanism is known, the processing speed of computers allows quite crude methods to be used to break these classical ciphers.

Nevertheless, someone receiving a radio signal from outer space has a very considerable problem in interpretation — the modulation system of the signal may be complex, the 'plaintext' language is unknown, the message may be enciphered in an unknown cipher algorithm and the encipherment may be controlled by an unknown key. The message may not be intelligible, even with immense computing power and unlimited time.

The cryptanalyst trying to obtain the meaning of a message couched in an unknown cipher with an unknown key faces a task which is nearly as daunting. Unless a very elementary cipher has been used, identification of the encipherment algorithm itself can present a severe problem. When the cipher is known, in most cases the key must be found before the message can be interpreted. Trying all possible values of the key to see if one of them fits is known as 'exhaustive search'. Finding the key by this method can be a monumental task when the number of possible values of the key is great.

Very often obtaining the key is a much more valuable prize than just the cracking of one message. It is likely that a whole stream of messages has been enciphered with this key and the whole stream becomes intelligible.

Classes of attack

In assessing the strength of a cipher system it is prudent to assume that the algorithm is known to the cryptanalyst and that the attempt to crack the enciphered message consists of trying to find the key. The cryptanalyst's task is most difficult when all he has to go on is ciphertext, with no knowledge whatsoever of the plaintext; this is the *ciphertext-only* attack. If there is no redundancy in the plaintext (for example, an arbitrary string of numbers) then it is impossible to find a key. For a given ciphertext, each key value gives for its plaintext an arbitrary string and all these strings are possible. Some known plaintext could resolve this problem, for example if the arbitrary message has a preamble in standard format. Keys can be tested until an instance of the preamble appears in the deciphered text; if the preamble is long enough then this may allow the key to be identified with confidence, since the preamble only appears for one key value. Otherwise it may be necessary to collect all possible candidate decipherments, thus identifying a set of possible keys. A knowledge of the representational code of a plaintext with, say, even parity, can give the cryptanalyst something to work on.

More favourable to the cryptanalyst is the situation where some plaintext and its matching ciphertext are available — the *known-plaintext* attack. The task is then to find a key corresponding to this match. If the length of the available text is sufficient, then the key can be identified with absolute confidence. Circumstances in which a plaintext–ciphertext pair is known are more common than one might imagine. For example, market changes may lead to predictable instructions being sent out to bankers or brokers. Sending enciphered press releases from a country to its embassy is another case in point. This is discouraged because of the working material it might afford a cryptanalyst seeking to break the diplomatic cipher. The remedy here is to ensure that the press releases are paraphrased before release.

The most favourable situation for the cryptanalyst arises from a *chosen-plaintext* attack. Here the cryptanalyst has somehow managed to introduce a plaintext of his own choosing into the encipherment process. The plaintext may be chosen to ease the task of key finding; for example, it may be helpful if the chosen plaintext could be 'all zeros'. The 'probable word' is an example of known plaintext, though its position in the text is unknown to the cryptanalyst. Sometimes the probable word is actually 'chosen plaintext'. An example of forcing a word into a cipher system occurred in the breaking of Enigma, where a bombing attack was made on a lightbuoy with the certainty that the comparatively rare word 'leuchttonne' would turn up in Enigma enciphered messages.[10]

In order to ensure the strength of a cipher it is best to make the most adverse assumptions about the information available to the enemy. The least that is assumed in practice is that the cipher algorithm is known, together with enough corresponding plain and ciphertext to determine the key, in other words that a known plaintext attack is possible. For a more critical view of its strength, chosen plaintext should be assumed. A similar criterion, less discussed but worth considering, is that of chosen ciphertext. For example, could the enemy insert his own text on the line and somehow get access to the plaintext? A good modern cipher should resist all these attacks.

If the method of attack is to be by exhaustive key search using a computer, given a known plaintext–ciphertext pair, then we can estimate the mean time it will take to find a key based on assumed times to test each key and the number of possible different keys. We assume for the sake of illustration key testing times of 1 μs and 1 ms. These times approximate to the performance one might expect from special-purpose, large-scale integration (LSI) hardware and a microprocessor system, respectively. We assume key domains of 24, 32, 40, 48, 56 and 64 bits. In the left-hand part of Figure 2.20 we show the times taken to find correct keys, assuming that one key is tested at a time and that on average the correct key is found after working through half the key space. Any results above and to the right of the thick line represent conditions where the cipher is very weak. Results expressed in days may indicate conditions to be avoided for very important messages, whilst results expressed in years may be strong enough for most purposes. The results for times greater than 100 years are not shown, being beyond the normal expectation of human life.

Imagine, however, that a cryptanalyst has at his disposal a more powerful machine which is able to test many keys in parallel at the same time. Our imaginary machine contains many distinct units, each of which is programmed to explore a particular part of the key domain; eventually one of the units identifies the key and the machine signals success in its key search. For the sake of illustration we shall assume that such a machine can be built and that it is capable of testing 1 million keys simultaneously. This changes the situation completely. The right-hand part of Figure 2.20 shows the times now achieved in key sear-

Key size bits	*Single test*		*10^6 tests in parallel*	
	1 ms	1 μs	1 ms	1 μs
24	2.33 h	8.4 s	8.4 ms	8.4 vs
32	24.9 d	35.8 m	2.15 s	2.15 ms
40	17.4 y	6.4 d	9.2 m	550 ms
48	> 100 y	4.46 y	1.63 d	2.35 m
56		> 100 y	1.14 y	10.0 h
64			> 100 y	107 d

Fig. 2.20 Times for exhaustive key search

ching. With the powerful key search machine implied here, even a 64-bit key may not be large enough to give adequate security.

2.7. THE STREAM CIPHER

Where a communication system needs to handle characters in a stream, as, for example, when a simple terminal without storage is connected, this implies certain constraints upon the choice of cipher system that may be used. Each character must be accepted as it becomes available, enciphered and then transmitted. Clearly some form of substitution cipher could be used, but not a cipher which transposes characters.

Plaintext data arriving in this fashion is normally handled by what is known as a 'stream cipher'. A stream cipher may be synchronous, when encipherment depends on the current plaintext character and its position in the stream (in addition to the encipherment key) and the ciphertext is not influenced by the particular plaintext characters which have gone before. Enigma produces this kind of cipher. If the positional information becomes inaccurate, then the result may be completely unintelligible. Other types of stream ciphers are self-synchronizing; decipherment then depends on the current plaintext character and on a limited number of previous ciphertext characters; synchronism may be lost temporarily, but recovered as more text is processed. Because of the dependence of the current ciphertext on the immediately recent ciphertext, each ciphertext character depends on the current plaintext character, its position in the plaintext stream and also on the whole of the preceding plaintext stream.

The Vigenère cipher is an example of a synchronous stream cipher. The modulo 26 addition of a key text into a plaintext naturally takes place character by character and depends on relative position of current plaintext and key text; correct registration of plaintext and key text is essential. In that context we pointed out that a truly random stream of characters used as key material would yield an uncrackable ciphertext. The idea of adding a random stream into a plaintext arose as a result of an important invention by Vernam in 1917.

The Vernam cipher

At that time Vernam was working with other employees of the American Telephone and Telegraph Company on methods of encipherment for the printing telegraph. Vernam's important contribution was the bit-by-bit combination of random characters with characters of the plaintext, the combination being carried out by modulo-2 addition (the XOR function); the principle is illustrated in Figure 2.21. Random characters, chosen by picking numbers out of a hat, were punched on paper tape; this tape was fed through one reader in synchronism with the plaintext tape fed through another reader. A new tape was punched in a tape punch, enciphered in this neat and elegant operation. The encipherment key was the random number stream punched on the key tape, sometimes called the 'key stream'.

Decipherment was equally simple; all that was necessary was to take the received enciphered tape and feed this through a tape reader, combining it by bit-by-bit modulo-2 addition with a copy of the random key stream. In other words,

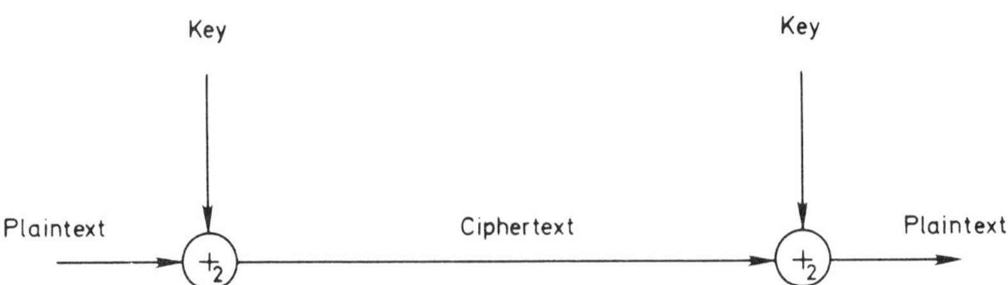

Fig. 2.21 The flow of data in the Vernam cipher

the process of decipherment used exactly the same mechanism and exactly the same key data as the process of encipherment. Figure 2.21 shows the data flows in the Vernam encipherment and decipherment. The modulo 2, or 'exclusive-OR', operation, used in the Vernam encipherment, has acquired great importance in designing modern cipher procedures. There is close similarity between Vernam's and Vigenère's methods except that one works with bits and the other with characters.

To operate Vernam's cipher it was necessary to provide sender and receiver with identical sequences of random numbers. Vernam's original idea was to use loops of tape, so that the random number stream was used over again as many times as necessary to encode a complete message. In order to generate longer number streams, two random number tape loops (with different random number streams) were used together, the tapes differing in length by one character. The two tape loops were run in synchronism and combined in an exclusive-OR operation, the result being combined with the plaintext. Thus, if one loop were 1000 characters long and the other 999, the number of combinations which would be produced before a repeat would be 999 000, but this is not a true indication of its complexity.

If tapes of truly random numbers can be provided at the two ends of a communication system using a Vernam cipher system, then the cipher is completely uncrackable. This is because, for any ciphertext with an unknown random key, all possible plaintexts of the same length are equally valid as decipherments. For very important communication channels it is worth going to the trouble of securely providing such random number tapes. Since the key is as long as the message, it is not generally useful, but there are situations needing great security in which the amount communicated is small but it is urgent and carrying bulk data as key is not difficult. This method, known as the 'one-time tape', has been used to protect channels such as the Washington–Moscow 'hot line'.

A similar method, operated by hand, is the 'one-time pad' which has been in the news more than once in connection with the unmasking of spies. Here the random number series is printed on tear-off pads, each sheet being used only once and then destroyed. Matching pads are held at sender and receiver locations.

As was pointed out in connection with the Vigenère cipher, so, with the Vernam cipher, it is important to avoid using the same key for more than one message, hence the use of the term 'one-time'. Use of the same key in encipherment for a number of plaintexts allows attacks either by a Kerckhoffs' superimposition or by elimination of the key by an exclusive-OR combination of the two ciphertexts.

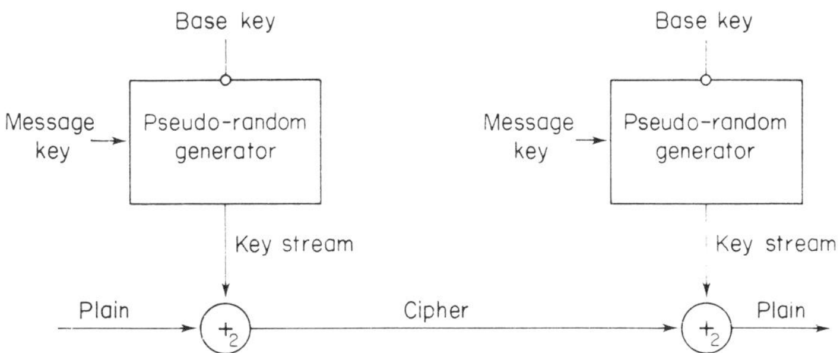

Fig. 2.22 Stream cipher with pseudo-random generators

Pseudo-random key streams

According to the Vernam principle the same random key stream is used for encipherment and decipherment. If a repetitive key stream is insecure and a completely random key stream is impractical, the compromise which is normally used is to employ a pseudo-random key stream generated by a finite state device at each end of the line as shown in Figure 2.22. Much of the ingenuity of cryptographers has gone into the development of these key stream generators. The key stream generator employs a secret key, the main cryptographic key, sometimes known in this context as the *base key*.

In order to synchronize the two ends of the line, the two pseudo-random key stream generators must begin with the same stored data, that is to say the same state of the machine. The number used to establish this state may be considered as part of the key, sometimes called the *message key* and it can be exchanged by a preamble to the main communication. Many systems allow the synchronizing data to be sent in clear through the network. Because synchronization can be lost due to a fault in the system or a bad communication line, practical systems which run automatically must be able to recover synchronism without human intervention. Effectively, this means that a portion of the channel capacity has to be given over to synchronizing information.

In some systems, the data being transmitted has considerable redundancy, perhaps because it is multiplexing a number of data streams. In this case, a loss of synchronism can be detected and both key stream generators stop their normal operation and enter into a resynchronizing sequence. In other systems, particularly those working at high speed or over networks with delay such as satellite links, the synchronizing information is sent continually and the two units resynchronize at a moment defined by a secret criterion based on the structure of the data being transmitted.

An example of this kind of key stream is given in chapter 4, section 4.4 — a method called output feedback.

2.8. THE BLOCK CIPHER

Block ciphers handle plaintext characters in blocks, operating on each block separately. The block size depends on the particular encipherment device in use;

the Data Encryption Standard (see chapter 3) is basically a block cipher operating on blocks of 64 binary bits.

An advantage of a block cipher is its ability in general to operate faster than a stream cipher. Data may be efficiently handled in blocks in applications such as file transfers. However, if there is no influence from other parts of a plaintext on the encipherment of each block this is a serious disadvantage. It means that if a block happens to occur more than once in a plaintext, then corresponding repeated ciphertext blocks will occur in the ciphertext, giving away information to an eavesdropper; this characteristic is very like that of material that has been encoded, as distinct from enciphered, when repeats may also be evident (indeed the simple block mode of using the Data Encryption Standard is usually known as 'electronic code book'). Even more seriously, an active line-tapper may be able to change the order of ciphertext blocks, delete selected blocks or duplicate particular blocks without risk of discovery.

Since the block cipher is effectively a substitution cipher acting on blocks instead of characters, if such a system is to be used with anything like adequate security, it is essential to use blocks of appreciable size in order to prevent a cryptanalysis based on frequency counts of block types.

Methods have been devised for using block ciphers as a basis for constructing stream ciphers. We shall see in chapter 4 how this has been done for the Data Encryption Standard.

2.9. MEASUREMENT OF CIPHER STRENGTH

Shannon's theory of secrecy systems

A most important contribution to the theory of encipherment systems was made by Claude Shannon, when he published his paper 'Communication theory of secrecy systems',[11] which followed soon after his better-known paper on the mathematical theory of communication. Shannon assumed that the cryptanalyst had only ciphertext to work on, which, from the point of view of cryptanalysis, is a more stringent assumption than the known or chosen plaintext assumption.

Shannon identified two significant classes of encipherment system — those that are unconditionally secure and those that are computationally secure. An *unconditionally secure* system is defined as one which defies cryptanalysis even though the cryptanalyst has unlimited computing power available to him. The only unconditionally secure cipher that is in common use is the one-time tape system, mentioned earlier, if used with a truly random key tape. As we have already seen all plaintexts with length equal to the ciphertext are decipherment candidates and there is no way of choosing between them, computational resources notwithstanding.

A particular cryptogram may be unconditionally secure if it is so short as not to contain sufficient material to lead to a unique solution. Even a monoalphabetic substitution ciphertext may fall into this class. Shannon defined the minimum length of text required to yield a unique solution as the 'unicity distance'. Stated simply, Shannon's criterion was that the redundancy in the plaintext must exceed the information contained in the key. For example, the key size for a monoalphabetic substitution cipher is 26! and $\log_2 26!$ is 88. If the redundancy of English-language text is 80 per cent, then each character contributes about 3.8 bits of redundancy. Therefore any ciphertext with more than $88/3.8 \simeq 23$ characters can be used to determine the key. The unicity distance is 23 characters. Since Shannon's estimate of the redundancy of English included spaces, cryptanalysis of text without spaces requires a little more text. Still, the result is close to the lower limit observed in cryptanalysis by traditional 'craft' methods. For a one-time tape cipher, the unicity distance is infinite.

A ciphertext which contains sufficient information for cryptanalysis is said to be *computationally secure* if the task of solution is so large as to be impossible of execution with the largest conceivable computational power. The 'one-way' functions that we shall meet in chapter 8 are of the class that are easy to compute in one 'direction', but are required to be computationally infeasible in the other.

There is no accepted definition of what is computationally infeasible, bearing in mind that extremes of parallel processing can be used, that time-memory trade-offs are possible and that some notion of the acceptable time for a solution must be obtained. Furthermore, processing and storage technology continually improve. It would generally be agreed that a calculation employing more than 10^{25} steps is not feasible today. Perhaps 'step' should be defined as 'that which an LSI chip can do in 1 μs'. To get a more absolute limit, going beyond our present technology, it is possible to set a thermodynamic limit.

Limits of computation

A reasonable assumption is that each logical step consumes energy kT, where k is Boltzmann's constant and T the absolute temperature. From a calculation of the amount of heat falling on the earth from the sun and assuming that the cryptanalytic computation must take place at a temperature of 100 K, it can be shown[12] that the number of elementary operations obtainable in 1000 years is about 3×10^{48}. The other significant consideration in a time-memory trade-off is the amount of available store for the computation. Let us assume that somehow a store can be constructed in which each bit requires 10 atoms of silicon; this store shall be 1 km high and cover all the land mass of the earth (or shall be placed in orbit as a new moon of equivalent mass). The number of bits that can be stored is about 10^{45}.

Being cautious in placing our limits of computational infeasibility, let us define these as 10^{50} operations in 1000 years and a store of 10^{50} bits. Any computations requiring more than this amount of computing power can be said to be totally impossible in the time scale stated. In practical terms, computations requiring very much less power than this are also likely to be impossible. A feel for the scale of the problem can be gained by looking at cryptanalysis of the Lucifer system[13] with its key of 128 bits; if this can only be attacked by exhaustive search for the key, given a known plaintext–ciphertext pair, then the number of keys that must be tested is about 3×10^{38}. If one Lucifer key can be tested every picosecond, then the time taken to find a solution is about 10^{19} years.

Key size and its effect on the time for exhaustive search set an upper limit on the strength of a cipher, but it is a great mistake to judge the strength of a cipher in this way. The history of ciphers is full of 'unbreakable ciphers' with large key spaces. Key search is no more than the universal method on which to fall back when all else fails. The strength of a cipher is a negative quality and depends on everyone's inability to find a feasible way of breaking it. It can never be guaranteed except for an unconditionally secure cipher. In chapter 8 we shall describe how 'complexity theory' might one day produce proof of the difficulty of breaking ciphers. At present this is just an aspiration. The best that can be done is to show that certain ciphers are as difficult to break as the difficulty of certain computational problems that mathematicians generally agree to be hard. The possibility of improved computational methods which make breaking feasible cannot be ruled out.

Looked at from the point of view of the theory of computational complexity, some systems can be proved to be unconditionally secure, but computational infeasibility of any cryptanalytic problem cannot be demonstrated. For this reason it is usual, when assessing the strength of a cipher, to postulate an attack under conditions which favour the cryptanalyst. This departs from Shannon's assumption of a ciphertext-only attack and makes the assumption of a known-ciphertext or chosen-ciphertext attack. It is on such assumptions that the strength of a modern cipher system is judged.

An application of Shannon's theory

An interesting application of Shannon's theory is to the solution of messages formed by adding two English texts. The operation can be modulo 2 addition of bits (Vernam) or modulo 26 addition of a letter code (Vigenère) — it makes little difference. From one point of view, each text is enciphering the other. Each text has a redundancy of about 80 per cent. The amount of key needed to solve it is provided by the other's redundancy. In other words, the 20 per cent equivocation in each text still leaves enough redundancy (60 per cent) to solve the problem. This, even if it were an exact argument, would not show that it could be done in practice because our knowledge of the structure of English is not perfect and long messages would be needed, with correspondingly long computation, to exploit all the redundancy which Shannon has estimated. In practice, by guessing first at one message and then the other the problem can often be solved. For a starting point, a probable word or phrase in one or other text is needed.

The same method of solution can be applied to an equivalent problem — solution where two texts have been additively enciphered with the same key stream. Such a solution gives both the texts and a sample of keystream for further analysis.

2.10. THREATS AGAINST A SECURE SYSTEM

Threats against any protected system, be it a communications system or a data storage system, can be classified conveniently into two categories — the passive and the active threat. A passive attack against a system is merely an attempt to get round the protective barriers and read what is being transferred or stored, but without attempting to alter it. This type of attack has been perpetrated against communication systems ever since the invention of the electric telegraph. In the days before multiplexing it was very simple indeed to introduce a wire tap onto a transmission line (Figure 2.23). Telephone tapping of a local connection is all too easy to achieve. The various forms of multiplexing now in operation and the

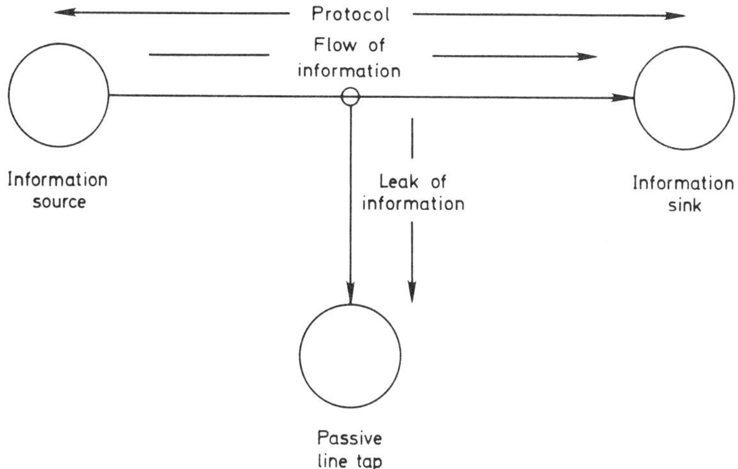

Fig. 2.23 A passive line tap

complex protocols in data networks demand greater sophistication of the intruding line-tapper, but this is well within the compass of the technologist criminal. It is often thought that multiplexing provides in itself a kind of safeguard against the intruder, but this is a dangerous and fallacious belief. It is well known that in the USA a foreign power made a practice of listening-in to microwave links (no physical line tap then being necessary) and selecting interesting telephone calls by recognition of the in-band dialling tones used to set up the calls.

A passive line tap need not break into the protocol controlling the flow of data. This simplifies the task of the passive line tapper enormously.

Detection of a passive line tap is feasible where a physical communication link is involved, based on precise measurements of line characteristics, but clearly this is impossible where a microwave link is tapped. Defence against passive line-tapping intrusion must therefore be based on protecting by encryption the data transmitted; it cannot hope to succeed by protecting the transmission line itself, except where the line is totally protected physically, an unusual state of affairs.

Active line taps

The active threat is potentially far more serious. Here the intruder seeks to alter the stored or transmitted data, and will hope to do this without discovery by the legitimate data owner or user. Alteration of stored data may take place either through misuse of the normal access channels or by re-writing data on exchangeable storage media. We are not so much concerned here with the access rules, though this is an important subject, but rather with the techniques for protecting data against attack via clandestine means. Use of encryption can protect against undetected alteration of the data by arranging that the encrypted data is structured in such a way that meaningful alteration cannot take place without cryptanalysis; we shall discuss this at greater length when we consider secure data formats using encryption. Where an intruder has illegal access to data-recording media, deletion of data cannot be prevented. Clearly this is a very

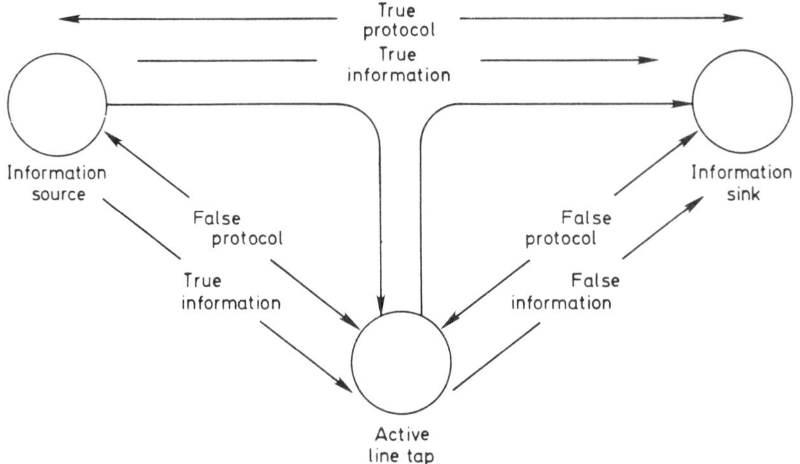

Fig. 2.24 An active line tap

serious matter and demands the creation of a secure physical environment for storage of tapes and discs, together with back-up copies at a different storage location.

An active line tap on a communication link is almost as difficult to prevent as is a passive line tap except in radio channels. Detection is much easier and may even be possible without encipherment or similar logical schemes. Successful alteration of data requires time to carry out and precise measurements of transmission times may indicate the presence of intrusion. Whereas a microwave link was easily tapped in passive mode, this type of link would present a severe challenge to an active line tap. If a physical communication link is under constant surveillance by its owner, with traffic constantly flowing and being monitored, then it may be almost impossible for the intruder to introduce his tapping connections without detection. The type of configuration that a line tapper would require is shown in Figure 2.24.

To operate an active line tap successfully is not a simple task. In Figure 2.24 communication between source and sink is controlled by a protocol, called in the figure 'true protocol'. The line tapper must interrupt this protocol and conduct a separate 'false protocol' with source and with sink, thereby deceiving each that they are still in direct communication. The true information flows from the source to the active line tap; false information flows from the tap to the information sink. The task of the line tapper is probably simplest when source and sink are allowed first to establish a genuine communication channel, to be taken over later by the line tapper. With some added complication the active tap takes part in the process of setting up the channel.

Methods of protection

Protection of transmitted data against an active line tap is based on the same principles as those mentioned earlier in connection with the protection of stored data. It is essential to prevent undetected alteration, addition, deletion and replay. In any one block of transmitted data the first three risks can be averted by arranging the encrypted data structure in a chain, so that any portion of the encrypted data is dependent on the corresponding plaintext data and also on the preceding plaintext; meaningful alteration cannot be carried out without finding the encryption key by cryptanalysis. We shall discuss chaining techniques in some detail in chapter 4 when we come to consider the modes of operation used with the US Data Encryption Standard.

Replay is the attack in which the intruder records an enciphered transmission and later sends it again from his line-tapping transmitter. For example, the aim might be to make money transfers to a bank account appear to take place more than once. Replay cannot be prevented, but it can be detected by including in the plaintext sufficient block identification data, a block number (either unique or cyclic with a large range) and, possibly, a time-stamp. The receiver would ignore any enciphered message whose decrypted plaintext did not contain valid identification. Detection of unauthorized changes is part of the subject called *authentication* and is treated in chapter 5.

2.11. THE ENCIPHERMENT KEY

In all but the most elementary of the systems we have discussed in this chapter, the role of the encipherment key is most significant; key-less encipherment algorithms, which always behave in a predictable manner, are only of academic interest; the security of such systems depends on keeping the design details secret and when the secret is lost a new design is needed to replace the lost cipher.

In keyed algorithms the encipherment process is carried out under the control of the key, which influences the behaviour of the encipherment algorithm. This property applies equally to substitution, transposition and product ciphers; it is particularly important when we come to machine ciphers. In the latter, capture of a machine may expose the detailed construction of a secret encipherment algorithm to an opponent. However, if the opponent does not know the relevant encipherment key, then knowledge of the algorithm alone should be totally insufficient to permit cryptanalysis of an enciphered message. It is a sound principle, much followed in modern practice, that security of a cipher system should not depend on the secrecy of the algorithm, but on the secrecy of the encipherment keys.

This principle places a considerable responsibility on the designer of the system under which keys are generated, distributed, used and destroyed after use. In the past the key material has been sent to encipherment users by human courier, but this is too slow and inconvenient for many of today's applications. Therefore, means must be devised for transporting keys more rapidly and efficiently. Often this implies sending the keys via the communication channel which is itself the subject of encipherment protection. Here we need an hierarchy of keys which distinguishes those keys which are used for data protection (and which may therefore be much in use) from those used solely for the protection of keys in transit. Keys used for encipherment of keys are used far less frequently and may be transported by less efficient means, such as a physical visit; the very fact that they are used less frequently means that they are less exposed to cryptanalytic attack.

It is important that the system designer gets the key management system right and does not leave loopholes exploitable by an intruder. We treat this subject in detail in chapter 6.

REFERENCES

1. Kahn, D. *The Codebreakers*, Macmillan, New York, 1967.
2. Gaines, H.F. *Cryptanalysis*, Dover, New York, 1956.
3. Friedman, W.F. *The Index of Coincidence and its Applications in Cryptography*, The Riverbank Publications, Aegean Park Press, Laguna, 1979.
4. Hinsley, F.H., Thomas, E.E., Ransom, C.F.G. and Knight, R.C. *British Intelligence in the Second World War, Vols I and II*, HMSO, London, 1979 and 1981.
5. Beker, H. and Piper, F. *Cipher Systems: The Protection of Communications*, Northwood Books, London, 1982.
6. Davies, D.W. 'The Siemens and Halske T52e cipher machine', *Cryptologia*, **6**, 4, 289, October 1982; 7, 3, 235, July 1983.
7. Gordon, J.A. and Retkin, H. 'Are big S-boxes best?', *Cryptography* (Proceedings, Burg Feuerstein, 1982, ed. T. Beth), Lecture Notes in Computer Science 149, Springer Verlag, Berlin, 1983.
8. Ayoub, F. *Some aspects of the design of secure encryption algorithms*, Hatfield Polytechnic, School of Engineering, Ph.D. Thesis, March 1983.
9. Brüer, J-O. 'On non-linear combinations of linear shift register sequences', *Proceedings, IEEE 1982 International Symposium on Information Theory, Les Arcs, France. Also* appeared as Internal Report LiTH-ISY-I-577, Linkoping University, March 1983.
10. Johnson, B. *The Secret War*, BBC Publications, London, 1978.
11. Shannon, C.E. 'Communication theory of secrecy systems', *Bell System Technical Journal*, **28**, 656, October 1949.
12. Davies, D.W. *Limits of computation*, NPL Internal Note, 1978.
13. Feistel, H. 'Cryptography and computer privacy', *Scientific American*, **228**, 5, 15, May 1973.

Chapter 3 THE DATA ENCRYPTION STANDARD

3.1. HISTORY OF THE DES

Why do we need a data encryption standard, since proprietary encipherment systems are available from many manufacturers? Users wishing to protect communication on a line need only purchase a pair of matching encipherment devices, install them, enter matching keys and they are ready to communicate securely. Whenever communication is internal to a particular group, protection of transmission by encipherment presents little in the way of organizational problems, such as reaching agreement on a particular method of encipherment. Until recently most data communication has been within single organizations, often carried on a private network of leased lines. In these conditions a standard is not needed.

When protected communication is required between users who belong to different organizations, this may not be so easy to arrange. The organization may have chosen encipherment devices from different manufacturers, in which case the devices are likely to be totally incompatible and unable to communicate either securely or otherwise. Furthermore encipherment must often be introduced in the context of standard communication protocols; this may place new restrictions on the means of encipherment. This kind of interworking is becoming increasingly common. Therefore, there is a case for considering standard encipherment algorithms for widespread use. Of necessity any standard encipherment algorithm must be public knowledge and this is a departure from previous cryptographic practice where users attempted to keep the nature of the encipherment algorithms in use a secret. The entire security of a system based on a public standard must rest on the strength of the algorithm and the secrecy of the encipherment key.

The initial move towards a public encipherment standard came from the US Federal Government, which, if it chose, could have dictated to its agencies involved in data processing what existing proprietary algorithm they should use. Encipherment had, of course, been in use by the US Government before this initiative; such encipherment was designed to protect data covered by the US National Security Act of 1947 or the Atomic Energy Act of 1954 and similar legislation. The new initiative was aimed at areas where the responsible authority deemed that the data in question, though not covered by legislation demanding encipherment (that is to say, not 'classified'), yet merited cryptographic protection; in general the application was, therefore, outside the military, diplomatic and other areas of national security.

The role of NBS

The US Federal department given the task of initiating this development was the Department of Commerce; within this department the responsible agency was the National Bureau of Standards (NBS). (It is now renamed the National Institute for Standards and Technology (NIST).) Under the Brooks Act (Public Law 89-306) NBS had the responsibility for setting Federal Standards for the effective and efficient use of computer systems; the relevant clause of PL 89-306 reads: 'The Secretary of Commerce is authorized . . . (2) to make appropriate recommendations to the President relating to the establishment of uniform Federal automated data processing standards.' The standards which have been produced as a result of this legislation are expressed in a substantial series of publications under the general title of Federal Information Processing Standards; they cover hardware, software, data, ADP operations, etc.

Under this general remit NBS initiated a study programme on computer security. One of the early actions in this programme was to express the need for a standard for data encryption. The Bureau took the view[1] that purchase of encipherment devices from commercial suppliers must be subject to the principle of 'caveat emptor' or 'buyer beware'. An NBS statement averred that 'the intricacies of relating key variations and working principles to the real strength of the encryption/decryption equipment were, and are, virtually unknown to almost all buyers, and informed decisions as to the right type of on-line, off-line, key generation, etc., which will meet buyers' security needs have been most difficult to make'.

The NBS's responsibility extended primarily to the protection of data as processed and transmitted by Federal agencies; however an important secondary area of responsibility as part of the Department of Commerce was to the general buyer of encipherment equipment. An interesting sidelight is thrown by the NBS's view that much of the proprietary (i.e. non-standard) encipherment equipment available came from outside the USA. Therefore the preparation of a standard for data encipherment would not damage the trade of US manufacturers to a serious degree.

In 1974 the US Congress passed a Privacy Act, which, strangely, did not address directly the question of data security. However, data security is an important tool in attaining in a satisfactory manner the requirements of legislation of this type, and the Office of Management and Budget assigned to NBS the responsibility of developing computer and data standards to meet the needs of this Act.

The NBS study programme got under way effectively in 1972. A search was begun for an appropriate encipherment algorithm which could form the basis of a Federal Information Processing Standard. The requirements for this were as follows:[2]

1. It must provide a high level of security.
2. It must be completely specified and easy to understand.
3. The security provided by the algorithm must not be based on the secrecy of the algorithm.
4. It must be available to all users and suppliers.
5. It must be adequate for use in diverse applications.
6. It must be economical to implement in electronic devices and be efficient to use.

7. It must be amenable to validation.
8. It must be exportable.

'Economical implementation' virtually demands that the algorithm be realizable in special purpose large-scale integration (LSI) hardware; mainframe computers are not at their best in carrying out the bit manipulations that form typical components of encipherment algorithms. Acting under the remit expressed in the eight points listed, the NBS team was not looking for ideas needing a lot of further development, only methods which were already well developed were sought.

In May 1973, NBS issued a call for ideas fitting these needs. The response was disappointing; many mathematicians sent them outlines of algorithms which needed development. One suggestion received was virtually a restatement of the 'one-time tape' method, based on the Vernam invention. None of the offerings came anywhere near filling the stated requirements. In August 1974 a second call for ideas was issued. In response, several algorithms were sent in; some of these were too specialized, some were not strong enough. However, one algorithm submitted showed great promise in meeting the requirements. This submission came from International Business Machines (IBM).

The IBM Lucifer cipher

The IBM submission to NBS was a development from an encipherment scheme invented in the early 1970s; this was the IBM 'Lucifer' encipherment algorithm, a 'product' cipher, one of the early applications of which is understood to have been in a bank automatic-teller terminal. As the eventual Data Encryption Standard (DES) was more or less directly developed from the Lucifer design it is worth examining Lucifer in a little detail. Lucifer certainly met the NBS requirement that the algorithm should be made fully public; it had been the subject of several articles quite freely available in various journals and reports.[3,4]

In chapter 2 we met various substitution and transposition ciphers and we saw that, used alone, many of these are very weak. However, as was pointed out by Shannon,[5] cipher operations which are weak in themselves can be combined together to form something much stronger; this is the concept of the product cipher. A product cipher may take a block of plaintext and transpose the order of its binary bits (called 'diffusion' by Shannon), following this with a systematic substitution of bit patterns with others according to look-up or substitution tables (called 'confusion' by Shannon). These two steps may be repeated a number of times. Figure 3.1 illustrates the process.

In Figure 3.1 the first logical step is a substitution, in which the bits of the plaintext block are taken in groups and replaced by other bit patterns of the same size according to substitution tables (S-boxes), with no expansion of data. The second step is a transposition, or re-ordering, of the bits of the whole output from the first step. The third step is a substitution, similar to the first step. This is followed by a further transposition and so forth. Thus transposition and substitution alternate until the ciphertext produced as a result of the total process is unpredictable from the plaintext without knowledge of the intervening transformations (the actual number of steps employed is set by the designer of the

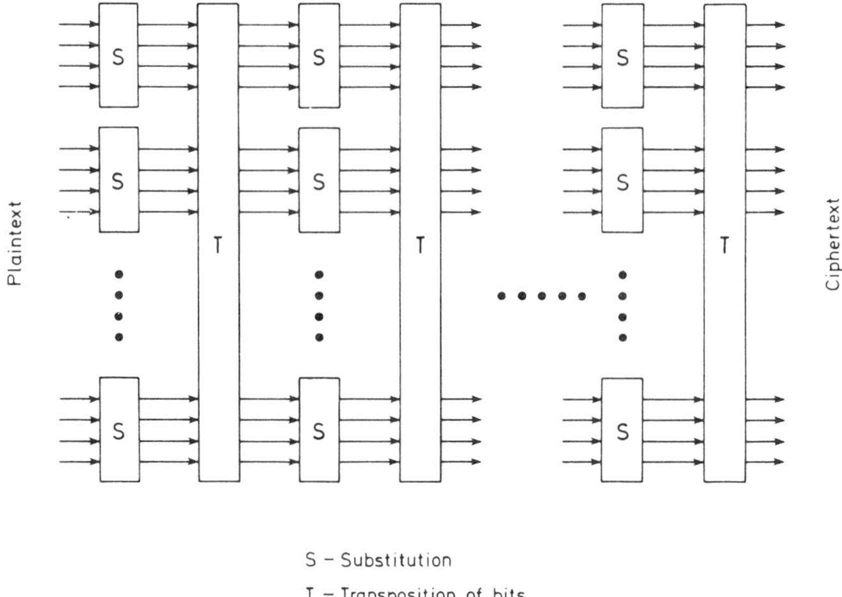

Fig. 3.1 A product cipher with transposition and substitution

encipherment). Decipherment by the possessor of this knowledge is carried out by running the ciphertext data back through the process, using the inverse of each S-box and transposition; without this knowledge a plaintext block should not be deducible from the ciphertext. Greatest strength of encipherment is achieved if the S-boxes used at each stage are different. If we say that the S-boxes are to carry out transformations on groups of 4 bits, then the total number of different S-boxes that can be specified is $(2^4)!$ or about 2×10^{13}, so there appears to be plenty of choice, although few of the choices are actually suitable.

The description of the transformations can be expressed in terms of a key, which should be shared knowledge between communicating parties and must, therefore, be readily transportable. The size of this key can be calculated precisely on certain assumptions. We have already postulated 4-bit S-boxes — now suppose that the transposition should be the same at all steps, and, therefore, need not be specified in the key. Take a plaintext block of 128 bits, the Lucifer block size, then at every stage thirty-two S-boxes will be required. An S-box usually comprises a read-only-memory (ROM) and in this case it stores sixteen words of 4 bits, but the information required to specify an *invertible* transformation of 4 bits is $\log_2(16!) = 44.2$ approximately. If there are 16 stages of transposition/substitution, then the total size of the specification of the transformations, i.e. the key size, will be $32 \times 45 \times 16$ (= 23 040) bits. This is clearly an inconveniently large key.

In the Lucifer cipher only two different S-boxes were specified; the selection of the S-boxes to be used at each substitution step was based on the digits of the encipherment key; the S-box specifications themselves were part of the algorithm and not affected by the encipherment key.

For a 16-cycle set of transformations the key size would need to be 32×16 (=512) bits. In Lucifer the key size was further cut down to 128 bits by using each key bit four times. The key was also involved in a process called 'interruption', giving a modulo-2 combination of selected key bits with the results of the substitution and transposition operations; the cycle followed the order, confusion, interruption, diffusion. The details of the Lucifer operation were public knowledge, as were the specifications of the transpositions and S-boxes. The whole of Lucifer's strength depended on the complex series of substitutions, with interposed transpositions, and on the secrecy of the encipherment key.

The Lucifer block size was 128 bits, with no data expansion in the encipherment process, and the key size was also 128 bits. Thus the number of different mappings of plaintext onto ciphertext under control of the encipherment key was 2^{128}, compared with the total number of ways of mapping 128 bits onto a 128-bit field with is $(2^{128})!$, a very much larger number.

The process of establishing the DES

The IBM reply to the second NBS solicitation for candidate encipherment standards was based on an algorithm of this type. During the months that followed the submission of the algorithm the proposal underwent government scrutiny to determine its suitability as a Federal standard. The National Security Agency (NSA) was called in by NBS to carry out an exhaustive technical analysis of the algorithm. During this time NBS and IBM negotiated the terms under which this piece of IBM intellectual property might be made available to others for manufacture, implementation and use. IBM agreed to grant non-exclusive, royalty-free licences for the manufacture or sale in the USA of DES devices where otherwise this would infringe IBM patents. The agreement was eventually recorded in the 13 May 1975 and 31 August 1976 issues of the Official Gazette of the United States Patent and Trademark Office. In March 1975 the description of the algorithm was issued for general public comment and in August 1975 a draft of the proposed standard was published.

The comments received showed a wide range of views. Many comments concerned the ways in which encipherment should be implemented in various computer architectures. The recommended modes of use of the standard algorithm are examined in chapter 4. Others were directed at the cryptographic strength of the algorithm. We shall return to this subject in the present chapter.

The process of comment, discussion and reply occupied about 18 months, following which the data encipherment algorithm was adopted as a Federal standard on 23 November 1976. It was published as FIPS Publication 46, entitled *Data Encryption Standard*, on 15 January 1977,[6] becoming mandatory 6 months later on 15 July 1977. After that date Federal agencies were expected to comply with its provisions, or show an acceptable reason for not doing so.

The abbreviation DES has achieved very wide currency for the Data Encryption Standard. The algorithm has also been adopted by the American National Standards Institute, where it is known as the Data Encryption Algorithm (DEA). We will now describe the way in which the algorithm carries out the encipherment.

3.2. THE ALGORITHM OF THE DATA ENCRYPTION STANDARD

A logical flow diagram of the DES is shown in Figure 3.2 which encompasses both encipherment and decipherment. The operation has two 64-bit inputs, the plaintext (or ciphertext) block and the encipherment key block, and one 64-bit output, the ciphertext (or plaintext) block. The algorithm transforms plaintext into ciphertext or vice versa, depending on the mode in which it operates. Of the 64 bits in the encipherment key block, only 56 enter directly into the algorithm. In the Federal standard the eight remaining bits take the values for odd parity in each 8-bit byte of the key block.

The building blocks of the algorithm are permutations, substitutions and modulo 2 additions (XOR). Permutations in the DES are of three kinds, straight permutations, an expanded permutation and permuted choices. Figures 3.3 (a), (b) and (c) illustrate the nature of these operations (the permutations actually used are defined in later figures). In a straight permutation bits are simply reordered. In the expanded permutation some bits are duplicated and the whole resulting field reordered. In permuted choices some bits are ignored and the remainder reordered.

Substitutions in the DES are known as S-boxes and are specified by eight different tables. In Lucifer the input and output of the S-boxes were each of 4-bits. In the DES the S-boxes have 6-bit inputs and 4-bit outputs.

The first operation to which the input data is subjected is the Initial Permutation, which has a highly regular character. The effect of this may be seen as an acceptance of data, an octet at a time, in eight shift registers (one bit to each register). As the successive input octets are accepted, their bits are distributed

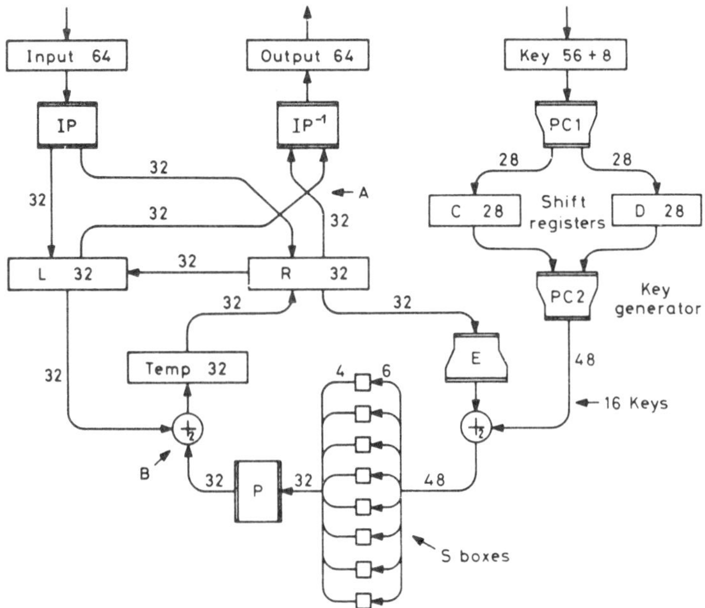

Fig. 3.2 The logical structure of the DES

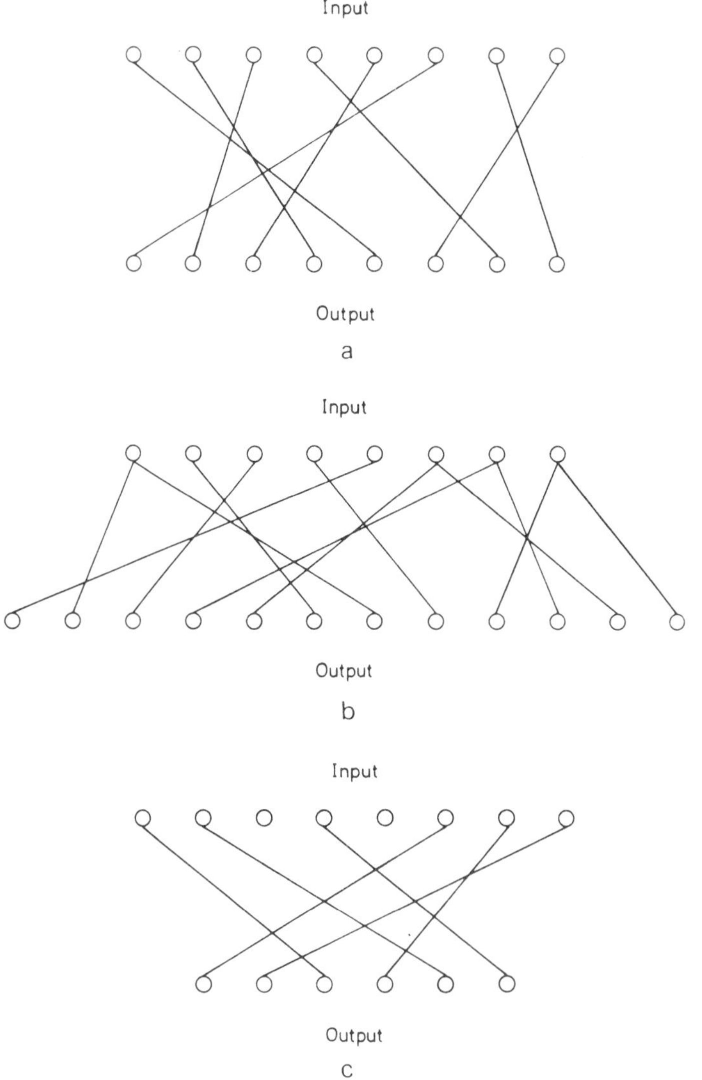

Fig. 3.3 Three types of permutation, a a straight permutation, b an expanded permutation, c a permuted choice

to the eight shift registers. When the input is complete and all 64 bits accepted, the contents of the shift registers are placed in sequence. The exact way in which the bits are permuted is shown in Figure 3.4. In this permutation input bit 1 emerges as output bit 40, input bit 2 emerges as output bit 8, etc.; output bit 1 corresponds to input bit 58, output bit 2 corresponds to input bit 50, etc.

One result of this permutation is that for a code such as American Standard Code for Information Interchange (ASCII), with every eighth bit representing parity, all the eight parity bits of the input block will be collected together into one octet at the beginning of the block, before transposition and substitution begins. There is, however, no evidence that this initial regularity introduces any

58	50	42	34	26	18	10	2
60	52	44	36	28	20	12	4
62	54	46	38	30	22	14	6
64	56	48	40	32	24	16	8
57	49	41	33	25	17	9	1
59	51	43	35	27	19	11	3
61	53	45	37	29	21	13	5
63	55	47	39	31	23	15	7

Fig. 3.4 The Initial Permutation of the DES

weakness into the encipherment operations; this is because of the powerful diffusion properties of the algorithm. It is understood that the algorithm includes this permutation for reasons of convenience in implementation; it has no cryptographic value.

Before the permuted and substituted data is available for reading from the output register at the end of the complete encipherment operation, it is subjected to the exact inverse of the Initial Permutation.

After the Initial Permutation, the 64-bit data field is split into two 32-bit fields which are sent to registers L and R, part of the main loop of the algorithm logic. Then the action of the main loop of the algorithm begins.

From register R the 32 bits are passed to the Expansion Permutation E, which is a regular transformation in which half the bits are replicated, so that the output of the permutation consists of 48 bits. For each of the four octets the first, fourth, fifth and eighth bits are replicated. The full permutation is expressed in Figure 3.5. In this permutation output bit 1 corresponds to input bit 32, output bit 2 corresponds to input bit 1, etc. The other output from register R is a direct 32-bit transfer to register L. We shall see later what happens to the displaced contents of register L.

The 48-bit output from the Expansion Permutation is next passed to an exclusive-OR operation in which it is combined with a selection of bits from the encipherment key. The 48-bit output from the exclusive-OR operation is split into eight sextets for input to the set of eight S-boxes, thus each S-box receives a 6-bit input, but produces only 4 bits as output. The individual S-box look-up tables are as shown in Figure 3.6. The layout of these tables requires further explanation. The

32	1	2	3	4	5
4	5	6	7	8	9
8	9	10	11	12	13
12	13	14	15	16	17
16	17	18	19	20	21
20	21	22	23	24	25
24	25	26	27	28	29
28	29	30	31	32	1

Fig. 3.5 The Expansion Permutation (E) of the DES

		Column															
		0	1	2	3	4	5	6	7	8	9	10	11	12	13	14	15
S1	0	14	4	13	1	2	15	11	8	3	10	6	12	5	9	0	7
	Row 1	0	15	7	4	14	2	13	1	10	6	12	11	9	5	3	8
	2	4	1	14	8	13	6	2	11	15	12	9	7	3	10	5	0
	3	15	12	8	2	4	9	1	7	5	11	3	14	10	0	6	13
S2		15	1	8	14	6	11	3	4	9	7	2	13	12	0	5	10
		3	13	4	7	15	2	8	14	12	0	1	10	6	9	11	5
		0	14	7	11	10	4	13	1	5	8	12	6	9	3	2	15
		13	8	10	1	3	15	4	2	11	6	7	12	0	5	14	9
S3		10	0	9	14	6	3	15	5	1	13	12	7	11	4	2	8
		13	7	0	9	3	4	6	10	2	8	5	14	12	11	15	1
		13	6	4	9	8	15	3	0	11	1	2	12	5	10	14	7
		1	10	13	0	6	9	8	7	4	15	14	3	11	5	2	12
S4		7	13	14	3	0	6	9	10	1	2	8	5	11	12	4	15
		13	8	11	5	6	15	0	3	4	7	2	12	1	10	14	9
		10	6	9	0	12	11	7	13	15	1	3	14	5	2	8	4
		3	15	0	6	10	1	13	8	9	4	5	11	12	7	2	12
S5		2	12	4	1	7	10	11	6	8	5	3	15	13	0	14	9
		14	11	2	12	4	7	13	1	5	0	15	10	3	9	8	6
		4	2	1	11	10	13	7	8	15	9	12	5	6	3	0	14
		11	8	12	7	1	14	2	13	6	15	0	9	10	4	5	3
S6		12	1	10	15	9	2	6	8	0	13	3	4	14	7	5	11
		10	15	4	2	7	12	9	5	6	1	13	14	0	11	3	8
		9	14	15	5	2	8	12	3	7	0	4	10	1	13	11	6
		4	3	2	12	9	5	15	10	11	14	1	7	6	0	8	13
S7		4	11	2	14	15	0	8	13	3	12	9	7	5	10	6	1
		13	0	11	7	4	9	1	10	14	3	5	12	2	15	8	6
		1	4	11	13	12	3	7	14	10	15	6	8	0	5	9	2
		6	11	13	8	1	4	10	7	9	5	0	15	14	2	3	12
S8		13	2	8	4	6	15	11	1	10	9	3	14	5	0	12	7
		1	15	13	8	10	3	7	4	12	5	6	11	0	14	9	2
		7	11	4	1	9	12	14	2	0	6	10	13	15	3	5	8
		2	1	14	7	4	10	8	13	15	12	9	0	3	5	6	11

Fig. 3.6 The S-box tables for the DES

input to each consists of 16 bits, of which the first and the last are taken together as a 2-bit number which has the effect of selecting the row in the S-box table; bits 2 to 5 of the input to each S-box, together giving values in the range 0 to 15, select the table element in the appropriate row (the left bit is most significant); each row of an S-box table represents a bijection from bits 2 to 5 of the S-box input. We shall return to the question of the structure of the S-boxes.

The 4-bit outputs from all the S-boxes are collected together as a 32-bit field for input to the permutation P, which is a rather irregular permutation, quite unlike the regular structures of IP and E. The details of the permutation P are as shown in Figure 3.7; this permutation neither replicates nor removes bits and therefore

16	7	20	21
29	12	28	17
1	15	23	26
5	18	31	10
2	8	24	14
32	27	3	9
19	13	30	6
22	11	4	25

Fig. 3.7 Permutation P of the DES

has a 32-bit output. In this permutation bit 1 of the output corresponds to bit 16 of the input, bit 2 of the output corresponds to bit 7 of the input, etc.

To complete the main loop of the algorithm the output from permutation P is combined with the erstwhile contents of register L in an exclusive-OR operation and the result moved into register R. In order to achieve this operation without a clash on register allocation, a temporary register TEMP is shown in Figure 3.2 between the exclusive-OR and register R.

The cycle of the main loop we have described here is repeated sixteen times before the contents of registers R and L are assembled in a 64-bit block *in the sequence R, L,* then subjected to the inverse of the Initial Permutation already mentioned and transferred to the output register. Note the final swap of registers R and L before the inverse Initial Permutation. The reason will appear shortly. Each cycle of the main loop is usually known as a 'round'.

It now remains to describe how the encipherment key participates in the process. We mentioned that the 48-bit output from the Expansion Permutation E is combined with selected bits from the encipherment key before input to the S-boxes. The selected bits are different for each of the sixteen rounds. At the right-hand side of Figure 3.2 the method of generating these bits is shown. After input of the key to the key register, the key is read into the Permuted Choice 1 (PC1) where every eighth bit is eliminated, otherwise the structure of this permuted choice is very similar to that of the Initial Permutation, probably signifying a similar kind of implementation; the structure of PC1 is given in Figure 3.8. Here bit 1 of the input to register C corresponds to bit 57 of the encipherment key and bit

C	57	49	41	33	25	17	9
	1	58	50	42	34	26	18
	10	2	59	51	43	35	27
	19	11	3	60	52	44	36
D	63	55	47	39	31	23	15
	7	62	54	46	38	30	22
	14	6	61	53	45	37	29
	21	13	5	28	20	12	4

Fig. 3.8 Permuted Choice 1 of the DES

1 of the input to register D corresponds to bit 63 of the encipherment key. The 56-bit field produced as the output from this operation is split into two 28-bit fields and loaded into registers C and D. The latter are cyclic shift registers; for encipherment they are cycled left by either one or two bit positions for each of the sixteen rounds of the algorithm. The schedule of left shifts for encipherment is given in Figure 3.9.

To produce the selection of 48 bits required for combination with the E output in the main loop of the algorithm, the combined contents of C and D are sent to a further Permuted Choice (PC2). This has a rather irregular structure, expressed in Figure 3.10. Bit 1 of the output corresponds to input bit 14 (of the combined C and D), output 2 corresponds to input 17, etc. The output from PC2 consists of 48 bits. For each round of the algorithm the output of PC2 forms the 'sub-key' used in the main loop.

Cycle	*K block*	*Number of left shifts before PC2*
1	K_1	1
2	K_2	1
3	K_3	2
4	K_4	2
5	K_5	2
6	K_6	2
7	K_7	2
8	K_8	2
9	K_9	1
10	K_{10}	2
11	K_{11}	2
12	K_{12}	2
13	K_{13}	2
14	K_{14}	2
15	K_{15}	2
16	K_{16}	1

Fig. 3.9 Schedule of left shifts for encipherment

14	17	11	24	1	5
3	28	15	6	21	10
23	19	12	4	26	8
16	7	27	20	13	2
41	52	31	37	47	55
30	40	51	45	33	48
44	49	39	56	34	53
46	42	50	36	29	32

Fig. 3.10 Permuted Choice 2 of the DES

Cycle	K block	Number of right shifts before PC2
1	K_1	0
2	K_2	1
3	K_3	2
4	K_4	2
5	K_5	2
6	K_6	2
7	K_7	2
8	K_8	2
9	K_9	1
10	K_{10}	2
11	K_{11}	2
12	K_{12}	2
13	K_{13}	2
14	K_{14}	2
15	K_{15}	2
16	K_{16}	1

Fig. 3.11 Schedule of right shifts for decipherment

This description of the structure of the encipherment algorithm is now complete. Decipherment is carried out very conveniently by a similar process, the only difference being in the generation of the sub-keys. For decipherment the cyclic shifting registers C and D are cycled to the right, the opposite sense from that used in encipherment. The schedule of right shifts for decryption is given in Figure 3.11. Note that in encipherment a shift takes place before the first sub-key emerges but not in decipherment. The result is that in decipherment the same set of sixteen sub-keys is produced, but in the opposite sequence.

The ladder diagram

The method of operation of the algorithm can be understood more easily if we consider the ladder diagram shown in Figure 3.12. This illustrates the relationship between the successive rounds of the algorithm. We see how the exchanges between registers R and L are made under the influence of the sub-keys, k_1 to k_{16}; the latter are the 48-bit patterns produced by operating on the main encipherment key; the action of the permutation E, the S-boxes and the permutation P are subsumed in the function F in this diagram. For the sake of economy of space only three of the sixteen rounds are shown.

The ladder diagram also enables us to illustrate quite neatly the relationship between encipherment and decipherment. In Figure 3.13 we show the process of encipherment followed immediately by the process of decipherment. The ladder has been 'untwisted' to make its nature more clear. The symbols R and L refer to the contents of the respective registers at each stage in the process. Note that the sub-keys are presented for decipherment in the opposite order from that in which they were used for encipherment. At each point in the ladder for

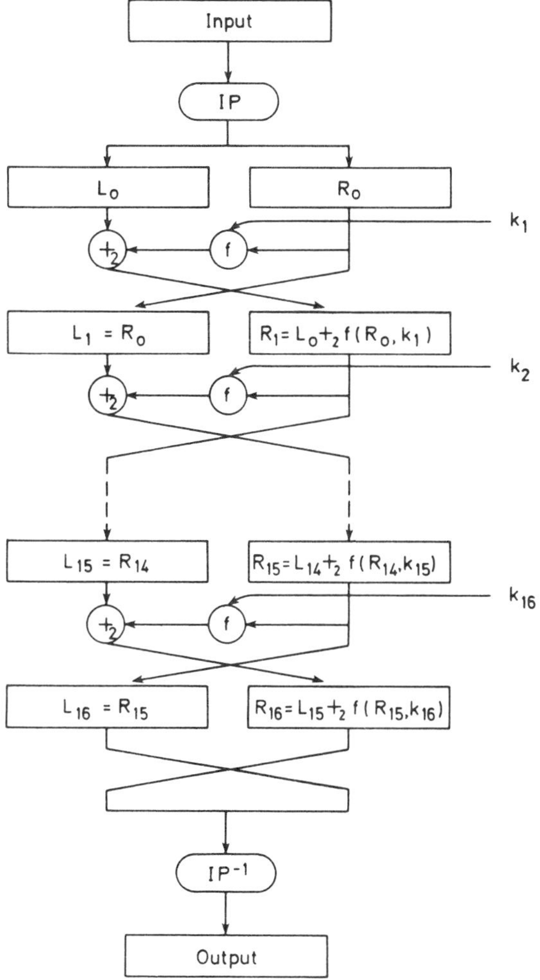

Fig. 3.12 Ladder diagram representation of the DES

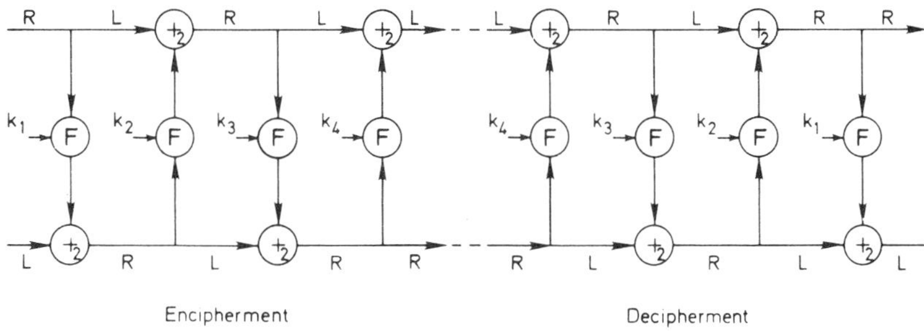

Fig. 3.13 Encipherment and decipherment mirrored in a ladder diagram

encipherment it is possible to establish that the state of the data is exactly the same as that of the corresponding state in the ladder for decipherment. The structure of the encipherment operation is exactly mirrored by the decipherment operation at their mutual boundary.

An algebraic representation

This result may also be expressed in algebraic form. We shall refer to the various rounds by the subscript j, and we shall call P the output of the permutation P. F has the same significance as in the ladder diagram. R and L refer to the contents of their respective registers and R_j, L_j to the contents after the jth round. The sub-key at the jth round is k_j.

During encipherment the following relationships are true

$$L_j = R_{j-1} \tag{1}$$

$$R_j = L_{j-1} +_2 P_j \tag{2}$$

$$P_j = F(R_{j-1}, k_j) \tag{3}$$

where the operator $+_2$ has the significance of exclusive-OR. The second and third of these equations may be combined

$$R_j = L_{j-1} +_2 F(R_{j-1}, k_j). \tag{4}$$

With equations (1) and (4) we now have a complete description of the jth state in terms of the $(j-1)$th state.

By simple rearrangement of equations (1) and (4) we can also write:

$$R_{j-1} = L_j \tag{5}$$

$$L_{j-1} = R_j +_2 F(R_{j-1}, k_j) \tag{6}$$

Substituting from (5) in (6) we get

$$L_{j-1} = R_j +_2 F(L_j, k_j). \tag{7}$$

Thus from equations (5) and (7) we have a complete description of the $(j-1)$th state in terms of the jth state. These equations can be understood as describing decipherment from L_{16}, R_{16} to L_0, R_0, using the sub-keys in the sequence $k_{16}, k_{15}, \ldots k_1$. Note that L and R are interchanged — this is the reason for the swap of L and R at the end of the sixteen rounds of encipherment or decipherment.

Each distinct round constitutes an involution. In decipherment the inverse operation is achieved by taking the same set of involutions in reverse order.

3.3. THE EFFECT OF THE DES ALGORITHM ON DATA

It is the aim of the designers of the DES algorithm that plaintext data processed by the algorithm shall be transformed in such a complicated way that it is not possible to find any correlation between ciphertext and plaintext, nor is it possible to show any systematic relationship between ciphertext and encipherment key. The effect of a change of one bit in an input plaintext block should ideally be to change the value of each individual bit in the output ciphertext block with a probability of one-half. There is little doubt that the designers have been successful in achieving this aim. To illustrate this assertion we shall now examine the way in which examples of plaintext blocks are enciphered by the algorithm. We propose to show that single bit changes of plaintext cause the ciphertext to be changed beyond recognition and that for the same plaintext single bit changes in the encipherment key produce totally different ciphertext.

In the figures that follow, plaintext, ciphertext and keys are all represented in hexadecimal notation. In Figure 3.14 a plaintext block is shown at the head of the table, with, later in the table, other selected blocks which each differ from the first by 1 bit only. All these blocks have then been enciphered under the same key (0123456789ABCDEF). A glance at the corresponding ciphertexts shows that the changes caused by the 1-bit differences in plaintext are large and random. To obtain some measure of the change the Hamming distance is listed of each of the subsequent ciphertext blocks from the first in the table; the mean Hamming distance from the first block measured over all the blocks is 31.06. This is very close to the expected value of 32.

In Figure 3.15 shows the result of enciphering the same plaintext block (ABCDEFABCDEFABCD) with a succession of different keys. At the top of the table we have encipherment with a chosen key, followed by encipherment with keys each of which differs from the first by 1 bit (parity correction has been included, so the appearance is given of a 2-bit difference). Against each subsequent ciphertext the Hamming distance from the first in the table is listed. Once again the mean Hamming distance of 32.88 is very close to expectation.

The DES transformation is thus seen to be efficient in obscuring similarities between plaintext. This can be illustrated even more graphically by examining the progression of the transformation within the algorithm. Figure 3.16 shows the result of encipherment of two plaintexts, which differ by 1 bit, under the same key (08192A3B4C5D6E7F). The contents of the L and R registers at the beginning of each of the sixteen rounds are listed (these quantities are called 'intermediate results'); the overall result of the encipherment transformation is also shown. The progression of the Hamming distance between the intermediate results at each stage can be traced (taking the combined L and R states within each encipherment process). For the first and second intermediate results the Hamming distance is low, but an appreciable distance soon develops and is maintained. By the completion of the fifth internal cycle there is very little correlation between the inputs and the intermediate results. There are many ways to examine this property and they all lead to the conclusion that five rounds would have been just sufficient to remove obvious regularities from the function. It seems that sixteen rounds are more than adequate.

3.4. KNOWN REGULARITIES IN THE DES ALGORITHM

Complementation

A notable feature of the DES algorithm is the property of complementation. If the complement of a plaintext block is taken and the complement of an encipherment key, then the result of a DES encipherment with these values is the complement of the original ciphertext.

Key 0 1 2 3 4 5 6 7 8 9 A B C D E F

	Plaintext	Ciphertext	Hamming distance
1	A B C D E F A B C D E F A B C D	C D E 8 7 2 D 4 A 4 7 1 3 4 6 F	
2	8 B C D E F A B C D E F A B C D	C D 3 D 0 A A 4 C 4 0 2 4 B 4 A	29
3	A 9 C D E F A B C D E F A B C D	8 0 1 F 8 A 2 9 6 8 B C 4 4 7 3	38
4	A B D D E F A B C D E F A B C D	5 D 9 8 C 4 7 D D D B A 6 F 3 0	36
5	A B C F E F A B C D E F A B C D	9 9 8 9 5 6 2 A 8 4 F 4 0 1 C 9	26
6	A B C D 6 F A B C D E F A B C D	6 7 C 2 6 9 F 2 5 4 2 7 9 1 F 9	30
7	A B C D E B A B C D E F A B C D	F 8 C 9 8 F 7 9 A D C 0 6 E A 4	33
8	A B C D F F D B C D E F A B C D	8 7 D 3 2 4 0 A B B F 4 4 0 7 4	34
9	A B C D E F A 9 C D E F A B C D	D B 9 9 8 B 6 7 0 4 6 C D C E 7	30
10	A B C D E F A B E D E F A B C D	2 F 6 E 5 4 7 0 E 4 E 3 5 1 A C	25
11	A B C D E F A B C C E F A B C D	B 5 3 E 4 2 D E 3 0 F 9 7 A D 0	29
12	A B C D E F A B C D 6 F A B C D	4 F 4 0 6 7 7 2 6 B 3 5 B 0 1 4	28
13	A B C D E F A B C D E 7 A B C D	A B 1 5 5 2 8 9 6 6 0 C 6 0 B 2	35
14	A B C D E F A B C D E F 2 B C D	5 B D A 9 3 F 7 D 4 2 7 B 8 D 2	30
15	A B C D E F A B C D E F A F C D	9 8 5 3 C 5 1 1 E D 5 6 8 8 7 E	34
16	A B C D E F A B C D E F A B D D	7 0 A A 2 4 0 7 9 5 9 F 0 4 B 1	34
17	A B C D E F A B C D E F A B C 5	8 9 2 B E C 4 7 C 9 7 1 2 B E 3	26

Mean Hamming distance 31.06

Fig. 3.14 *DES encipherment of a series of data blocks each differing by 1 bit from a chosen block*

Plaintext A B C D E F A B C D E F A B C D

	Key	Ciphertext	Hamming distance
1	0 1 2 3 4 5 6 7 8 9 A B C D E F	C D E 8 7 2 D 4 A 4 7 1 3 4 6 F	
2	8 0 2 3 4 5 6 7 8 9 A B C D E F	1 B 7 3 F E 8 B C 0 B 8 8 6 0 6	35
3	0 2 2 3 4 5 6 7 8 9 A B C D E F	0 F 9 2 F 6 0 D 2 F D 4 D 8 B 7	32
4	0 1 6 2 4 5 6 7 8 9 A B C D E F	3 1 A F D 8 C 5 4 F B F 4 B C D	37
5	0 1 2 6 4 5 6 7 8 9 A B C D E F	C 7 9 F 5 9 6 3 D 4 6 5 A 7 E E	29
6	0 1 2 3 0 4 6 7 8 9 A B C D E F	3 6 5 2 9 C C 1 0 7 1 7 A 3 8 9	39
7	0 1 2 3 4 6 6 7 8 9 A B C D E F	7 F 3 5 F 7 E 6 C E C 5 7 E E 3	30
8	0 1 2 3 4 5 4 6 8 9 A B C D E F	C 9 F 3 F D 9 2 6 0 C 6 8 1 8 A	27
9	0 1 2 3 4 5 6 4 8 9 A B C D E F	E 6 9 2 8 3 2 2 E E 8 B 9 A 6 9	36
10	0 1 2 3 4 5 6 7 A 8 A B C D E F	0 9 9 7 A 5 A F 6 E 4 E 1 4 6 0	37
11	0 1 2 3 4 5 6 7 8 A A B C D E F	C 7 B 4 1 C 4 F 3 8 D 9 A F 7 A	31
12	0 1 2 3 4 5 6 7 8 9 E A C D E F	4 B 4 A A 2 0 A B 2 1 4 4 D D D	30
13	0 1 2 3 4 5 6 7 8 9 A 8 C D E F	1 2 9 9 7 E E 8 0 0 1 B D 2 7 C	32
14	0 1 2 3 4 5 6 7 8 9 A B 4 C E F	4 2 D 1 7 B 7 D F 5 3 4 3 B 7 9	28
15	0 1 2 3 4 5 6 7 8 9 A B C E E F	8 7 F 6 4 9 3 C 4 C 8 9 8 3 2 7	33
16	0 1 2 3 4 5 6 7 8 9 A B C D 6 E	F C 3 0 F 6 F 7 6 B D 4 2 5 9 2	31
17	0 1 2 3 4 5 6 7 8 9 A B C D E C	1 C B D E C 4 B 7 9 B C A 7 A 1	39

Mean Hamming distance 32.88

Fig. 3.15 DES encipherment of a data block with a series of keys each differing by 1 bit from a chosen key

Encipherment with Key 08192A3B4C5D6E7F

Plaintext 1 0000000000000000 Plaintext 2 0000000000000001

Round	Register L	Register R	Register L	Register R	Hamming distance
1	00000000	00000000	00000080	00000000	1
2	00000000	AF0D68FD	00000000	AF0D687D	1
3	AF0D68FD	CE0A36EA	AF0D687D	CE3A32E2	5
4	CE0A36EA	0BDCC5FE	CE3A32E2	81BDED5F	15
5	0BDCC5FE	5D181CC3	81BDED5F	F8CA39B2	26
6	5D181CC3	1744B978	F8CA39B2	A994B918	26
7	1744B978	9B1CB0D8	A994B918	3E9D05D8	22
8	9B1CB0D8	7AE8C7E0	3E9D05D8	23F48DFF	26
9	7AE8C7E0	A2AC7B3F	23F48DFF	9D58DCFB	34
10	A2AC7B3F	580E51EF	9D58DCFB	3F076303	34
11	580E51EF	2865DBD4	3F076303	97AE4AE3	35
12	2865DBD4	BF68F77C	97AE4AE3	68ABB612	37
13	BF68F77C	8F57F629	68ABB612	09D9C398	32
14	8F57F629	0827B240	09D9C398	44B0435C	31
15	0827B240	F2DEBFAC	44B0435C	319FD4B8	29
16	F2DEBFAC	16CD0EB8	319FD4B8	42684FF9	24

Ciphertext 1 25DDAC3E96176467 Ciphertext 2 1BDD183F1626FB43 22

Fig. 3.16 DES decipherment of a block with two keys differing by 1 bit, showing the intermediate results before each round of the algorithm

Thus if

$$y = Ek(x),$$

then

$$\bar{y} = E\bar{k}(\bar{x}).$$

This result may seem surprising, especially when the non-linearity introduced by the S-box transformations is considered. The effect is entirely due to the presence of two exclusive-OR operations, one of which precedes the S-boxes in the logical flow of the algorithm and the other follows the permutation P. If the two input fields to an exclusive-OR operation are complemented, then the resulting output is exactly the same as that obtained for the uncomplemented inputs. Complementation of the plaintext and key inputs to the DES algorithm results respectively in complementation of the output of the expansion permutation E and the output from the sub-key selection procedure (following PC2); these two data fields are the inputs to the exclusive-OR before the S-boxes, so the S-boxes received unchanged data. Their unchanged output enters the exclusive-OR gate which transfers the (complemented) L register to the (complemented) R register.

The complementation property of the DES algorithm does not represent a serious weakness in the security of systems using the DES; there is no reduction in the time taken for an exhaustive key search, given a known plaintext–ciphertext pair. If a chosen plaintext attack can be mounted, then, by exploiting the complementation property, the time taken to exhaust the key domain by exhaustive

```
01 01 01 01   01 01 01 01
FE FE FE FE   FE FE FE FE
1F 1F 1F 1F   0E 0E 0E 0E
E0 E0 E0 E0   F1 F1 F1 F1
```

Fig. 3.17 The weak keys of the DES

search may be reduced by a factor of two, given the plaintext x and the values of $y_1 = \mathrm{E}k(x)$ and $\bar{y}_2 = \mathrm{E}k(\bar{x})$, so that $y_2 = \mathrm{E}\bar{k}(x)$. If all values of x are searched to find if the $\mathrm{E}k(x)$ value equals either y_1 or y_2, then each test covers the two key values k and $\bar{k}$. If this is a matter for serious concern it might be said that steps should be taken to frustrate a chosen plaintext attack, probably by controlling access to the DES function. A system should not enable an opponent to discover both $\mathrm{E}k(x)$ and $\mathrm{E}k(\bar{x})$. From another point of view, it does not appear that any attempt has been made to exploit the complementation property in the design of applications of the DES.

The weak keys

A further regularity in the operation of the DES algorithm is due to the way in which bits are selected from the encipherment key to form the individual sub-keys, k_1 to k_{16}. Clearly, if the encipherment key bits are either all ones or all zeros, then the output from PC2 at each cycle will be the same for all rounds. Therefore, presentation of the sub-keys in the order k_1 to k_{16} will be identical with the order k_{16} to k_1. This means that encipherment and decipherment are exactly equivalent. Thus the effect of the algorithm on a plaintext block with a key of all zeros or all ones will be the same whether encipherment or decipherment is invoked. The result of this is that if a block is enciphered with either of these keys and then re-ciphered with the same key, the original plaintext will be restored; the function is an involution for these values of key. Futhermore, because of the way in which the C and D registers are segregated, having zeros in C and ones in D (or vice versa) will also produce a set of constant sub-keys. For this reason four encipherment keys which have this property can be specified. These are known as 'weak' keys[7] and are listed in Figure 3.17. They should be avoided when encipherment keys are being selected for critical situations, like master keys.

The semi-weak keys

In addition to the weak keys it can be shown that a further set of encipherment keys have a somewhat analogous property. These are the 'semi-weak' keys[8] which occur in pairs. For each pair of semi-weak keys the sets of sub-keys are identical but in the reverse sequence. Thus k_1 for the one encipherment key is equal to k_{16} for the other key of the pair, and so on. The result of encipherment with one key of a pair, followed by encipherment with the other key of the pair, is to restore the original plaintext. There are six pairs of semi-weak keys, which are listed in Figure 3.18. In the semi-weak keys, one or other of the C and D registers contains a pattern of alternate ones and zeros . . . 0101 . . . Their property is due to the symmetry of the shift patterns shown in Figures 3.10 and 3.11.

01	FE	01	FE	01	FE	01	FE
FE	01	FE	01	FE	01	FE	01
1F	E0	1F	E0	0E	F1	0E	F1
E0	1F	E0	1F	F1	0E	F1	0E
01	E0	01	E0	01	F1	01	F1
E0	01	E0	01	F1	01	F1	01
1F	FE	1F	FE	0E	FE	0E	FE
FE	1F	FE	1F	FE	0E	FE	0E
01	1F	01	1F	01	0E	01	0E
1F	01	1F	01	0E	01	0E	01
E0	FE	E0	FE	F1	FE	F1	FE
FE	E0	FE	E0	FE	F1	FE	F1

Fig. 3.18 The semi-weak keys of the DES

The semi-weak keys should also be avoided when selecting encipherment keys, though it is unlikely that they represent such a security risk as do the weak keys. Note that NBS uses the term 'dual keys' to embrace both weak and semi-weak keys.[2]

Further regularities in cycle keys are possible, in which some sub-keys are repeated in the sequence. In 1983 an account[7] was published in the journal *Science* of an alleged analysis claiming to show that patterns arising in the cycle keys led to serious deficiency in the strength of the DES algorithm. Here the term 'weak' was extended to include the keys we have called weak and semi-weak, together with other keys exhibiting regularity of cycle key pattern. It was asserted that cryptanalysis of the algorithm with any key was possible in 8 hours and that with the strongest of a general class of weak keys cryptanalysis should be possible in 4 hours (the 'worst-case' from the cryptanalyst's point of view). The 8-hour assertion is not dissimilar to that made by Diffie and Hellman and referred to later in section 3.5; note that this assertion demands the existence of a very large and powerful cryptanalytic machine. The size of the general class of weak keys was not defined in the published article, but it was said to include 25 different categories. Claims such as were made in the *Science* article would certainly be very damaging to any confidence that users might have in the security of the DES algorithm. We have seen no evidence to suggest that the claims made can be substantiated. In section 3.6 we discuss some of the academic studies that have been applied to the DES; none of the reported results makes claims remotely like those of the *Science* article. Indeed the paper by Moore and Simmons, discussed in section 3.6, gives an account of a detailed analysis of the significance of regularities in the cycle keys.

Hamiltonian cycles in the DES

There remains to be described a regular feature in the permutations of the DES algorithm.[8] We have seen how the strength of the whole algorithm depends on the complexity of the function resulting from the iteration round the loop (R)ESP(R), where (R) represents the contents of the R register, E the expansion permutation, S the S-boxes and P the permutation of that name.

Consider now the transposition component of the functional loop, that represented by the E and P permutations. Since an arbitrary renumbering of the bits of both L and R is unimportant, it is not the individual permutations P and E that are significant but their product, the operation P followed by E (taken in this order because the path from S-box back to S-box takes them in this order).

The identification of the inputs and outputs of an individual S-box is also mostly insignificant. All the outputs are of equivalent status. The inputs, due to the nature of E, divide naturally into three classes. Let us denote the six input bits of an S-box as ABCDEF. The EF values of one S-box equal the AB of its neighbour to the right, for example in the case of S-boxes 1 and 2, E1 = A2 and F1 = B2. For our purpose, input bits A and B are not distinguishable because a permutation of the bits of the L and R registers could interchange any pair of these, interchanging also the corresponding EF pair.

The significant feature of the product of the P and E permutations is therefore not the bits they interconnect but the groups AB, CD and EF which they join. These can be tabulated as in Figure 3.19, showing for each of the S-box outputs which S-box input it affects, by way of the PE permutation. For example S-box output bit 1 of S1 affects F2 and B3 on boxes S2 and S3 respectively.

The way in which the outputs have been arranged in the DES obscured the pattern which is already evident, namely that of the four S-box output bits, two go round the loop to C or D inputs and two go to A, B, E or F inputs. The pattern can be shown best as a graph with directed arcs pointing from S-box output to S-box input. Figure 3.20 shows graphs separately for the CD and EF inputs. The nodes of the graphs are the eight S-boxes.

The graph in Figure 3.20 for CD can be expressed as two Hamiltonian cycles (87654321) and (84725163). The graph for EF has an identical form reflected about a vertical line. Both are very regular graphs which seem likely to have been a design intention. The apparent randomness of the permutation P is due only to the way that S-box inputs and outputs have been identified.

If the corresponding graph for the AB inputs is drawn on a pattern similar to those for CD and EF, then Figure 3.21(a) is produced which is much less regular than Figure 3.20. It can, however, be redrawn to show (Figure 3.21(b)), a less-confused structure.

The permutations in DES are a fixed feature, not subject to the key, and their purpose is to diffuse the effect of the substitutions so that a single bit change, if it expands to four potential bit changes after the first round, will (potentially)

S1				S2				S3				S4			
f2	f4	d6	d8	f3	e7	c1	c5	e6	e4	c8	c2	c7	e5	c3	f8
b3	b5			b4	a8			a7	a5				a6		b1

S5				S6				S7				S8			
e2	c4	f6	d1	e1	f7	d3	d5	e8	e3	c6	d2	f1	d7	d4	f5
a3		b7		a2	b8			a1	a3			b2			b6

Fig. 3.19 Interconnections between S-box outputs and inputs

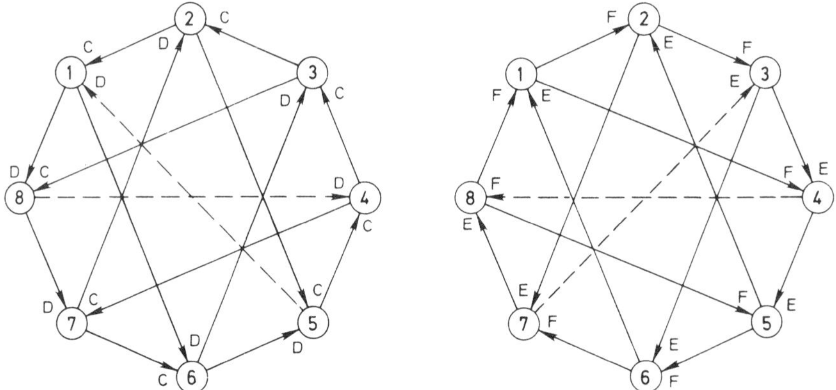

Fig. 3.20 Directed arcs for S-box CD inputs and EF inputs

affect the inputs to four S-boxes in the next round. We say (potentially) because it would be wrong to expect all the S-box outputs to change value — approximately half of them should do so. Also, some bit changes in the S-box outputs will affect two S-boxes on the next round, due to the expansion permutation. The aim of the permutation design (as reflected in our graphs) should be to spread the bit changes as rapidly as possible to all boxes. In this respect the CD and EF graphs are excellent, the AB graph less so. We can speculate that such considerations as these led to the remarkable regularity of two of the graphs and that the third graph is an unavoidable consequence of the first. Since they represent 'spreading' processes which operate in parallel, the good structure of two of the graphs is not impaired in any way by the third one. We understand that these permutating structures, though they were the consequence of a quest for good spreading, were not designed as regular structures except perhaps for the outer Hamiltonian cycles which are fairly obvious. Note also that our principle of ignoring the identity of S-box inputs and outputs, though it cannot be questioned for outputs, does cut across the stated design of the S-boxes as four separate bijections.

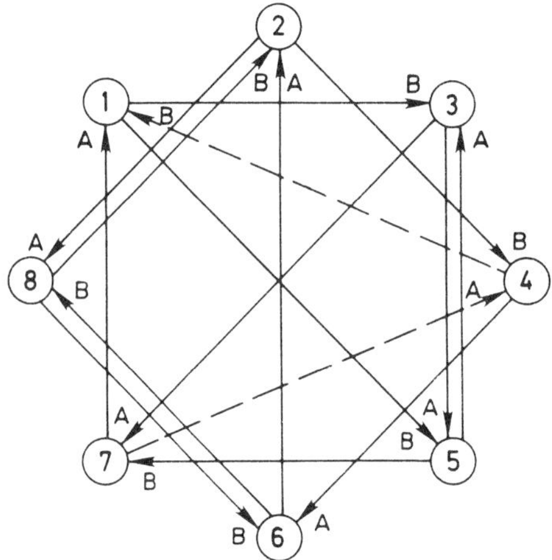

Fig. 3.21a Directed arcs for S-box AB inputs

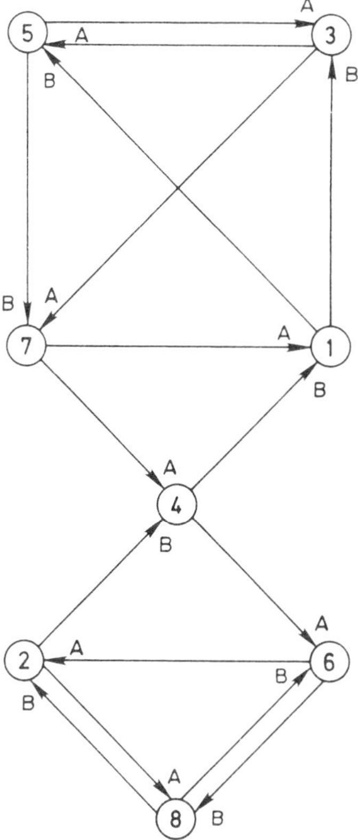

Fig. 3.21b Figure 3.21a re-drawn

3.5. ARGUMENT OVER THE SECURITY OF THE DES

When the DES algorithm was first published in the spring of 1975, a considerable amount of criticism was rapidly generated.[9] This was based on two premises — the length of the encipherment key was considered by some to be inadequate and it was alleged that 'trapdoors' or deliberate weaknesses might have been built into the S-box structures. These two criticisms are considered in turn.

Exhaustive search for a DES key

The length of the encipherment key is important when the security of the system against exhaustive search for the key is evaluated. For exhaustive search the cryptanalyst needs corresponding pairs of plaintext and ciphertext blocks; these may have been obtained by some breach of security in handling legitimate texts (a known-plaintext attack) or by deliberate introduction of text into the encipherment device by the opponent (a chosen-plaintext attack). If partial knowledge of the plaintext is all that is available to the cryptanalyst, perhaps in the form of knowing that the plaintext characters enciphered have odd or even parity, then his task is greater. In this case the search must list all the keys which yield possible plaintexts obeying the parity constraints; it is then necessary to repeat the operation with other ciphertext blocks until the candidate keys have been narrowed down to one only.

Proceeding then on the assumption that the attacker has somehow obtained matching blocks of plaintext and ciphertext, the next step is to test all possible keys by enciphering the plaintext in turn with each and comparing the result with the known ciphertext; the process is known as a 'key search'. When a match is obtained between the produced and known ciphertext, then the current key value is that which is being sought. Clearly the time taken to exhaust the whole of the key domain will depend on the time taken to carry out the DES encipherment. If we assume that the time taken for the encipherment is 100 ms (the speed of the slowest microprocessor implementation known to us), that other processing times may be neglected, and that only one device is used, then we can calculate the time required to test all possible keys to be 7.2×10^{15} s, or about 228 million years. If we assume that the encipherment device is somewhat faster, with an operation time of 5 μs (a fraction slower than the fastest LSI implementation widely available), then the total time required to exhaust the key domain is just over 11 000 years. Clearly these times are totally impracticable for a cryptanalyst.

The exhaustive search times only become practicable when we postulate key searching which uses many DES devices in parallel, each of which is searching a different part of the key domain. Diffie and Hellman[10] have considered in some detail the design of a hypothetical machine constructed for this purpose. They assume that the operation time for each DES device is as low as 1 μs; this is justified on the dual grounds of advancing technology and economy of input/output. Since each special-purpose DES device need only be loaded once with the known matching pair of plaintext and ciphertext and with the encipherment key, then the overheads normally associated with input/output operations need not be taken into account. The only output from the device occurs when the required key has been found, when the fact of its recovery and its value will need to be output. Diffie and Hellman further assume that 1 million such DES devices can be assembled into one machine.

On these assumptions the time taken to exhaust the key domain can be recalculated; the time is reduced to about 72 000 s or just over 20 h. On average a key is found after testing half of the key domain, so the average time to find a key should be about 10 h. Evidently this time is low enough for users of the DES to be worried about the security of their data. We must, therefore, consider whether the assumptions made in arriving at this figure were reasonable.

Concern about the suggestion that an exhaustive key search machine was possible led NBS to hold a workshop 'on estimation of significant advances in computer technology' in August 1976.[11] This brought together experts who produced a range of estimates for the cost of the machine. The outcome from this meeting gave it a cost of $72 million and an expected availability date of 1990.

It was alleged that the mean time between failure of LSI devices is such that one could not run the machine without failure for more than a very few hours. Diffie and Hellman counter this by proposing a system of self-checking with an automatic switch-over to a reserve machine segment when failure is discovered. The cost of the machine has been estimated by Diffie and Hellman to be about $20 million, based on a unit cost for each DES device of about $10; IBM have put the cost at more like $200 million, based on a different unit cost. Diffie and

Hellman quote the total power consumption of the machine to be about 2 MW; another estimate that has been given is 12 MW.

On the assumption that the cost of the machine is amortized over a period of 5 years, Diffie and Hellman estimate that the cost per solution is about $5 000.

It is evident that the construction of an exhaustive-search machine is a very large undertaking, requiring very large resources. It has been conjectured[12] that the reason that the key length chosen was no more than 56 bits was in order that the key search machine should be viable. If the key length were 64 bits, an increase in key search times of a factor of 256 would be obtained; the parity bits do not seem to serve a particularly useful purpose. Against this it has been argued that excessive security is unnecessary for the intended field of application (unclassified, commercial, etc.) and that an increase in key length would increase the cost of the DES chips.

Multiple DES encipherment

It has at various times been suggested[13] that multiple-block encipherment under different keys would increase the security of enciphered data. It might be thought that an effective key length of 112 bits could be obtained by enciphering twice with different keys. However, a time-memory trade-off reduces the effective key length to 57 bits.[14] This is based on the assumption of a known plaintext attack in which the known plaintext is enciphered under all possible keys, then the known ciphertext is deciphered under all possible keys, giving two sets of data. These data are sorted to order and then compared one against the other, so that candidate key pairs can be identified. Note that the data unit for the sorting process consists of two elements — the result of the DES operation (encipherment or decipherment) and the key with which the result was obtained; this amounts to 120 bits per item. Since it is likely that a large number of candidate key pairs will be found (the theoretical number is 2^{48} false key pairs), it is necessary to check the candidate pairs against other pairs of plaintext and ciphertext blocks. Given a second known plaintext block and its ciphertext, the probability of finding a wrong key pair in addition to the correct pair is theoretically 2^{-16}. This attack has been called 'meet in the middle'.

Successive use of two encipherment keys can be made more secure by applying them in a more complex way. IBM use encipherment with key 1, followed by decipherment with key 2, followed by encipherment with key 1.[15] This is very much stronger and is used by IBM for protection of master keys in some of their cryptographic systems. A convenient feature of this system is that compatibility with single-key encipherment can be achieved by making the two keys have the same value. Even the degree of protection given by this triple operation has its critics,[14] Merkle suggests that a chosen plaintext attack will yield the two keys with about 2^{56} operations and 2^{56} words of memory. Note that this attack requires a chosen plaintext, whereas the attack on simple double encipherment with two keys requires only a known plaintext. Indeed the amount of chosen plaintext material required corresponds to 2^{56} DES operations; availability of this amount of material is so unlikely that the method is not practical.

Merkle suggests that a worthwhile increase in security can be obtained by triple use of the algorithm with three different keys (triple encipherment was put forward as early as 1976 by Diffie and Hellman); this involves encipherment with the first key, decipherment with the second and encipherment with the third — if compatibility with the IBM two-key scheme is required, then this can be achieved by making the first and second keys of equal value.

Trapdoors in the DES?

The criticism of the DES algorithm has also concerned the choice of S-boxes. Hellman and others[16] have investigated the S-box structure and have shown that the security of a DES-like algorithm can be reduced by careful choice of S-boxes. By replacing the DES S-boxes by others of their own design, they have shown that it is possible to weaken the security of the encipherment while concealing the weak S-box structure to some extent. The assertion is then made that the actual DES S-boxes may themselves contain deliberate weaknesses put there in order to make the algorithm subject to a 'trap-door' attack by those in possession of the design information. On the other hand Hellman and his colleagues have not been able to demonstrate weaknesses in the DES S-boxes, though they point to certain unexplained systematic structure.

The permutation regularity which we discussed in the previous section of this chapter is one of construction. It does not, to our knowledge, result in any regularity of the DES function. The purpose of the permutations is to spread any change among the S-boxes in as few rounds as possible. For this purpose the EF and CD graphs are well chosen. For example a change in the input of S1 spreads to 6 and 8, and, from then on, the S-boxes affected in successive rounds are 3457, then 12345678, by which time all S-box inputs have been affected — on three rounds only. Since only half of the 64 bits circulating in the L and R registers are changed at each round, in practice six rounds are needed to spread a single bit change completely. The chosen graphs CD and EF are very efficient in this respect. Reference to Figure 3.16 where the intermediate data are shown for each round confirms this effect.

If the DES designers at IBM and NBS were able to set out their design criteria, then the kind of criticism which has been outlined in this section might well be allayed. However, at the request of NSA the design criteria have been withheld on the grounds that the designers came across principles of cryptographic design which were considered to be of importance to national security. The same ban on publication has been applied to the results of an IBM study of cryptanalysis of the DES, said to have occupied seventeen man-years.

Senate investigation of the DES

Because of these bans on publication, the controversy of criticism of the DES has been fuelled, rather than subdued. It has been alleged that NSA has exerted undue influence in the matter. Allegations of this kind led the US Senate Select Committee on Intelligence to investigate the role played by various government bodies during the design and development of the DES. The report of this committee is unavailable as classified material, but an unclassified summary[17] of its findings was issued in 1979. NSA was exonerated from tampering with the design of the DES in any way. It was said to have convinced IBM that a shorter key was adequate, to have indirectly assisted in the development of the S-box structures and to have certified that the final DES algorithm was, to the best of their knowledge, free of any statistical or mathematical weakness. The Senate committee reported that the overwhelming majority of scientists consulted felt that the security of the DES was adequate for the time span for which its use was planned.

3.6. RECENT ACADEMIC STUDIES OF THE DES AND ITS PROPERTIES

During the public existence of the DES there has been a steady level of academic interest in its security and general properties; this has led to publication of a number of interesting results, some of which have only emerged relatively recently. A research group at the University of Louvain has published a series of papers,[18,19] giving an account of several investigations. In 1983 it was shown, using various equivalent representations of the DES algorithm, that is was possible to specify a totally iterative representation of the DES. It was also shown that the algorithm could be described with one fewer internal table and that it was related to the mixing transformation described by Shannon in his classical paper; a link was also established to nonlinear feedback shift register theory and to permutation networks. A successor paper in 1984 studied the degree of dependence of the output of the algorithm on its input, with particular attention being paid to small avalanche characteristics. The 1984 paper drew attention to some newly discovered properties of the S-boxes; the studies included tests with some S-box input bits fixed and observations made of the effect of varying the other inputs, whilst another test included complementation of certain input bits. Attention was also paid to the key scheduling part of the algorithm, reducing the number of internal cycles in order to simplify the study. Links between the S-box structure and the performance of the algorithm were identified. The paper is full of minor results, but the authors make no claim to have made cryptanalysis of the algorithm any easier; they do assert that a better understanding of the algorithm has been achieved.

Two papers[20,21] published in 1985 by the MIT group led by Ronald Rivest gave particularly interesting results. The general aim of the work was to determine whether the DES algorithm represented a group. The basis for this aim was the suggestion that if the set of permutations describing the DES transformation was closed under functional composition, then the DES would be vulnerable to a known-plaintext attack with 2^{28} steps on average, a possibly serious weakness. The results obtained by Rivest and his colleagues show, with an acceptable level of confidence, that the DES is not a group, therefore the basic supposition of DES vulnerability on this score is not substantiated. However, in the course of experiments carried out in this connection, a really surprising result was noted. The particular work related to cycling tests on the DES for keys selected in various ways, random keys and alternate encipherment with all zeros and all 1s as keys. A cycling test is designed to discover when, after a series of operations of the algorithm beginning from a recorded starting value, an output value is reached which has the same value as the start point. Encipherment with alternate keys of all zeros and all 1s produced a cycle length of rather less than 2^{33} operations; this was far less than that predicted according to analysis. It was estimated that the probability of such an occurrence was less than 1 in 10^9 if the permutation was chosen at random from the symmetric group on message M. The result was very striking and caused much comment. The true explanation of the phenomenon was given by Coppersmith,[22] who showed that, given a weak key, there were 2^{32} fixed points for the DES; a fixed point represents a plaintext which is not altered by a cryptographic transformation — the ciphertext output of the algorithm is the same as the plaintext input. Each weak DES key has 2^{32} associated fixed points. If one pictures a cycling test as tracing a path through message space, it is immediately apparent that a fixed point causes this path to be reversed. It is the reversal of the path through message space that causes the start point to be reached much earlier than would be expected. Again, this result does not aid the task of the cryptanalyst. The discovery of the fixed points makes the case for avoiding weak keys in practical applications all the stronger.

Another interesting empirical result was reported by Shamir[23] in his paper given at Crypto '85; this showed some remarkable structural patterns in the DES S-boxes. If one divides the S-box outputs into those containing an even number of 1s and those containing an odd number of 1s (in other words a classification based on the parity of the outputs), the distribution of this classification in the S-box rows is very striking. In almost all S-box rows there is a bias of output parity either to the right or left half of the row; in most cases this bias is very marked, the phenomenon being most apparent in S-boxes 1, 5 and 8. The input element which selects the right or left half of an S-box row is the 'B' bit, using the notation expressed in section 3.4. The conclusion is that the 'B' input bit has a very strong influence on the parity of the output pattern. Whilst Shamir did not explain the significance of the phenomenon in his paper, he did consider that the unexpected result demonstrated the deficiencies of current certification techniques, arising from a lack of deep understanding of the nature of the DES algorithm. Having discussed the result with Brickell and Coppersmith, Shamir suggests that the observed properties of the S-boxes could be an unintentional consequence of some of the design criteria. Shamir also notes that the phenomenon was also reported independently by Franklin in a Berkeley MSc dissertation.

Further illustration was thrown on the effect of regularities in the DES cycle keys by Moore and Simmons[24] in 1987. Their study was prompted in part by the 1985 results of Rivest and his co-workers, followed by the interpretation put by Coppersmith on these results. Moore and Simmons confirm the explanation advanced by Coppersmith; they identify a class of cycles of the kind noted by Rivest and call these 'Coppersmith cycles'. A formal derivation is provided of the set of inverse DES key pairs that comprises the weak and semi-weak keys as defined earlier in this chapter. The related phenomenon of palindromic and anti-palindromic keys is investigated. Palindromic keys are those which give the same sub-key sequence whether they are read from sub-key 1 to sub-key 16 or from sub-key 16 to sub-key 1, in other words sub-key 1 is equal to sub-key 16, sub-key 2 to sub-key 15 and so on; only the weak keys have this property. Anti-palindromic cycle keys are those where sub-key 1 is the complement of key 16, sub-key 2 is the complement of sub-key 15 and so on; four of the twelve semi-weak keys have this property. Another interesting result is a proof of the assertion that each of the weak keys has exactly 2^{32} fixed points; similarly it is proved that the four anti-palindromic semi-weak keys have exactly 2^{32} 'anti-fixed' points, where the result of encryption is the exact complement of the plaintext.

The second part of the Moore and Simmons' paper is an extensive analysis of nine classes of cycles; it is beyond the scope of the present account to cover this further work, but it can be recommended to the interested reader. The main feature from this paper that we stress is that, apart from the specification of the fixed and anti-fixed points for eight specific keys, none of the results of the study seems to be of any use to a cryptanalyst faced with the task of deciphering a ciphertext which has been enciphered with an unknown key.

One of us (DWD) in an unpublished work has analysed an attack on the DES which exploits the correlation between two bits of S-box input in neighbouring S-boxes, due to the E permutation. In principle this provides inferences concerning the bits of the key if enough plaintext–ciphertext pairs are given. The information available from this method would amount to 16 bits of key and greatly weaken the cipher. The analysis of the effect shows that the quantity of plaintext–ciphertext pairs needed is comparable with 2^{56}, so the method does not improve on a systematic key search. There is slight evidence that the S-boxes might have been constructed to reduce this effect, but this is not conclusive. The analysis was extended to include three adjacent S-boxes, with the same result.

3.7. IMPLEMENTATIONS OF THE DATA ENCRYPTION STANDARD

Since the announcement and official launch of the DES in 1975 a number of manufacturers of microprocessors and LSI devices have designed, made and marketed devices embodying the DES. The performance spectrum has been extremely wide, ranging from inexpensive microprocessor implementations capable of carrying out the algorithm in 100 milliseconds to very fast LSI implementations which perform the algorithm in less than 5 microseconds. Some of the devices simply implement the algorithm in its basic block mode, whilst others also offer the more complex modes of operation which we shall encounter in our next chapter. The physical protection given to the devices also covers a very wide range, from simple chips with no protection to tamper-resistant boxes designed to withstand determined mechanical and/or electronic attack. There has therefore been available a very wide spectrum of equipment embodying the DES algorithm.

Not all the DES devices announced by manufacturers are still available. There is an observable tendency for system designers to choose fast LSI devices rather than the slower microprocessor implementations; the cost differential is no longer as great as it once was. Some of the slower implementations may not now be available as they are no longer considered commercially viable and have been withdrawn. Equally there is little doubt that some DES implementations have been extremely popular and these are found in many commercial systems for ensuring data security.

3.8. THE IBM CRYPTOGRAPHIC SCHEME

International Business Machines have developed a sophisticated data security system, using the DES, which caters for data in transit between terminals and hosts (and between hosts) and in storage within hosts. This system, which is particularly noteworthy for its elaborate key management facilities, is discussed in chapter 6. It is an integral part of Systems Network Architecture (SNA) and may be used within the operating systems of IBM 360 and 370 computers. Within the host computers the DES capability is provided either by the Programmed Cryptographic Facility Program Product (i.e. in software) or by the 3848 Cryptographic Unit (a hardware device); the 3848 requires the Cryptographic Unit Support Program Product to control it. At the terminals the DES is implemented in special hardware.

The maximum data transfer rate of the 3848 is 1.5 Mbytes/s. The unit enciphers in CBC mode, using 64-bit data blocks. The encipherment operation is said to be faster than the input/output channel can handle data, so it appears that the equivalent block encipherment time is less than 5.3 μs. Since the unit is not intended for use on insecure sites, it is not heavily armoured. However, the master key storage is arranged to delete the keys if attempts are made to access this illegally.

A two-level key hierarchy is operated; the keys are classified as master and operational. Entry of master keys into the 3848 is carried out using the same design of key entry module as that used for the 3845 and 3846 tamper-resistant boxes. Entry of the master key into the 3848 requires operation of a physical lock. The master key is double length, 128 bits (112 effective), in order to enhance the protection given by master key encipherment to the operational keys. Operational keys are protected by multiple DES operations under the control of the two halves of the master key; if we call the latter ka and kb, each of normal length, then the encipherment of operational keys follows the sequence

Encipher under ka
Decipher under kb
Encipher under ka.

Compatibility of the 3848 device with single-key encipherment is obtained by giving identical values to ka and kb. Examination of the encipherment sequence just quoted shows that this has the effect of encipherment once with the identical key value; thus the security given to the operational keys is reduced.

The same scheme of treble encipherment with two keys has been adopted in the standards ANSI X9.17 and ISO 8732 described in chapter 6.

It is recommended that ka and kb be chosen by some random method such as coin tossing. The keys entered must have correct parity, as the 3848 is programmed to reject keys with incorrect parity; calculation of parity depends on the person generating the keys. As a further degree of protection the unit is programmed to reject any of the weak keys should these be input by the operator. As the IBM key management system uses variants of the master key, calculated by inverting selected bits, the unit also rejects variants of the weak keys. Tests for the semi-weak keys are not carried out.

3.9. CURRENT STATUS OF THE DES

We have discussed in this chapter the early development of DES in the field of US federal and national standardization. Before we leave the subject it is appropriate to consider the current status of the algorithm.

The algorithm has been reviewed and re-certified by the US federal authorities for use in the US private commercial sector every five years since its official launch in 1977.

A recent change of US policy towards the algorithm can be traced back to the issue in 1984 of US National Security Decision Directive 145 (NSDD-145). Under this directive a Systems Security Steering Group, with members from some of the major US departments of government, was enjoined by the US President to encourage, advise and assist the US private sector in achieving protection of some communications. The executive agent for this group is the Director, National Security Agency (NSA), acting as National Manager for Telecommunications Security and Automated Information Systems Security.

Arising directly from NSDD-145 a Commercial COMSEC Endorsement Program (CCEP), already implemented by NSA in the context of protection of government information, was made available to the private sector. Under this program communication security modules have been created that embody cryptographic algorithms designed by or for NSA; mechanisms have been set up for supplier approval. These NSA algorithms come in two types, I and II. Type I is used for the protection of classified government information. Type II was originally intended only for the protection of unclassified but sensitive national security information, but its application is now extended to cover the commercial private sector. Neither Type I nor Type II algorithms will be published, nor will either be available for export from the USA.

One may now ask where this leaves the DES and its users. It was originally intended that new equipment using the DES would not receive endorsement for government purposes after 1 January 1988; existing endorsed equipment was to be allowed to continue in use for an indefinite period. When this policy was announced it was perceived that it might cause problems for the US Treasury program for certification of electronic funds transfer authentication equipment. Indeed the US banks were known to be using DES very widely for protection of funds transfers of all kinds, some of which involved international traffic. Arising from concerns of this kind, discussions took place that resulted in a revision of policy regarding the DES. On 22 January 1988 the US Department of Commerce issued a press release in which the National Bureau of Standards reaffirmed use of the Data Encryption Standard for another five years, saying that it continued to be a sound method of protecting computerized data. However, for federal non-financial applications the DES must now be regarded as superseded by the algorithms developed under CCEP.

Fuller discussions of these developments can be found in Newman and Pickholtz[25] and in Howe and Rosenberg.[26] We shall consider the wider impact of this US government policy on the international use of DES in chapter 11 on security standards.

REFERENCES

1. Davis, R.M. 'The Data Encryption Standard in perspective', *Computer Security and the Data Encryption Standard,* National Bureau of Standards Special Publication 500-27, February 1978.
2. National Bureau of Standards *Guidelines for implementing and using the NBS Data Encryption Standard,* Federal Information Processing Standards Publication 74, April 1981.

3. Feistel, H. 'Cryptography and computer privacy', *Scientific American*, **228**, 5, 15, May 1973.
4. Smith, J.L., Notz, W.A. and Osseck, P.R. 'An experimental application of cryptography to a remotely accessed data system', *Proceedings of the ACM Annual Conference*, 282, August 1972.
5. Shannon, C.E. 'Communication theory of secrecy systems', *Bell System Technical Journal*, **28**, 656, October 1949.
6. National Bureau of Standards *Data Encryption Standard*, Federal Information Processing Standards Publication 46, January 1977.
7. Kolata, G. 'Flaws found in popular code', *Science*, 219, 369, 28 Januiary 1983.
8. Davies, D.W. *'Some regular properties of the 'Data Encryption Standard' algorithm', Advances in Cryptology. Proc. Crypto '82*, 39, August 1982.
9. Branstad, D., Gait, J. and Katzke, S. *Report of the Workshop on Cryptography in Support of Computer Security*, National Bureau of Standards, Report NBSIR 77-1291, September 1977.
10. Diffie, W. and Hellman, M.E. 'Exhaustive cryptanalysis of the NBS Data Encryption Standard', *Computer*, **10**, 6, 74, June 1977.
11. Meissner, P. (Ed.) *Report of the Workshop on Estimation of Significant Advances in Computer Technology*, National Bureau of Standards, Report NBSIR 76-1189, December 1976.
12. Hellman, M.E. 'DES will be totally insecure within ten years', IEEE Spectrum, **16**, 7, 32, July 1979.
13. Diffie, W. and Hellman, M.E. 'Privacy and authentication: an introduction to cryptography', *Proc. IEEE*, **67**, 3, 397, March 1979.
14. Merkle, R.C. and Hellman, M.E. 'On the security of multiple encryption', *Comm. ACM*, **24**, 7, 465, July 1981.
15. Tuchman, W. 'Hellman presents no shortcut solutions to the DES', *IEEE Spectrum*, **16**, 7, 40, July 1979.
16. Hellman, M., Merkle, R., Schroeppel, R., Washington, L., Diffie, W., Pohlig, S. and Schweitzer, P. *Results of an initial attempt to cryptanalyze the NBS Data Encryption Standard*, Stanford University, Centre for Systems Research, Report SEL 76-042, November 1976.
17. 'Senate Select Committee on Intelligence' *New York Times*, 13 April 1978; *Computerworld*, 17 April 1978.
18. Davio, M., Desmedt, Y., Fossprez, M., Govaerts, R., Hulsbosch, J., Neutjens, P., Piret, P., Quisqater, J-J., Vandewalle, J. and Wouters, P. 'Analytical characteristics of the DES'. *Advances in Cryptology. Proc. Crypto '83*, 171, August 1983.
19. Desmedt, Y., Quisquater, J-J. and Davio, M. 'Dependence of output on input in DES: small avalanche characteristics', *Advances in Cryptology. Proc. Crypto '84*, 359, August 1984.
20. Kaliski, B.S., Rivest, R.L. and Sherman, A.T. 'Is the Data Encryption Standard a group?', *Advances in Cryptology. Proc. Eurocrypt '85*, 81, April 1985.
21. Kaliski, B.S., Rivest, R.L. and Sherman, A.T. 'Is the DES a pure cipher? (results of more cycling experiments on DES)', *Advances in Cryptology. Proc. Crypto '85*, 212, August 1985.
22. Coppersmith, D. 'The real reason for Rivest's phenomenon', *Advances in Cryptology. Proc. Crypto '85*, 535, August 1985.
23. Shamir, A. 'On the security of the DES', *Advances in Cryptology. Proc. Crypto '85*, 280, August 1985.
24. Moore, J.H. and Simmons, G.J. 'Cycle structure of the DES for keys having palindromic (or antipalindromic) sequences of round keys', *IEEE Trans. on Software Engg*, **SE-13**, 2, 262, February 1987.
25. Newman, D.B. and Pickholtz, R.L. 'Cryptography in the private sector', *IEEE Communications Magazine*, **24**, 8, 7, August 1986.
26. Howe, C.L. and Rosenberg, R. 'Government plans for data security spill over to civilian networks', *Data Communications*, **136**, March 1987.

Chapter 4 USING A BLOCK CIPHER IN PRACTICE

4.1. METHODS FOR USING A BLOCK CIPHER

Block encipherment operates on blocks of data of fixed size but a message to be enciphered can be of any size. Encipherment is sometimes applied to a stream of data, where the length of the stream may not be known when encipherment begins. So a block encipherment algorithm seems, at first, to have some limitations. Perhaps a stream encipherment algorithm might be closer to the real needs?

This reaction is probably wrong. A block encipherment method like the Data Encryption Standard (DES) can be used in various 'methods of operation' which cover all the practical requirements. Instead of being a limitation, the fact that the DES operates on fixed-length blocks makes it a convenient starting point for at least four different methods of operation which offer a very versatile set of 'tools' for encipherment.

Given a plaintext block x and a key k, the encipherment algorithm enables us to compute the ciphertext block $y = Ek(x)$. This, in itself, is a 'method of operation' sometimes called the 'native mode'. It can be applied to any source of plaintext in 64-bit blocks. This is actually a frequent requirement, for example when a DES key has to be enciphered. The DES key in its 64-bit form can conveniently be enciphered using the native mode of the DES algorithm.

A more general case is the encipherment of a message of arbitrary length perhaps made up of many 64-bit blocks. We can think of such a message as having been generated in a processor and stored ready for encipherment. We need a method for enciphering such a multi-block message which is convenient and secure. For this purpose there is a method of operation called 'cipher block chaining'.

The third case is the encipherment of a string of characters such as might come from a teleprinter which is being operated continuously. For such a stream we need a method of operation which enables each character to be enciphered as it arrives and does not have to wait until several characters have arrived before block encipherment can be carried out. This is a form of stream cipher and the related method of operation which will be described is called 'cipher feedback'.

There is a fourth requirement, which will emerge later in this chapter, for which another method of operation is used, called 'output feedback'.

The native mode of operation is called 'electronic codebook' because, for a given key, it associates with every value of the input block a unique value of the output block and vice versa. Because of the large number of such values, 2^{64}, the codebook in question is a very large one, never likely to be enumerated in full.

These four methods of operation have been found to cover most of the practical requirements for the use of encipherment in computer and network systems. They are usually called by the initial letters of their names and they can be summarized thus:

1. *Electronic Codebook (ECB)* the straightforward use of the algorithm to encipher one block.
2. *Cipher Block Chaining (CBC)* the repeated use of the algorithm to encipher a message consisting of many blocks.
3. *Cipher Feedback (CFB)* the use of the algorithm to encipher a stream of characters, dealing with each character as it comes.
4. *Output Feedback (OFB)* which is another method of stream encipherment with its own properties, to be described later.

These four methods of operation are the result of some careful study since the DES first became available. Other methods are possible but these four seem to meet most requirements. They are defined in a US Federal Information Processing Standard[1] and an international standard.[2]

The four methods can be used with any block cipher. In the descriptions which follow, the DES will be used because it is easier to describe the methods by a well-known example. It will be obvious how the methods can be used for any block cipher, changing only the details according to the block size.

The limitations of the electronic codebook mode

It might seem that a message comprising many blocks could be enciphered simply by dividing it into 64-bit blocks and enciphering each one separately. This is a simple application of ECB encipherment — the native mode. Decipherment is equally obvious. But, in general, ECB encipherment should not be used in this way because of some severe security weaknesses which arise from the structure of typical messages.

In all practical applications, fragments of messages tend to repeat. One message may have bit sequences in common with another. Messages generated by computers have their own kind of structure. The need for formal definition of protocols and the desire for generality in their definition has made large parts of the formats used by computer messages very repetitive. They may contain strings of zeros or spaces. The protocol designer provides parameters which in practice are rarely used and, therefore, take constant values. Computers often generate messages in a fixed format, having the important data always in the same place.

If the ECB type of encipherment is used with this kind of traffic, an alert enemy who intercepts collections of these messages will be able to detect the repetitions. Figure 4.1 illustrates how ECB encipherment fails to conceal this structure.

Repeated phrases will not show through the ECB encipherment unless they happen at the same phase relative to the block size. Cryptanalysts will search for and exploit the occasional visible regularities. The commonest repeats, such as strings of zeros, show up easily. The biggest danger arises when the significant parts of the messages change very little and appear in fixed locations. Analysing these parts becomes a 'codebook' exercise in which the number of code values is small.

Suppose now that a set of messages has been analysed and some partial

Block	Plaintext	Ciphertext
1	T H E Y C A N	60 99 46 42 52 82 22 49
2	H A V E S E	FF BF BC 77 8B BB F2 06
3	V E R A L A C	0D 4D 86 DE B6 CD 92 5D
4	T I V E P E R	99 63 A8 0F 32 D3 E7 E9
5	M A N E N T V	10 49 1F 3B DE 67 21 B7
6	I R T U A L C	BD 2D 6D 61 42 08 C7 B8
7	I R C U I T S	19 F1 01 A4 89 6A AE 4C
8	A N D / O R V	84 DB CC EC 35 18 58 9C
9	I R T U A L C	BD 2D 6D 61 42 08 C7 B8
10	A L L S A T	D4 3C D4 5A 9E 0B A5 ED
11	T H E S A M E	84 52 01 AC 2D FE 98 3A
12	T I M E .	89 F1 89 E9 DB CC CB BB
	Key	01 23 45 67 89 AB CD EF

Fig. 4.1 A weakness in ECB encipherment

understanding has been obtained of these parts of the messages. An enemy can re-construct messages from the parts of the message he has intercepted. The ECB mode of operation has not linked these blocks together, so they can be reassembled in any order. For example, a bank payment message might have the payment amount or the payee account number in fixed positions. Even if the enemy does not know the precise amount, he might be able to deduce the payee. He may also be able to find which are the large payments. He could exploit his knowledge of the enciphered blocks by forging large payments for chosen accounts, simply by reconstructing messages from known pieces.

The vulnerability of systems to 'codebook' analysis is greatest at the beginning of messages, where formal headers appear which contain the destination, serial number, time etc. The problem of the 'stereotyped beginning' of messages has to be considered carefully in all cipher systems.

The weakness of the ECB method lies in the fact that it does not connect the blocks together. By enciphering each block separately it leaves them as separate pieces which the cryptanalyst can analyse and assemble for his own benefit. What we need is a method of enciphering successive blocks which makes the cipher meaningless except in the given sequence. This is called 'chaining'.

4.2. CIPHER BLOCK CHAINING

Cipher block chaining uses the output of one encipherment step to modify the input of the next, so that each cipher block is dependent, not just on the plaintext block from which it immediately came, but on all the previous plaintext blocks.

Figure 4.2 illustrates how the method operates. All the operations in the figure work with 64 bits in parallel. In a real system, some of these transfers might be serial but the principle is best explained as for parallel operation throughout.

The operation of addition in the figure is the 'modulo 2 addition' function and it operates in parallel on the 64 bits of the blocks. The encipherment method, shown on the left of the figure, has a feedback path from the ciphertext output back to plaintext input. Except for the first block, each successive block, before encipherment, is added to the ciphertext of the previous one and this makes the nth ciphertext block C_n a function of all the plaintext blocks $P_1, P_2, \ldots, P_n$.

The figure shows on the right-hand side how this process is reversed for decipherment. After decipherment of the ciphertext, a correction is made which corresponds exactly to the operation carried out at the beginning, namely the addition, modulo 2 of the *previous* ciphertext block.

The words 'addition, modulo 2' do not adequately express the function shown by the circles in the figure. In fact, each bit of the input block P_n is added, modulo 2, to the corresponding bit of the previous ciphertext C_{n-1}. In actual implementations, the operations may be serial, or be carried out on one octet in parallel, according to the speed required or the DES implementation employed. The figure shows the operations as though they take place with 64 bits in parallel.

In order to make quite clear the sequence of operations, the ciphertext block is shown at both ends as entering a register where it remains between successive operations. This clarifies the point that, at both ends, the quantity entering the modulo 2 addition is *the ciphertext of the block which preceded the one now being processed.*

The operation of the cipher block chaining can be expressed as follows

encipherment:

$$C_n = \mathrm{E}k(P_n +_2 C_{n-1})$$

decipherment:

$$Q_n = \mathrm{D}k(C_n) +_2 C_{n-1}$$

In order to prove that the result of decipherment, Q_n, is indeed the plaintext, apply the operation Dk to the first equation, giving

$$\mathrm{D}k(C_n) = P_n +_2 C_{n-1}.$$

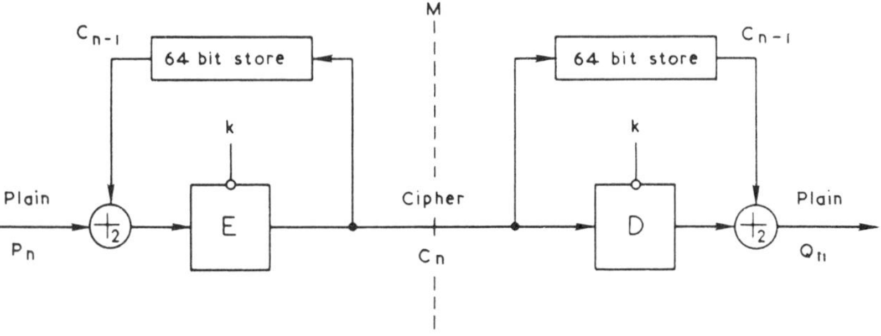

Fig. 4.2 Cipher block chaining

Substituting this value in the second equation

$$Q_n = P_n +_2 C_{n-1} +_2 C_{n-1} = P_n.$$

There is a more graphic way to demonstrate this inverse property which derives from the symmetry of Figure 4.2. It has 'mirror' symmetry about the broken line marked M, except for the data flows. The flow of data through the encipherment/decipherment operations is consistent with a symmetric pattern of data on the lines, noting that modulo 2 addition is naturally symmetric because the flow of data on the horizontal line can be reversed without changing the data values. It follows that the right- and left-hand sides are symmetric about the line M in the data values carried by the lines. From this it follows that the input and output data are the same. This kind of argument is very useful for establishing quickly the plausibility of systems employing a block encipherment algorithm, though it should be checked by a more rigorous analysis and the timing or sequencing of the operations should be checked. The 'mirror' argument is an appealing one because it can be seen directly in a figure and we shall use it again.

The first and last blocks

To complete the definition of CBC operation, both the start and the finish of the procedure must be detailed. At the beginning, the 'previous ciphertext block' is not available so the initial contents of the registers in Figure 4.2 must be specified otherwise. This is the quantity shown in the following equations as I. It is known variously as the 'initializing variable' or 'initialization vector'. We shall call it the *initializing variable (IV)*. The choice of this value is important to the security of the operation and it must be the same for the sender and receiver. The equations for expressing encipherment and decipherment must be replaced, for $n=1$, by

$$C_1 = \mathrm{E}k(P_1 +_2 I)$$
$$Q_1 = \mathrm{D}k(C_1) +_2 I$$

and the equations given earlier hold for $n=2,3, \ldots$.

When cipher block chaining is used in a communication system it is normal to keep the IV secret. This can be done by transmitting the IV encipherment in ECB mode using the same key that will be employed for data encipherment. More important than the *secrecy* of the IV is its *integrity*. We want to prevent an enemy from being able to make purposeful changes to the plaintext which emerges from the encipherment system. With CBC encipherment this cannot be done by manipulating the ciphertext because changes to the ciphertext will produce an apparently random effect in the plaintext output. But changes to the initializing variable would have the effect of making corresponding bit changes to the first block of plaintext output. In this way the whole of the first block is vulnerable to purposeful changes by an enemy. The problem is avoided by sending the IV in enciphered form.

Given a message of arbitrary length, it may not divide neatly into 64-bit blocks but leave something at the end. We can deal with the short end block in several ways but the simplest is to pad out the message by adding some extra bits until the block reaches the correct size. These padding bits could be zeros, because the addition of the previous ciphertext block before encipherment prevents them being used for codebook analysis. On the other hand, the prudent designer might consider padding out with random digits just to make the analyst's job harder.

The use of padding gives a new problem to the system designer because the amount of padding has to be indicated somewhere so that the receiver can remove it. One convention is that the last octet of padding indicates by a 'padding indicator' (PI) the number of padding octets, as shown in Figure 4.3. This method requires that the existence of padding be revealed earlier in the message format. In a format that has been designed with encipherment in mind, special provision for a padding indicator might be made.

For many applications the system designer must have a method of handling the short end block which does not expand the message size. This would be needed, for example, when a stored plaintext is being enciphered and put back into the same store. For this purpose the scheme shown in Figure 4.4 is useful.

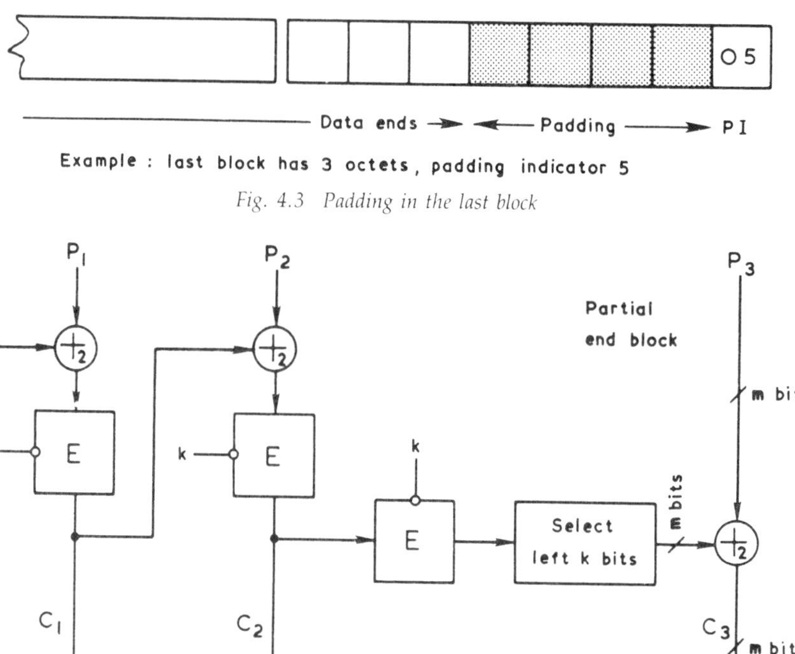

Fig. 4.3 Padding in the last block

Fig. 4.4 A method for enciphering the end block

This figure shows the same basic procedure as Figure 4.2, but the treatment of each block of the message is shown separately. It shows the case of a message of two complete blocks followed by one partial block which has *m* bits. The final ciphertext block from the cipher block chaining process is enciphered once more and used by modulo 2 addition to treat the last, short block. Its value is sufficiently random for this purpose. The first *m* bits of the generated pattern are used to add, modulo 2, to the short block for transmission. A similar process at the receiving end generates the same pattern of *m* bits which deciphers the short block by modulo 2 addition.

There is a weakness in any such additive encipherment. Although the enemy may not know what is being enciphered, he can change it systematically. He can choose individual bits in the last plaintext block and change their values simply by changing corresponding bits in the ciphertext. This is not dangerous unless the part of the message we are talking about contains essential information and fortunately this is not usually the case. The end of a message usually carries sum checks which cannot usefully be changed. When used with care, this method of dealing with the short end block can be effective.

Transmission errors in CBC encipherment

The CBC method of operation can be characterized as *feedback* of ciphertext at the sending end and *feedforward* of ciphertext at the receiving end. The feedback process has the property of extending indefinitely the effect of any small change in the input so that a single bit error in the plaintext message will affect the ciphertext from the block in which it occurs and every succeeding ciphertext block until the chain ends. This is not as important as it may seem because the decipherment process restores the correct plaintext except for the error.

Errors in the ciphertext are more important and may easily result from a noisy communication path or a malfunction in a storage medium. The effect of a single bit ciphertext error on the received plaintext is shown in Figure 4.5.

The first effect is to change the input to the DES algorithm at the receiver. The output of the algorithm is then, for all practical purposes, random, because of the good cipher properties of the algorithm. At the time of the next block, the ciphertext error emerges from the register in the feedforward path and a single

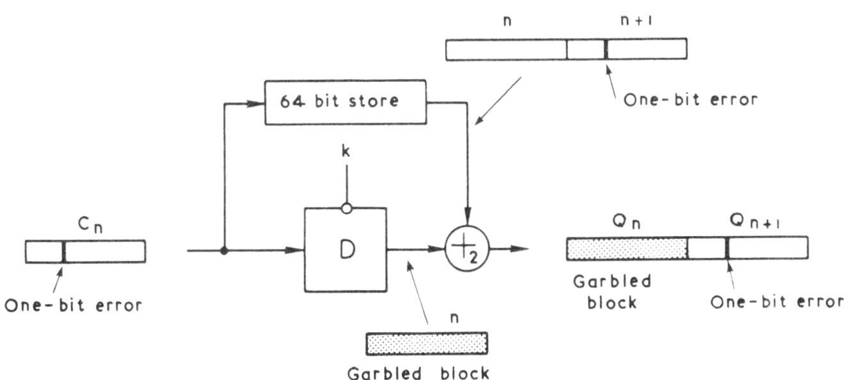

Fig. 4.5 One-bit error in cipher block chaining

bit error in the ciphertext causes a single bit error in the plaintext, in a corresponding position of this second block.

Blocks after the second are not affected by the ciphertext error, so in this respect the cipher block chaining procedure is self-recovering. That is to say: two blocks in the plaintext output will be affected, but the system recovers and continues to work correctly thereafter.

In contrast to the rapid recovery from bit errors, the system does not recover well from synchronization errors. If it is possible for a bit to be lost or gained in transmission so that blocks are now shifted one bit out of position, then the receiving system will generate garbage indefinitely. The method of using cipher block chaining must ensure that the framing of blocks cannot be lost indefinitely. This will generally be achieved when it is used for data in storage or in conjunction with data-link protocols that use framing.

The property of converting brief errors in transmission into lengthier errors of plaintext output is called 'error extension' or 'garble extension'. It is a nuisance in communication systems and may play havoc with a nicely tuned error-control procedure that depends on the statistics of error occurrences. For this reason, error control should be applied directly to the ciphertext stream, not around the larger plaintext loop which includes encipherment and decipherment. Figure 4.6 shows these two alternative locations of error control.

The error properties of CBC carry a security risk if they are not handled correctly. Errors randomize one block but they cause controlled changes to the *next* block. A system which ignored the errors in one block (or allowed them to be recovered later) and accepted the succeeding block would be vulnerable to calculated changes being made by an enemy to the ciphertext.

For this reason, although CBC recovers fairly quickly from line errors, the way in which it is used should abort any chain in which errors have occurred. Error recovery procedure should be designed to repeat the whole chain and to accept it only when it is all delivered correctly. For this reason, it is normal to provide a sum check on the plaintext to detect these errors at the receiver. Any enemy trying to modify bits in the chain would probably cause a random sum check and this would be detected. In chapter 5 this is treated more fully.

Choice of the initializing variable

We described cipher block chaining as a method by which the blocks could be linked together to prevent codebook analysis. This is achieved for all the blocks after the first, but the encipherment of the first block cannot depend on earlier blocks. If the initializing variable is kept constant for a series of chains, the repetition of patterns in the beginnings of the chains will show through. Using the same IV and key, two messages which have the same plaintext string at the beginning will have the same ciphertext until the point at which the two messages diverge. This degree of transparency would be very unwise, so we must contrive that either the beginning of the message or else the value of the IV differs on each occasion.

For communication systems it is customary to use as chain identifier a serial number at the start of the chain which does not repeat during the currency of a key. On the other hand, there are applications of cipher block chaining where this message expansion is not acceptable, such as data enciphered *in situ* in storage. For these, it is the IV that must be made variable. For example, the IV can be derived from the index used for looking up the data in storage.

4.3. CIPHER FEEDBACK

Data are handled in many forms, as complete messages, or sequences of frames, blocks, 8-bit characters or binary digits. When messages must be treated character by character or bit by bit, another kind of chained encipherment is used which is known as *cipher feedback.*

When we treat data as a stream of bits or characters we are at a low level in the hierarchy of protocols which makes up the computer network architecture. This is a level at which transparency is important. We must introduce encipherment in a way which disturbs the existing system as little as possible. For example, consider a character terminal which sends out a stream of characters in the ISO 7-bit code with an added parity bit. Each of these octets from the terminal must be transmitted through a communication network at once and encipherment must not cause a delay. So the encipherment method, although it chains these octets together, must operate on each octet as it arrives and deliver it, enciphered, as soon as possible. The same requirement for immediacy applies to the decipherment of the octet stream coming into the terminal.

The cipher feedback method is illustrated in Figure 4.6. Superficially it appears similar to cipher block chaining but the important difference is that the DES block encipherment operation takes place in the feedback path at the transmitting terminal and in the feedforward path at the receiving terminal.

The encipherment of the character stream consists of adding, modulo-2, the character derived from the output of the DES algorithm to the plaintext character to form the ciphertext character and then at the receiving end adding the same data into the ciphertext character to restore the plaintext. The design of the system makes this added data stream pseudo-random and yet provides the same stream at the sending and receiving ends.

Whereas cipher block chaining operates on whole blocks, cipher feedback

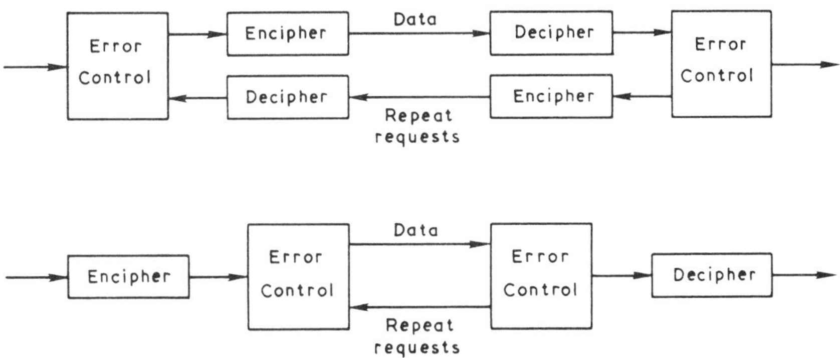

Fig. 4.6 Alternative locations for error control

operates on one character at a time, and the character length *m* can be chosen as a parameter of the design. It is known as '*m*-bit cipher feedback'.

The smallest character length is 1 bit, resulting in '1-bit cipher feedback'. In principle, any value of *m* could be used up to 64. The most usual choice of *m* is the character size used in the communication system. (In older systems this can be 5 or 6 bits and in modern systems 7 or 8.) Probably the most frequent choice of character size today is the octet and Figure 4.7 is illustrated with this character width.

The eight bits used to encipher the character stream by modulo 2 addition are derived from bits 1 to 8 of the output of the DES algorithm. It is important to realize that the DES algorithm is performing an *encipherment* at both ends of the line. The input of this algorithm comes from a 64-bit shift register which contains the most recent 64 bits transmitted as ciphertext. Every character which goes out on the line at the sending end is shifted into the high-numbered bits of the shift register, displacing a similar number of bits from the other end. The same thing is done to the received ciphertext at the other end of the line. The argument of symmetry which was useful in understanding cipher block chaining can be applied to Figure 4.7, but the layout of the figure is intended to make clear the numbering of bits into and out of the DES algorithm and this rather obscures its symmetry.

If the ciphertext is transmitted without error, the contents of the shift register are the same at both ends. Therefore, the output of the DES algorithm is the same at both ends and the same eight bits are added to the data stream. This ensures that encipherment is correctly reversed by decipherment at the receiving end.

The sequence of operations as each character passes through the system is as follows. The incoming plaintext character has added to it, modulo 2, the bits coming from the DES algorithm. By this addition the character is enciphered and made ready for transmission. At the same time as it is transmitted it is loaded into the shift register, displacing the same number of bits from the left-hand or low-numbered end. Now the shift register contains a new pattern of bits and the DES algorithm is invoked to produce a new output. This output is ready for encipher-

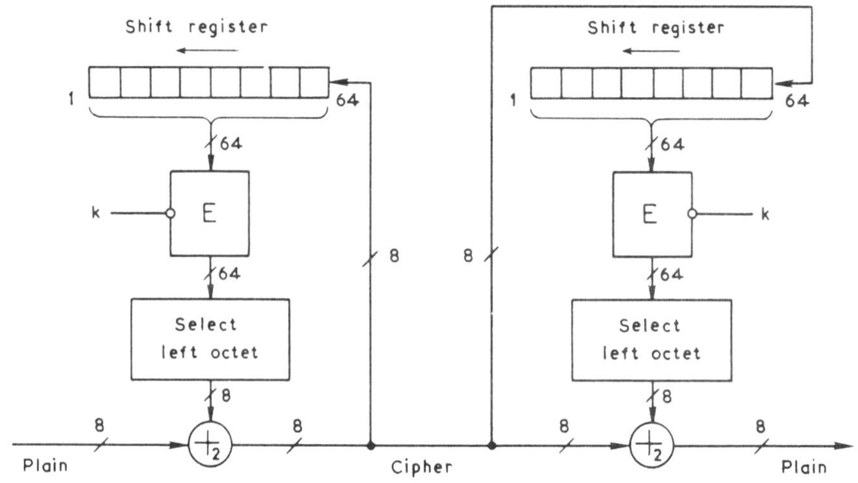

Fig. 4.7 8-bit cipher feedback

ing the next character. A similar procedure takes place at the destination where the incoming characters are deciphered (by adding bits from the DES output) and also fed into the shift register to produce the same stored pattern as at the sending end.

Each passage of one character through encipherment or decipherment invokes the data encipherment algorithm once. This contrasts with cipher block chaining where each operation of the algorithm enciphers 64 bits of users' data. For this reason, cipher feedback throughput could be limited by the speed of the DES operation. Fortunately, the lower levels of network hierarchy, at which cipher feedback is used, are often quite slow in operation so this speed limitation is not important.

Like cipher block chaining, cipher feedback chains the characters together, making the ciphertext a function of all the preceding plaintext.

Error extension in cipher feedback

CFB operation is like cipher block chaining in having a feedback path at the sending end and a feedforward path at the receiving end. Correspondingly, changes in the plaintext affect all the succeeding ciphertext, and errors on the communication line are subject to 'error extension'. We can follow the effect of a single bit transmission error in Figure 4.8. The first effect is to alter the corresponding bit in the plaintext output of that character. The error also enters the shift register, where it remains until the shift process pushes it off from the left of the register. While the error is in the shift register, the output is garbled. In the case of 8-bit cipher feedback which is illustrated, nine plaintext characters are affected by a single error. The error properties are similar to those of CBC except that the 'controlled' change takes place at the beginning of the error pattern instead of the

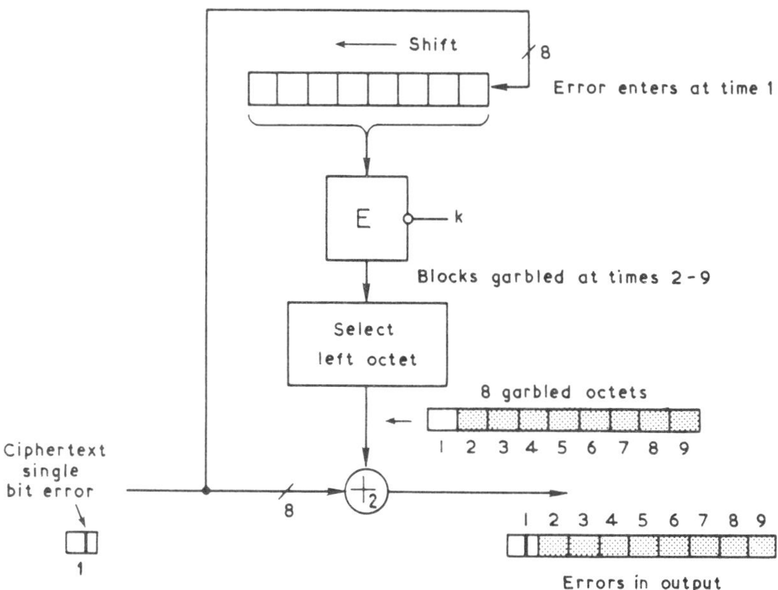

Fig. 4.8 One-bit error in 8-bit cipher feedback

end. An enemy could change selected bits of a character in the plaintext while completely garbling the following eight characters.

When the purpose of encipherment is simply to avoid disclosure this error pattern is a nuisance but no more. For an active attack it could be a weakness because the first character of the error pattern can be modified purposefully. Usually this trick would be detected because of the following garble, which a plaintext sum check would reveal as an error. A dangerous case occurs with the last character of a message, where the garbled characters never come. This matters little for 1-bit CFB and is unlikely to matter for 8-bit CFB since the 8 bits in question are probably part of the sum check, but it could be serious with 64-bit CFB and, for example, a 16-bit sum check. The enemy could alter the last 48 bits of the message and correct the sum check to match. Of course, 64-bit CFB is unusual and CBC is likely to be used instead.

Initializing with CFB

To start up the process of cipher feedback, the shift registers must be loaded with an IV. It is customary to use a different IV for every chain during the lifetime of a key, in order to mask any repetitions at the beginnings of the chains. These IV values can be transmitted in clear form because they enter the process through the encipherment algorithm.

In cipher block chaining the transmission of the IV is a procedure quite distinct from normal message transmission because the IV is enciphered. The plaintext IV which is sent at the start of CFB operation can be regarded as a preamble to the message. It enters the shift register at both ends of the line and, when the registers are loaded with the same data, secure transmission begins. This method of start-up is characterized by random data being produced at the sending end (to conceal any regularities in the starts of messages) and different random data being produced at the plaintext output of the receiving end. In effect, the self-correcting properties of this cipher method are being used for starting up.

More formal procedures for initializing are described in chapter 11, in which a randomly chosen IV is sent before transmission of data begins. Each chain incurs this transmission overhead. There are polling systems which use very short messages, each message sent to a different station on a multi-station line. If CFB encipherment is used with a different key for each station, then each of these short mssages must be preceded by an IV. To avoid the full-length, 64-bit IV, it is customary to send a shorter-length value and pad it on the left (most significant) end with zeros to make up the 64-bit pattern in the shift register. Of course the transmitted value should not be too short or the danger of code book analysis of the starts of messages returns. For example, if an 8-bit IV is transmitted the enemy has to construct 256 versions of the codebook. Suppose that the polling message consisted of an address which was fixed for the station followed by a command with only a few possible values, then it would not take long to analyse these command values for a given station.

Encipherment of an arbitrary character set

Cipher feedback is used for enciphering a stream of characters, where each character is represented by m bits. In such a character set a character is represented by a number in the range 0, 1, ... $n-1$ where $n=2^m$. Both the plaintext and the ciphertext characters are in the same range.

Some character codes do not occupy a range of 2^m values. For example, we might want to use the digits 0, 1, ... 9. Some character sets must avoid those binary values that represent control functions. For example the ISO 7-bit coded character set[3] (which is similar to the ASCII code) includes character values which mean: start of heading, start of text, end of text, acknowledgement, data link escape etc. In many communication systems, if we tried to send such a character in a ciphertext stream it would have an unfortunate effect on the network's procedure, so we have to restrict the transmitted characters to a 'safe' set. In the ISO case this usually means the binary values representing the numbers 32 to 127, ninety-six values in all.

We could consider an arbitrary character set which does not even use a consecutive set of character values. Then the first step in encipherment would be to map these characters into the consecutive values 0, 1, 2, ... $n-1$ where there are n different character values in the allowable set. If $n=2^m$, normal CFB operation can be used. Otherwise m bits are needed to encode the character where m is the smallest value for which $n<2^m$.

Figure 4.9 shows how the cipher feedback method of operation can be modified for the case of an arbitrary character set.[4] The process of encipherment in CFB is the addition of a stream of m-bit characters coming from the least significant m-bit positions of the DES output into the plaintext character stream. For dealing with the arbitrary character set modulo 2 addition of bits is replaced by modulo n addition of character values. Both these methods of encipherment are well established in classical ciphers, where binary addition, modulo 2, characterizes the Vernam method and modulo n addition characterizes the Vigenère method.

Cipher feedback has the useful property that the method of deriving the character stream for encipherment can be chosen almost at will, provided the same method is used by the receiver for decipherment. We must now show how a character in the range 0, 1, 2, ... $n-1$ can be produced from the output of the DES encipherment.

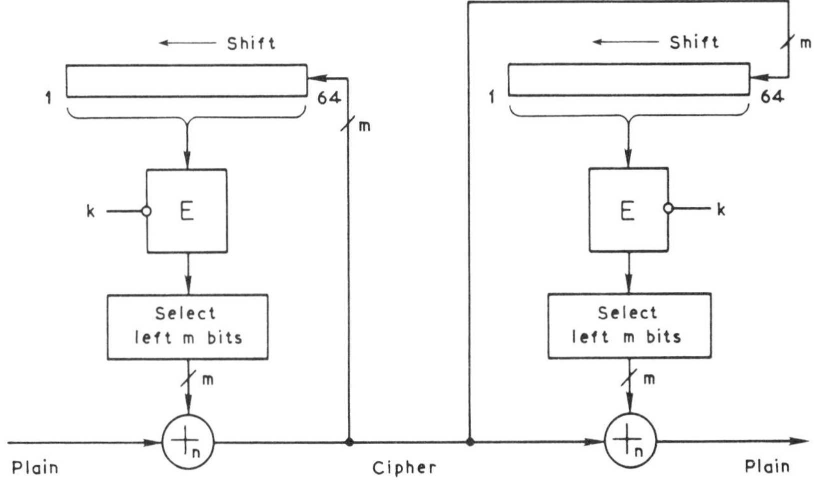

Fig. 4.9 Modulo n cipher feedback, $n<2^m$

The simplest proposal is to use the least significant m bits of the DES output where

$$2^{m-1} < n < 2^{m}.$$

The character value produced by these m bits could be added, modulo n to the plaintext to generate the ciphertext.

Figure 4.10 illustrates the procedure for deriving the ten character values 0, 1, ... 9 from a 4-bit output taken from the DES. This method is not recommended because the values 0, 1, 2, 3, 4 and 5 will occur twice as often as the values 6, 7, 8 and 9. If the digits are not equiprobable in the plaintext there will be a distorted distribution in the ciphertext, which can be exploited by a cryptanalyst. For general use we need the additive character stream to give all the values 0, 1, ... $n-1$ with equal probability.

A useful method is illustrated in Figure 4.11 for the example of generating decimal digits. The 64-bit output of the DES is divided into sixteen blocks of 4 bits each. Starting from the least significant end, the first four bits are examined and if they lie in the permitted range 0 to 9 this value is used for the additive encipherment. If not, the next block is tested in the same way. Since there are sixteen possibilities, it is highly probable that one of them will generate a satisfactory character and the first one to be reached is the one used for encipherment. If, unfortunately, all the values happen to lie in the range 10 to 15 then we choose some arbitrary prearranged value to add to the character for encipherment, say 5 for example. The probability that this arbitrary value will have to be used is $(6/16)^{16}$ which is approximately 1.53×10^{-7}. Clearly the method is not perfect since the probability of one particular character value (5) is very slightly larger than the rest. To exploit this flaw, even if the plaintext stream had one very highly probable character, the cryptanalyst would need to collect a very large sample of text. For most purposes the method is adequate.

This slight flaw could be removed, at some expense. Instead of the default condition when all sixteen tests fail, the DES output could be enciphered again to

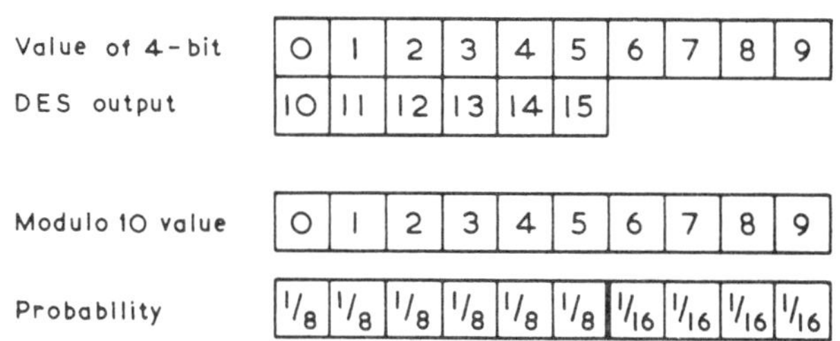

Fig. 4.10 The modulo 10 value — an uneven distribution

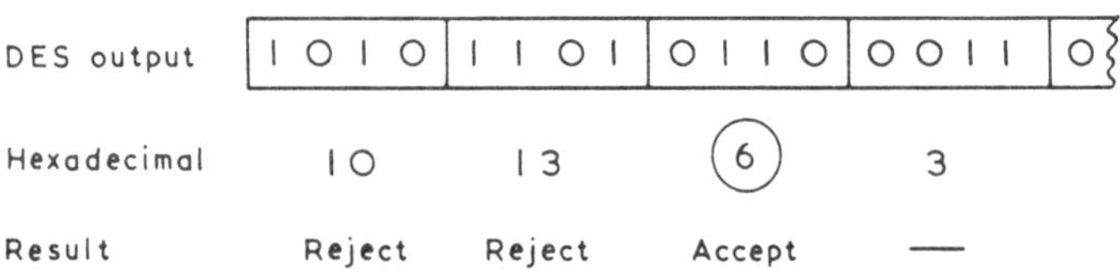

Fig. 4.11 An improved rule for modulo 10

generate a new pseudo-random block of 64 bits and the tests could begin again from the least significant end. If the second block fails, encipherment can be used to generate a third, and so on. In practice, these multiple encipherments will be rare. The result of this process would be, as far as we know, a set which is equiprobable in the range 0, 1, . . . 9. But in some applications the extra encipherment cannot be allowed because of the extra time it would take. Whereas the sixteen tests can be made rapidly, some systems could not face the possibility of an indefinitely long process to produce a satisfactory character, however rare that event might be. For most practical purposes, sixteen tests followed by a default procedure is adequate.

The probability of the default condition should be checked to make sure that it does not weaken the system. Take another example, which is the 96-bit ISO graphic set. Since this is a 7-bit code, there will be nine useful blocks in the DES output and the probability of a default condition is $(32/128)^9$ which is approximately 3.8×10^{-6}.

4.4. OUTPUT FEEDBACK

Of the three methods of operation that have been described, ECB, the 'native mode', is for limited use and the two others CBC and CFB are general-purpose methods. The fourth, output feedback, is intended for applications in which the error-extension properties of the two general-purpose methods are troublesome.

Coded speech or video channels can withstand a small amount of random noise in transmission or storage, basically because they are highly redundant. The listener or viewer is not bothered by occasional clicks or spots due to faulty speech samples or picture elements. If we required it, an automatic process could detect and correct these errors, in the same way that blemishes in old films can be corrected electronically. We do not want these occasional, isolated errors to be magnified by the error-extension properties of a cipher method. For example a standard speech channel with pulse code modulation (PCM) uses 8-bit speech samples. A single error would be extended by encipherment to damage about eight further samples. This could make a bad noise in the channel or, with multiplexing, cause clicks in eight other channels.

Output feedback uses a cipher of the kind invented by Vernam; only the pseudo-random source is new. In one respect it resembles CFB operation because it can be applied to streams of m-bit characters for any value of m between 1 and 64. It has the property of all additive stream ciphers (Vernam type) that errors in the ciphertext are simply transferred to corresponding bits of the plaintext output.

Figure 4.12 shows OFB encipherment, working with m-bit characters and an m-bit wide feedback path. It resembles CFB operation in all respects except the place from which the feedback is taken. The pseudo-random stream is the result of feedback applied to the DES cipher.

The non-error-extension property of additive stream encipherment carries disadvantages with it. An active attack on the channel can produce controlled changes to the plaintext. Since OFB systems usually run continuously on a synchronous channel, the enemy may have no knowledge of where each message starts and ends, so the usefulness of this trick to the enemy might be limited.

Synchronization of the pseudo-random number generators at the two ends of

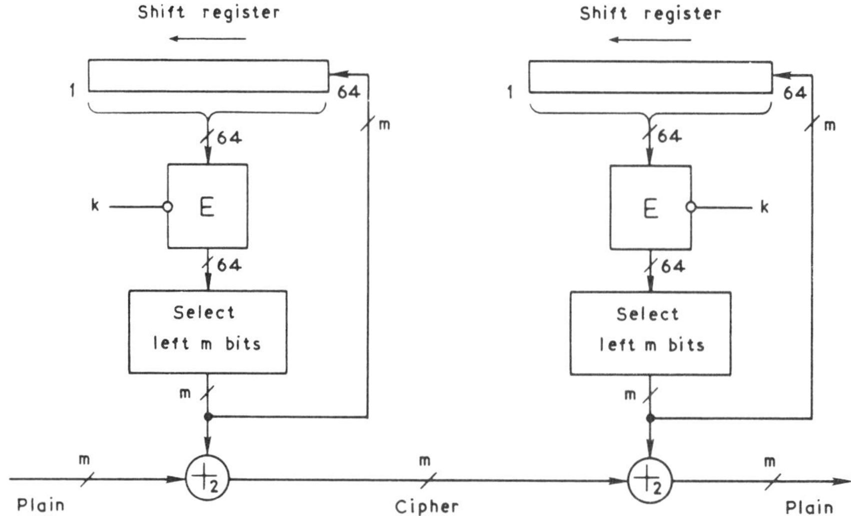

Fig. 4.12 m-bit output feedback

the line is critical. If they get out of step the 'plaintext' output at the receiver will be random. Whereas CFB operation can recover when character synchronization is restored, OFB will not if characters have been gained or lost. A system employing OFB must have a method of re-synchronization after such a failure. This is the same as restarting with a new IV.

To start up or resynchronize OFB, the shift registers must be loaded with identical values. The IV can be sent in clear form because knowledge of it does not help the enemy to construct the pseudo-random stream. In the same way as for CFB operation, the full 64-bit field need not be used. The IV sent over the line can be padded on the left with zeros to fill the shift register. There is the same trade off between IV overhead and security as in CFB initialization.

Key stream repetition

The pseudo-random stream which is added (modulo 2) to encipher and decipher is called the *key stream*. If ever the key stream repeats it weakens the cipher. If X_i is the key stream and it is used for two different plaintexts A_i and B_i, the two ciphertexts are: A_i+X_i and B_i+X_i (modulo 2 addition). Adding these, modulo 2, produces A_i+B_i, eliminating the key stream. The result is that which would have been obtained by enciphering A_i with B_i as a 'running key' cipher, or vice versa. Solution of such ciphers is sometimes easy. The enemy may not know, at first, where the key-stream repetitions occur, but if the messages are in natural language there are statistical tests which distinguish their sum from the random stream caused by X_i. Digitally coded voice and video have even stronger statistical properties. Sliding the enciphered messages and testing their sum will detect the key repetition. Such repetitions must be avoided.

Output feedback, like any other pseudo-random generator, repeats its output after a certain time, that is, it enters a cycle. The key-stream generator can be regarded as a finite state machine with 2^{64} states, each state corresponding to one value x in the shift register of Figure 4.12. More than 2^{64} states per cycle are impossible.

Output feedback with $m=64$ iterates the function $x_{i+1}=Ek(x_i)$, which is a one-to-one mapping of x on to y. It permutes the 2^{64} values of the state variable x. When such a generator re-cycles, it returns to its initial state. Permutations can be represented by the cycles they produce and it is well known that when all the $n!$ permutations are written out as cycles, the total length of all the cycles of any length c is exactly $n!$, whatever the value of c. In other words, starting from an arbitrary state x_0, the length of cycle is equally likely to take any one of the values 0, 1, 2 ... n. In the case of OFB, where $n=2^{64}$, the mean cycle length (averaging over all keys and starting values — or with randomly chosen starting values) is $2^{63}+\frac{1}{2}$. This result depends on a significant assumption, which is that the encipherment functions $Ek(x)$ are effectively random choices among the $(2^{64})!$ permutations.

A cycle length which is random and uniformly distributed is not as safe as a cycle of known length, but at least the average length is high and the probability of a very low cycle length is small. OFB with $m=64$ is satisfactory.

The analysis of OFB with other values of m is more difficult. The function $x_{i+1}=f(x_i)$ relating each state x_{i+1} to the previous state x_i is not one-to-one. A simplifying assumption can be made, according to which it is a *random* function so that, until the cycle repeats, each new x value is effectively chosen at random. A cycle begins when any x value repeats one that has already occurred. Although this is not a good description of the OFB operation, it has been found experimentally to give the correct cycle behaviour.[5] The problem of finding cycle length is similar to the solution of the famous 'Birthday problem' which is of such importance in the analysis of data security that it is described in an appendix to chapter 9, beginning on page 279. The solution of that problem indicates that coincidence of two x values is likely to happen after about $\sqrt{n}$ iterations, where $n=2^{64}$ is the number of values from which the x values are randomly chosen. A more careful analysis described in reference 5 shows that the average cycle length for a random function is $(\pi n/8)^{1/2}$, which is close to 2^{31}. The OFB function is, however, not random and by a closer analysis it was possible to estimate the way that the average depends on m.

The largest effect of non-randomness occurs for $m=1$ and $m=63$ and it increases the average cycle length by a factor $\sqrt{2}$. The effect decreases for m = 2, 3 ... and m = 62, 61 ... so that the average $(\pi n/8)^{1/2}$ is accurate for most values of m. The average cycle length is always in the range 2^{31} to 2^{32}.

An average cycle length of this magnitude is not satisfactory in an encipherment scheme intended for general use. For example, a digitized speech channel at 2^{16} bit/s rate would recycle on average after 2^{16} s or 18 hours. There is a further reason why a random function is not suitable for generating a key stream. The average number of distinct cycles for such a function is very small, namely $\frac{1}{2}\ln n$, which for 2^{64} system states is only 22. The conclusion from these calculations is that OFB should be used only with full 64-bit feedback. The other feedback possibilities for OFB in Federal Information Processing Standard (FIPS) publication 81 should be avoided. The effect of this on standards is described in chapter 11 (page 349).

4.5. STANDARD AND NON-STANDARD METHODS OF OPERATION

We can now compare the properties of the four 'standard' methods of operation.

The electronic codebook method (the simple use of a block cipher) is not recommended for messages larger than one block because of the danger of 'codebook' analysis of repeated block values and the possibility of reassembling messages from known blocks. When ECB is used to encipher a block of data, the way it is used should prevent repetitions or make them very unlikely.

The most usual application of the ECB method is to encipher a key for transmission. Since a key contains 56 bits of random data, the building up of a codebook is impractical. Of course, the danger of a replay of old keys must also be considered when the security is analysed.

If very short messages (such as acknowledgements) have to be sent which fit into one block, codebook analysis should be defeated by making the block sufficiently variable. There are several ways in which the variability can be achieved. One is to add an initializing variable, effectively using cipher block chaining. Instead of sending an IV, which must be protected from manipulation, the added variable can be a serial number, incremented for each transmission. If the message is short enough, the serial number can be incorporated in it, which is more convenient for checking and resynchronizing.

An alternative is to put a date and time in the message which can be checked to detect a replay. Since a compact form of date and time, down to the second, occupies 48 bits, there will be space for 16 bits of message. Another method is to include with the message a random number. This prevents codebook analysis but does not easily guard against replays. Methods like these can be 'tailor-made' for a particular application. Compared with general-purpose protocols these methods can be more efficient and justify the special software that has to be written. In most circumstances it is better to use general methods based on CBC or CFB operation and accept their overheads.

Cipher block chaining is the recommended method for messages of more than one block. This method avoids codebook analysis generally but not at the start of the chain. Communication systems generally use chain formats which begin with a serial number so that the first block differs for all chains using a given key. In other applications the IV may be varied.

This method extends single bit errors in the ciphertext to affect two successive blocks at the plaintext output. An enemy with control of the ciphertext can introduce controlled errors into the second of these blocks, therefore the method of error recovery should not trust chains in which errors are detected.

Cipher feedback is recommended for enciphering streams of characters when the characters must be treated individually. Like CBC it operates on chains — in this case chains of characters. The problem of codebook analysis at the starts of chains is similar but in communication by CFB operation the variability of the start of the chain is usually provided by varying the IV. Error extension is present also in CFB but the security implications are different because it is the first character of the error pattern in the plaintext output which is under the control of the enemy, if he can introduce ciphertext errors. The chain format should not allow important information to appear in the last character. This requirement is met, for example, if the last character is part of the sum check and cannot usefully be altered by the enemy.

Both CBC and CFB methods recover from bit errors, which is another way of saying that their error extension is finite. Loss of block synchronism in CBC or byte synchronism in CFB causes an indefinite garble. The case of 1-bit CFB is an exception, with no synchronization requirement beyond bit synchronism. It is often regarded, in consequence, as the most 'transparent' of the methods. Fortunately, cipher feedback is often applied with asynchronous, start-stop operation which restores the synchronism of characters immediately. Cipher block chaining is often applied to data which is in storage or is transmitted by a link procedure, such as HDLC, which restores synchronism at least in the next frame after a synchronization error. So the normal error recovery procedures can deal with the problem.

Output feedback is needed when error extension is undesirable. In this method of operation, synchronization errors are not recovered. Therefore loss of synchronism must be detected and notified in a signal back to the sender which causes a new IV to be sent, in order to restore synchronism. As an alternative where a return message is impossible, synchronization can be re-established so frequently that the loss due to an error is very brief.

The four methods of operation are versatile enough for nearly all applications. In order to guard against active threats the designer has to choose carefully the methods of forming the IV and transmitting it. He may also need to provide variability at the beginning of a chain by means of a serial number, a time and date or a random number. The four modes of operation are a set of tools, extending the versatility of the block encipherment algorithm but still requiring to be applied with some thought.

These methods do not exhaust the possibilities. We can construct other methods for using the block cipher on chains of blocks. For an example, Figure 4.13 shows how, operating on 64-bit blocks, encipherment can be incorporated into both the feedforward and the feedback paths. The symmetry argument helps us to generate such schemes. This one can be expressed by the equations

encipherment:

$$C_n = Ek_1[P_n +_2 Ek_2(C_{n-1})]$$

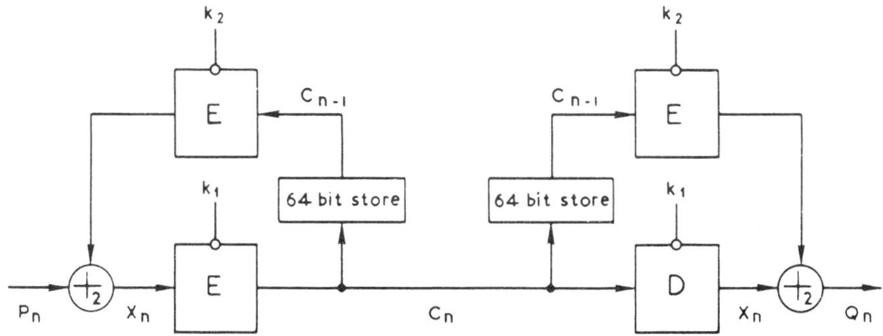

Fig. 4.13 A non-standard chaining method

decipherment:

$$\begin{aligned} Q_n &= Dk_1(C_n) +_2 Ek_2(C_{n-1}) \\ &= P_n. \end{aligned}$$

In error performance, synchronization etc., this method is like CBC but it uses two keys k_1 and k_2. To some extent these provide extra security, rather like the effect of double encipherment. We claim no special virtues for this method.

4.6. SECURITY SERVICES IN OPEN SYSTEMS INTERCONNECTION

The International Standards Organization has produced an architectural model for networking which is defined in ISO 7498 entitled 'Open Systems Interconnection — Basic Reference Model'. This OSI model is the framework for all dependent ISO standards. Unfortunately, it is not universal because the IBM *systems network architecture* (SNA) is the basis for many of today's networks. Among other computer systems suppliers, OSI is gradually achieving a dominant status. Though there is a general similarity in the approach taken by SNA and OSI, there are basic problems in interworking between these two systems. In this book we will follow the development of a systematic architecture for security services in OSI. The model is contained in part 2 of ISO 7498.[6]

Open systems interconnection is structured in the well-known seven layers from the physical layer at the bottom to the application layer at the top. The lower three layers, physical, data link and network form the communications structure and their standards have been put forward initially as CCITT recommendations. The higher layers, transport, session, presentation and application employ the communications facilities of the network layer. These upper layers can be considered as operating from end to end through the communications system. In principle, security features could be built into any layer of the OSI structure but the security architecture is more specific than this and defines, for each security service, those particular layers which can provide it.

The OSI security architecture makes a clear distinction between security *services* and security *mechanisms*. For example, encipherment is a *mechanism* which can be used as part of a number of services, of which the most obvious is that of confidentiality. The emphasis in the OSI document is on the services rather than the mechanisms, since the latter form part of dependent standards detailing how these services can be provided at the various layers of the architecture.

Definition of security services

The security services fall into five groups: confidentiality, authentication, integrity, non-repudiation and access control. Some of these can be broken down into sub-types according to the method of communication and the details of the service required. Definitions of the terms can be found in the Glossary starting on page 360.

Data confidentiality ensures that information is not made available or disclosed to unauthorized individuals, entities or processes. It can be applied either to the whole of the message or to specific fields within a message, which is called *selec-*

tive field confidentiality. Data confidentiality can also be distinguished according to whether the communication protocols used are connectionless or employ established connections between peer entities.

The word *authentication* is used in OSI in a specialized sense which is different from the way it has been used in the past. Authentication in OSI refers to the certainty that the data received comes from the supposed origin. In the context of connectionless service the term used is *data origin authentication* and is defined as 'corroboration that the source of data received is as claimed'. Where an association has been set up between peer entities the meaning of authentication can be strengthened and is contained in the term *peer-entity authentication* meaning 'corroboration that a peer-entity in an association is the one claimed'. The term authentication is not therefore extended to include the integrity of the data which are being transmitted.

Data integrity is treated as a separate concept, meaning the property that data have not been altered or destroyed in an unauthorized manner. Outside the context of OSI, this form of data integrity is sometimes called authentication but the OSI distinction can be useful and will be maintained in the remainder of this chapter.

In practice data integrity and data authentication are linked in two ways, they usually employ the same mechanisms and they are usually both required together.

Concerning the need for both services, if it can be assured that the data received come from the claimed source, this is not useful in practice if the data may have been changed in an unauthorized way. Conversely, if we know the data have been transmitted without any unauthorized changes, this is not useful unless we know that the data came from the claimed source and not from an imposter. Therefore, in practice, data integrity and authentication are required together rather than separately.

The mechanisms by which these services are obtained are the subject of chapter 5 and depend on cryptographic processes using a secret key shared between the communicating parties. In many cases, the possession of this key establishes the authenticity of the data source and at the same time allows the integrity of the data to be checked. Not all authentication mechanisms employ cryptography but the more secure forms of authentication always do.

Data integrity services can vary according to whether the communication is connectionless or employs a peer-entity association. Where an association exists the data integrity mechanism can be provided with a form of recovery mechanism. Data integrity services can be applied to the whole of the message or to selective fields, as was the case for confidentiality.

Repudiation is defined as 'the denial by one of the entities involved in a communication of having participated in all or part of the communication'. Non-repudiation is a service which can be provided in the OSI architecture in two forms, either with proof of origin or with proof of delivery. The concept of non-repudiation will be understood better in the context of digital signatures, which are the subject of chapter 9. Digital signature is one of the mechanisms by which they can be provided, the other is notarization by a trusted third party.

In the case of non-repudiation with proof of origin, the recipient of data is provided with proof of the origin of the data which will protect against any attempt by the sender to falsely deny sending the data or to deny its contents. A digital signature can provide this service in such a way that the message and signature can be taken to a third party who can verify for himself the origin of the data. Where notarization is employed, the authority of the notarization service must be called upon to provide this proof. Non-repudiation with proof of delivery is a similar service to protect against any subsequent attempt by the recipient to falsely deny receiving the data or to deny its contents. Essentially this requires the recipient to provide a signed receipt for the message or else to send that receipt to a trusted notarization authority.

The final category of security services recognized in OSI is *access control* and this can be applied either near the source of a communication, at an intermediate point or near the destination. Access control may be needed in order to protect a network against deliberate saturation by an opponent, but normally it protects an information service against unauthorized access and is therefore frequently used at the application layer or above. Within the network layer, an access control service can provide protection for a sub-network, allowing only authorized entities to establish communication within it. The transport layer can provide a similar service.

A special form of confidentiality is *traffic flow confidentiality* which protects against traffic analysis, namely the inference of information from the observation of traffic flows. It is difficult to obtain complete protection against traffic analysis and a number of mechanisms may be required simultaneously, such as encipherment at the physical layer together with traffic padding to confuse observation of traffic levels.

The mechanisms employed to provide these services include encipherment with both symmetric and asymmetric algorithms, digital signature, integrity and authentication mechanisms, traffic padding, routing control and notarization. There are also *pervasive mechanisms* which can be applied at all levels of the OSI architecture and three of these are distinguished. The first is *trusted functionality* which means that the parts of the system which carry out security functions must be both protected and trusted. The second is called *event detection and handling* and includes the detection of attempts to break into the system and the way in which this information is acted upon. The third category of pervasive mechanism is the *security audit trail* which provides a record of what has happened and enables the security of the network to be reviewed and action taken when trouble is found.

Security services in relation to the OSI layers

Many of the services envisaged in the OSI architecture can be provided as alternatives at different layers. Table 4.1 shows, for each of the defined services, in which layers it may be provided. The session layer provides no security services so this has been omitted from the table. Many of the services provided by the application layer will actually employ security mechanisms within the presentation layer. For example, if the presentation layer changes the coding or format of some messages this can only be done prior to encipherment; therefore encipherment to provide confidentiality at the application layer may, in practice, be carried out by the presentation layer after the recoding or reformatting has been done. In this table we have combined the two forms of authentication, peer-entity

Table 4.1 Location of security services in OSI layers

	Layer of OSI					
	1	2	3	4	6	7
Confidentiality						
Connection	Y	Y	Y	Y	Y	Y
Connectionless	—	Y	Y	Y	Y	Y
Selective field	—	—	—	—	Y	Y
Traffic flow confidentiality	Y	—	Y	—	—	Y
Authentication (both kinds)	—	—	Y	Y	—	Y
Integrity						
Without recovery	—	—	Y	Y	—	Y
With recovery	—	—	—	Y	—	Y
Selective field	—	—	—	—	—	Y
Non-repudiation (both kinds)	—	—	—	—	—	Y
Access control	—	—	Y	Y	—	Y

OSI layers: 1, Physical; 2, Data link; 3, Network; 4, Transport; 6, Presentation; 7, Application.

and data origin authentication, and the two forms of non-repudiation service, with proof of origin and with proof of delivery.

The letter Y in the table means that the service should be incorporated in the standards for the layer so that the provider of this layer can include the security services as an option. Where there is no Y the service cannot be provided.

For each of the OSI layers, work is proceeding to incorporate the security services as an option. Some of the progress to date is described in chapter 11.

4.7. THE PLACE OF ENCIPHERMENT IN NETWORK ARCHITECTURE

Even at a casual glance, a computer network can be seen to have a complex structure. It consists of switching centres, often called nodes, connected by communication links and joined to multiplexers or concentrators which provide the paths to the network's host computers and terminals. In such a structure, encipherment could be employed in many different ways. A more careful study of network structure reveals that the complexity goes beyond its communication system. It has further levels of structure incorporated in the host computers and the terminals.

When encipherment is used in a network, the security properties it gives are dependent on the layer at which it is applied. The OSI model shows that potentially it can be applied at any of six layers and the effect is has depends on the network structure and on the responsibility for operating different parts of the network. For example, in a private network the nodes may sometimes be regarded as trusted facilities whereas they would not be trusted in the shared use of a public network.

The OSI view of network structure encompasses both the communications system and the intercommunicating computers, terminals etc. For the present, we take a less-detailed view and look at the main alternatives for using encipher-

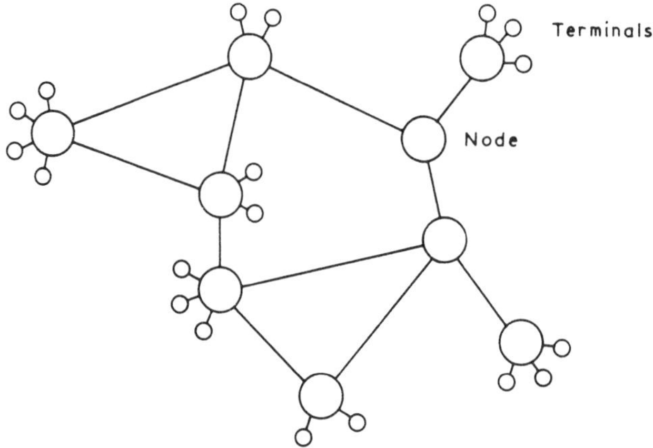

Fig. 4.14 Network of nodes — switches, multiplexors, etc.

ment in the communications part of a network. For this purpose we shall describe three ways in which encipherment can be employed, called, for the present purpose 'link-by-link', 'end-to-end' and 'node-by-node'.

Figure 4.14 shows a network as a set of switching centres or 'nodes' connected by communication lines.

The simplest place to apply encipherment is on the lines between the nodes. These communication lines are the most vulnerable part of the system. This is called encipherment 'link-by-link'. Since the lines are used for carrying bits or characters from one node to another, the encipherment in principle need take no account of the structure of the messages, such as the presence of headers. Therefore some forms of this encipherment are very simple. Information must be deciphered as it enters the nodes, because these nodes look inside the structure of the messages and must see the information in clear form. The information is completely vulnerable to attack while it is in the node: in the OSI model this form of encipherment is in layers 1 or 2.

Many users will require a different form of encipherment, which allows the content of their data to remain enciphered whenever it is in the network, even in the nodes. This can be achieved with 'end-to-end encipherment' in which the communication network is regarded as a single entity which is transparent to the enciphered message: in the OSI model this is encipherment in any of layers 3 to 7. The message content is enciphered at the point of entry to the network and deciphered at the point of exit. This makes the data as secure inside the nodes as elsewhere in the network.

We shall also consider another method of making data secure inside the nodes which is called 'node-by-node encipherment'.

Link-by-link encipherment

The line is a path of communication betwen nodes and uses a protocol or procedure which operates over this line. To be more precise, both the *physical* and *data link* layers of the architecture may be involved. Figure 4.15 illustrates link-by-link encipherment.

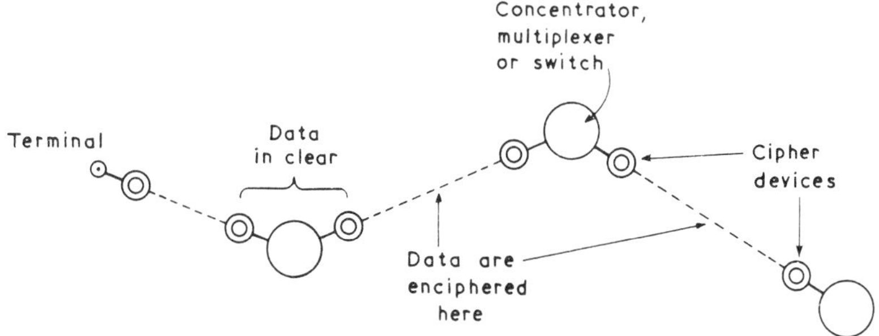

Fig. 4.15 Link-by-link encipherment

If we encipher at the lowest level, treating what passes over the link as a sequence of characters or bits, then an enemy who taps the line will see nothing of the structure of the information. He will therefore be unaware of the sources and destinations of messages and may even be unaware whether messages are passing at all. This provides 'traffic-flow security' whch means that we conceal from the enemy, not only the information, but also the knowledge of where the information is flowing and how much is flowing. By using the lowest level of encipherment, traffic-flow security can be obtained.

Consider a synchronous communication line, carrying a stream of bits. This stream can be enciphered using 1-bit cipher feedback. After the initialization, the process can continue to run as long as the line is open, recovering automatically from bit or synchronization errors. An enemy tapping the line will see only a random sequence of bits, which does not reveal whether data messages are passing or not. The designer might have to consider whether the 'infill' between actual messages should be a repetitive pattern or random. Having no knowledge of the beginning or end of messages, the enemy could not attempt any useful active attack. Supposing that he did change some of the information, this would appear as unpredictable changes in the plaintext. Detection of these changes at the lowest level of protocol would be difficult, but they should be detectable at a higher level if the error control mechanism is well chosen.

The situation is less clear when asynchronous, start-stop communication is used. In this case, also, the entire message content can be enciphered by 1-bit cipher feedback or by character-wide cipher feedback, leaving the start and stop elements unchanged. There remains one clue for the enemy, which is the rate at which characters are passing. If this traffic information must be concealed, the system requires an extra provision which sends dummy messages during quiet periods. These dummies can be generated and absorbed within nodes. In packet switching networks a logical channel could be devoted to this purpose and made to fill up a part of the unused capacity of the line.

What we have described belongs in the *physical* layer of the architecture. The *data link* layer protocols, such as HDLC or BSC, define 'frames' which pass between the nodes. If the encipherment procedure is applied to the data content of these frames, some more traffic information is given to an enemy tapping the line. This illustrates that raising the level at which encipherment is carried reduces the traffic-flow security. More details of the precise methods used for encipherment at layer 1, the physical layer, are contained in chapter 11.

Key distribution for encipherment at these low levels is straightforward. The exchange of the keys concerns only the two nodes connected by the line and is a local matter. The key can be changed at local discretion without affecting the rest of the network.

The limitations of this form of encipherment appear when we look at the higher-level functions of the network. The information contained in the nodes (the equipment, other than the lines or the link protocols) is in clear form. This could be tolerated only in a private network in which the degree of care in physical protection, personnel selection, maintenance etc. is just as high in the nodes as it is at the terminals or in the operation of host computers. Network errors, accidental program bugs or deliberate bugs introduced by an enemy might redirect a message to the wrong destination. If all the users of the network trust each other and have the same level of security clearance, this need not matter. However, even in a private network, information owned by one department may have to be withheld from another, so the kind of accident which might redirect messages should be avoided. In a public network, few users could accept the possibility of redirection of messages or the presence of clear messages in the switching or concentrator nodes.

To summarize, the use of encipherment at a low level has the advantage of providing traffic-flow security but the great weakness of protecting data only on the communication lines, and not providing adequate separation between one network user and another.

We therefore argue that the use of link-by-link encipherment should exploit its advantages and the weaknesses should be dealt with by encipherment at a higher level. When link-by-link encipherment is used in this way it should preferably be done at the physical layer, that is, applied to the individual bits or characters and without regard to the structure imposed by the link protocol. The best kind of low-level encipherment is at the lowest level, when it is supplemented by encipherment at a higher level.

End-to-end encipherment

Figure 4.16 depicts end-to-end encipherment, with an encipherment device interposed between each terminal and the network. The encipherment process must take into account the protocol of communication between the terminal and the network. The purpose of a call from one terminal to another is to set up a

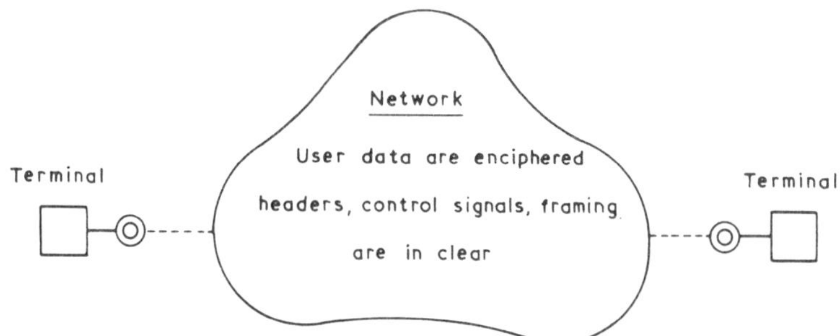

Fig. 4.16 End-to-end encipherment

circuit or virtual circuit for communicating data. Some of the data passing over the line between terminal and network are for use by the network, not the terminal, and these data cannot be enciphered because the network does not possess the key. Therefore, the encipherment device must distinguish between the 'user data', which are destined for the distant terminal, and all the rest.

To give an example from packet switching, we can see how this relates to the use of an X.25 interface. The lowest layer of the X.25 interface is the physical layer with, immediately above it, the data link layer that provides a HDLC mechanism for communicating packets. None of this changes when end-to-end encipherment is used. All that happens is that these mechanisms carry some enciphered data. At layer 3, in which packets are formed, there are about twenty-six different packet types. Most of these are concerned with call set up and clearing calls, with flow control, reset and restart. None of these packets is affected when enciphered data is handled. Only the two packet types that carry users' data need be affected, the DCE data and DTE data packets. In these, the header includes a logical channel number and the send and receive sequence numbers, forming part of the mechanism operated by the network itself to separate one virtual circuit from another and to control flow. This header must remain in clear. The user data which follows it is enciphered. The DTE and DCE interrupt packets also contain user data which is not examined by the network so that this information can be enciphered if the user wishes.

If encipherment is carried out by a unit interposed between the terminal and the network, as Figure 4.16 suggests, this unit must interpret some part of the X.25 protocol in order to find these two kinds of users' data fields, which are the subject of encipherment. In practice, encipherment might be carried out on the data before they enter the X.25 mechanism of the terminal, so avoiding much of the complexity. It is also likely that the various virtual calls carried by a single X.25 interface would have different keys, since they probably concern different users.

In order to specify exactly how encipherment of data over a virtual circuit would take place, additional protocol features are needed, if only to let the destination terminal know that a call is enciphered. In practice the protocol features associated with security are moderately complex. They can be provided either by a special security protocol or as an amendment or enhancement of one of the higher protocol layers.

All the operations of the network which set up a virtual circuit are unprotected by encipherment. The 'call request' and 'call accepted' packets can be observed passing through the network and so the progress of each call and the quantity of information carried can be monitored. There is no traffic-flow security in end-to-end encipherment. The best overall approach is to use encipherment twice, at a high level and again at a low level, because the virtues of high- and low-level encipherment are complementary. End-to-end encipherment protects the users' data everywhere in the network and link-by-link encipherment protects traffic flow information on the lines.

It might seem unreasonable to ask for traffic-flow information to be protected within the nodes, because this is where routing decisions are made, and they require traffic information to be exposed. But there are ways in which even this degree of protection can be achieved. It is done by providing in the network secure 'rerouting centres' which share encipherment keys with the users. The message hops from one re-routing centre to another, eventually finding the right destination. There can also be some dummy traffic created by the rerouting centres themselves, to conceal the flow. Such a scheme[7] requires the users to trust the rerouting centres with whom they share the keys necessary to determine the successive destinations of the messages. This scheme has been studied theoretically but seems likely to have only very specialized use, when extreme traffic security is the prime consideration.

The key distribution problem for end-to-end encipherment

The encipherment and decipherment functions at both of the terminals in Figure 4.16 must have the same key. Therefore each pair of terminals that wishes to exchange messages must somehow obtain a key for that purpose. If there are N terminals, potentially there will be $N(N-1)/2$ possible keys. In chapter 6 we shall see how session keys can be distributed using only N master keys, so the problem is not as bad as it seems. Nevertheless, key distribution for end-to-end encipherment has a degree of complexity which was not present in line encipherment, where the keys for each line were the concern of only the two nodes at the ends. The multiplicity of keys is a consequence of end-to-end encipherment's big advantage — it separates each call from all the others. If a packet goes astray it arrives at a place where it cannot be deciphered and little harm is done. We shall return to the key distribution problem in chapter 6.

Node-by-node encipherment

A hybrid method of enciphering in networks has been described under the name 'node-by-node' encipherment.[8] This shares with link-by-link encipherment the property that each key belongs to a particular line, so the key-distribution problem is made easier. This has the corresponding effect that an error in the system could deliver messages to the wrong destination. But if ease of key distribution is an overriding factor, node-by-node encipherment might be attractive. It is used in some EFT networks.

In this form of encipherment (unlike link-by-link encipherment) only the users' data are enciphered, not the headers. This enables the switches, multiplexers or concentrators to operate on the messages as though they were in clear form, but without having access to the users' data.

There remains the problem of changing the encipherment from one key to another as data come in on one line and go out on another. This function is carried out in a physically secure module which is contained in the node but given special protection. The secure module contains all the keys which the node must use, one for each of its lines. Figure 4.17 shows the principle. In order to carry out the re-encipherment the secure module must be told the identity of the input and output lines, after the node has made its routing decision. With these data, the module extracts the necessary keys, deciphers the data stream with the incoming line's key and re-enciphers it with the outgoing line's key. Clear users' data never exists in the less-protected part of the node and the risk associated with line encipherment is avoided.

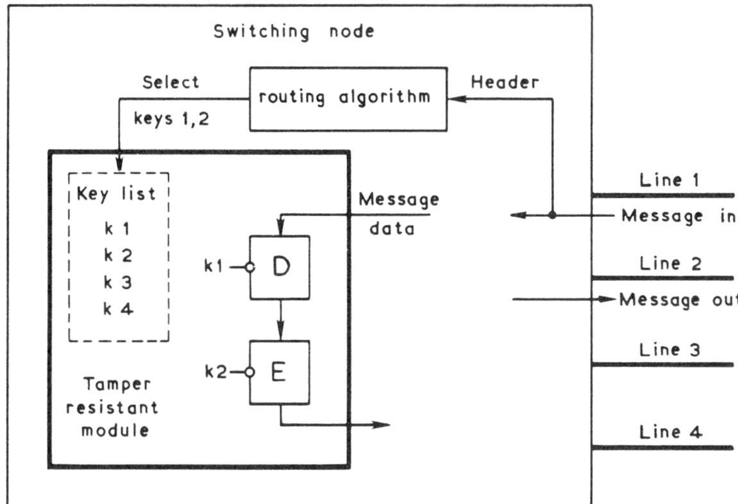

Fig. 4.17 Node-by-node encipherment

But in removing one of the risks of link-by-link encipherment, we have lost its main virtue. All the headers and routing data are sent through the network in clear. Therefore node-by-node encipherment achieves very little since it has neither the main virtue of link-by-link encipherment, which is traffic security, nor that of end-to-end encipherment, which is separation of the users' data.

A best place for encipherment in network architecture?

A rule of thumb could be formed from the preceding arguments, saying that encipherment is most useful at either a very low or a high level in network architecture. The low level, possibly in or just above the physical layer, provides traffic-flow concealment. The higher level, being one of the end-to-end protocol layers, separately protects data flows. But there are other possibilities. For example if link-by-link protection against active attack is needed encipherment at layer 2 may be preferred. For many commercial purposes, traffic flow security is not important and an end-to-end encipherment method will be all that is required.

The backbone of data security will remain end-to-end encipherment, in spite of the key-distribution problem.

REFERENCES

1. *DES modes of operation*, Federal Information Processing Standards Publication 81, National Bureau of Standards, US Department of Commerce, December 1980.
2. *Information processing — modes of operation for a 64-bit block cipher algorithm*, International Standard ISO 8372, International Organization for Standardization, Geneva, 1987.
3. *7-bit coded character set for information processing interchange*, International Standard ISO 646, International Organization for Standardization, Geneva, 1973.
4. *Guidelines for implementing and using the NBS data encryption standard*, Federal Information Processing Standards Publication 74, National Bureau of Standards, US Department of Commerce, April 1981.

5. Davies, D.W. and Parkin, G.I. 'The average cycle size of the keystream in output feedback encipherment', *Lecture Notes in Computer Science* No. 149, Cryptography, (Proceedings, Burg Feuerstein 1982), Springer Verlag, 263–279, 1983.
6. *Information processing — OSI Reference model — Part 2: Security architecture*, International Standard ISO 7498/2, International Organization for Standardization, Geneva, 1988.
7. Chaum, D.L. 'Untraceable electronic mail, return addresses and digital pseudonyms', *Comm. ACM*, **24**, No. 2, 84–88, February 1981.
8. Campbell, C.M. 'Design and specification of cryptographic capabilities', *Commun. Soc. Magazine*, **16**, No. 6, 15–19, November 1978.

Chapter 5 AUTHENTICATION AND INTEGRITY

5.1. INTRODUCTION

A piece of data, a message, file or document is said to be 'authentic' when it is genuine and came from its reputed source. In a broader sense we may also mean that the source had the authority to issue it. We might ask whether a payment is authentic in the sense that it corresponds to a genuine invoice. The concept of authentication is therefore very wide. For the purpose of this chapter we shall limit the concept to the following kinds of situation.

1. A message passes from A to B through a communication network. A and B may variously be persons, procedures or 'entities' in a hierarchy of protocols. For a general term we will call them *entities*. We want B to know that the received message came from A and has not been changed since it left A, in other words that it is genuinely A's message. It is not reasonable to separate the authenticity of the message from the identity of A, if we believe there may be enemies about. A message from A that has been undetectably modified is not different in its risks from an unmodified message that seemed to come from A but did not.

2. A message that has been stored by A is later read from the store by A or by B, a different entity. We want the reader of the data to know that it was stored by A and was not changed by others before it was read. Since stored data can be changed legitimately it may be necessary for the reader to know the originator and all those entities which subsequently altered the data. We shall see that a 'version number' is sometimes needed in stored data to keep track of alterations.

3. Two entities A and B undertake a transaction in which messages pass between them, alternately from A to B and B to A. For the transaction to be authentic, A must verify the identity of B, B must verify the identity of A, and the contents of each message in the interaction must be genuine as received — each after the first must be a genuine reply to the last message sent.

Authentication is the procedure for ensuring authenticity in any of these situations. The word *authenticator* is used for the specially constructed items of data used for this purpose. The question of secrecy of information can often be separated conceptually from its authentication. Obviously encipherment protects against passive attack (eavesdropping) whereas authentication protects against active attack (falsification of data and transactions).

In the terminology of Open Systems Interconnection a distinction is made between *authentication* and *data integrity*. Authentication refers to the means by which the parties to a communication verify each others' identities. If it is a two-way conversation with an established association, this is known as *peer-entity authentication*. For a single message which is not part of a connection the corresponding service is *data origin authentication*. The security service which prevents unauthorized modification or destruction of data is known as *data integrity*. We have already noted that authentication and data integrity are both required to make communications safe against active attack. One of these by itself provides little useful help. Furthermore, it is quite usual for the same mechanism to provide both authentication and data integrity. It is quite common to use the word *authentication* in the more general sense of providing all the protection needed against an active attack. We shall therefore adopt the usage of Open Systems Interconnection in the OSI context, but otherwise we shall use the term authentication in the more general sense, to include integrity.

Active attack is much more complex than passive since there are many different ways of altering data. Among the threats to be considered are; (a) alteration, deletion, addition; (b) changing the apparent origin; (c) changing the actual destination; (d) altering the sequence of blocks or items; (e) using previously transmitted or stored data again; and (f) falsifying an acknowledgement.

It is noteworthy that encipherment, by itself, only protects against inserting *new* data. Most of the attacks listed above can be carried out, in principle, using old data that is available already in enciphered form. We shall detail in section 5.7 below the extra precautions needed for full protection. Since it is difficult to protect a widespread system from enemy interference the most that can be ensured is that these active attacks are detected.

In this chapter we shall first consider methods of protecting data against accidental errors in preparation and transmission and then broaden the discussion to examine methods of authenticating data both by encipherment and by special authentication functions.

Papers and signed documents are an important part of business transactions, usually setting out instructions or promises to perform actions. Because of this reliance upon documents their integrity has assumed great importance. On the other side, criminals have found much to gain from the successful falsification of business documents.

The advent of digital communication networks is causing a revolution in commercial practice. Transactions of a type which have in the past involved exchange of paper documents are now conducted by network messages; this trend can be expected to increase until the transport of messages by data networks dominates commercial practice. The accuracy and integrity of transmitted messages are therefore of very great significance.

Recognition of a valid document has customarily relied upon the written signatures of the parties to an agreement. The signature serves at least two purposes — it indicates the assent of the signatory to the terms of the document and it vouches for the accuracy of the document as signed. Indication of assent by means of a signature is a kind of proof of origin for the document; we shall return to this in detail in chapter 9. Here we are concerned with the proof of accuracy of the document, which is one of the aspects of 'authentication'.

In the following sections we treat in turn defences against the various kinds of error, leading up to *message authentication* which aims to provide data integrity through the ability to detect deliberately introduced changes.

5.2. PROTECTION AGAINST ERRORS IN DATA PREPARATION

A first step in ensuring the integrity of stored or transmitted documents and messages is to check that they have been captured correctly at the preparation stage. There are many ways of achieving this, such as double keyboarding (preferably with different operators) with automatic comparison. An alternative method, particularly relevant in the preparation of financial documents, is to use a check digit or digits — a *check field*.

It is well known that keyboarding errors usually fall into well-defined classes, such as single characters typed wrongly, adjacent characters transposed and characters repeated or omitted. Where trailing zeros are involved the number of zeros may be wrong. A check field provides a simple function of the numerical characters in the important fields (date, amount, currency, account identity, transaction serial number etc.) of a document in such a way that it is highly probable that mistakes of these kinds will be detected.

Before the data entry (keyboard) operation begins, the data to be entered must be provided with a check field. Sometimes this is permanently attached; thus part numbers or account numbers can be constructed in this way — usually with a single check digit. After data entry the computer checks the correctness of the check field and rejects those which indicate errors. In the case of entirely new data the check field is provided by human effort, but this is not obviously better than having two independent operators enter the data. Use of a check field is not a substitute for the security provision of having at least two 'independent' people involved in data entry. An operator could maliciously enter wrong data as part of a fraud attempt, while giving the correct check field, since no secret is used in forming it. The principle of 'dual control' is applied by having one person enter the data and another check it, perhaps an inspector belonging to another department. The value of a check field in this case is that it reduces the number of genuine errors to be picked up by the inspector. It can also be argued that the presence of a large number of errors to be corrected increases the opportunity for fraudulent changes.

An excellent source of good check-field schemes is the International Organisation for Standardization (ISO) Draft Standard 7064[1] entitled *Data processing — check characters*. The kinds of error that occur in data entry have been measured, for example 80 per cent of all errors are single character errors and 6 per cent are adjacent transpositions. Check digits to detect errors in numerical fields are best calculated in modulo 11 arithmetic (since 11 is prime) and with different weights in each column. One method using modulo 11 arithmetic is in turn to add each digit, then multiply by 2. The process starts from the left with the initial value 0 and the check digit is chosen so that the final value of 'balance' is 1. Here is an example with the given number 530128; in the example the Roman symbol X is used to represent the value 10 in modulo 11 arithmetic.

		check	
digit d	5 3 0 1 2 8	4	
digit c	0 X 4 8 7 7	8	$c_i = 2y_{i-1}$
digit y	5 2 4 9 9 4	1	$y_i = c_i + d_i$.

The starting value of c is 0. The c digit is added to the given number's digit to produce y, in the first column this is 5. To get the next c value this is multiplied by 2, giving X in our notation. Then X+3=2, modulo 11, etc. Finally, the c value 8 requires a check digit 4 to produce the 'balance' 1. The effect of this calculation is a weighted sum, modulo 11 with the following weights, starting from the rightmost digit (the check digit):

1, 2, 4, 8, 5, X, 9, 7, 3, 6, 1, 2, ... repeating.

These values are 2^i, modulo 11, for $i=0, 1, \ldots$ and it can be seen that $2^{10}=1$, so the weights repeat after 10 columns. The appendix to chapter 8 (page 245) explains the mathematical basis for methods like this.

It is very important to note that use of a check field does not provide any protection against deliberate fraud. It is quite easy to generate a fraudulent version of a document which yields the same value of check digits as the genuine version. It is equally easy to create an entirely false document and generate a corresponding value for its check field, because the method of generating the check field is not kept secret. Nevertheless the check field has a useful role to play in handling data — to reduce genuine errors.

5.3. PROTECTION AGAINST ACCIDENTAL ERRORS IN DATA TRANSMISSION

In any communication system, users need assurance that their messages are delivered with neither accidental nor deliberate alteration. Accidental alteration of a message transmitted over a noisy channel is so frequent that it must be detected, and corrected either by retransmission or by application of a forward error-correcting code.

To detect accidental errors in a communication system a range of techniques is available; use of the eighth bit for parity in a 7-bit character code is perhaps the best-known and simplest example. On a slightly more sophisticated level, 'longitudinal parity' may be applied to a message, providing check digits on groups of characters by modulo 2 addition of successive words of 8 or sometimes 16 bits. Parity checks suffer from the serious defect that they may fail to detect the occurrence of a group of errors. For this reason methods have been devised to cope with a wider range of error patterns.

Cyclic redundancy checks

The best known error detection method is the cyclic redundancy check (CRC), which operates by treating the binary patterns as polynomial forms, 'dividing' a block of data at the transmitter by a predetermined polynomial and taking the 'remainder' of this arithmetic step as the check field which is appended to the data and transmitted with it. At the receiver, the received data is subjected to the same arithmetic process and the remainder from this operation compared with that received as the transmitted check field. Agreement of the two shows that it is highly unlikely that transmission errors have occurred. Recommendation V.41 of the International Telephone and Telegraph Consultative Committee (CCITT)[2] defines a standard version of this process providing a 16-bit CRC for use on communication lines; this is applied in line protocols such as HDLC. The CCITT V.41

CRC detects all single and double bit errors, all errors with an odd number of bits, all burst errors of length 16 or less, 99.997% of 17-bit error bursts and 99.998% of 18-bit and longer bursts. An extension of CRC can be applied to forward error correction, in which certain classes of errors can not only be detected but also corrected without calling for retransmission of the faulty block. This application is somewhat cumbersome and is not often found in practice. The more powerful the degree of forward error correction, the greater the amount of data expansion, and the overheads involved may be appreciable. An interesting proposal from British Telecom[3] brings forward error correction into play after it has been discovered that a particular channel is subject to serious noise, thus avoiding unnecessary overheads when a channel is performing satisfactorily.

Though checks of this kind provide strong protection against the effects of accidental error, they are not sufficient to protect transmitted data against active attack because the method of creating the check data, including the parameters of the CRC polynomial, is public knowledge. The active line tapper has only to divert the transmitted message, make whatever alteration he desires and recalculate the contents of the check field. The reconstituted message can then be sent on its way to the intended receiver where it passes the prescribed check and, in the absence of other evidence to the contrary, would be accepted as authentic. A more effective method is required in order to detect deliberate alteration.

5.4. DATA INTEGRITY USING SECRET PARAMETERS

The weakness of the methods we have discussed so far, from the point of view of secure authentication, is that they are based on parameters which are not secret — anybody can forge the data in the check field accompanying a message. Methods are needed which depend on a secret key known to sender and receiver only, information which is denied to any third parties. The distribution of this secret key represents problems like those of encipherment keys.

Figure 5.1 shows the principle of the method. It is based on a public algorithm represented by the function A(k,M). The key k is known to sender and receiver, but not to the enemy. The message M to be authenticated may be of any length and the function A(k,M) must depend on every bit of the message. The function A is the *authenticator* that accompanies the message to its destination. The receiver

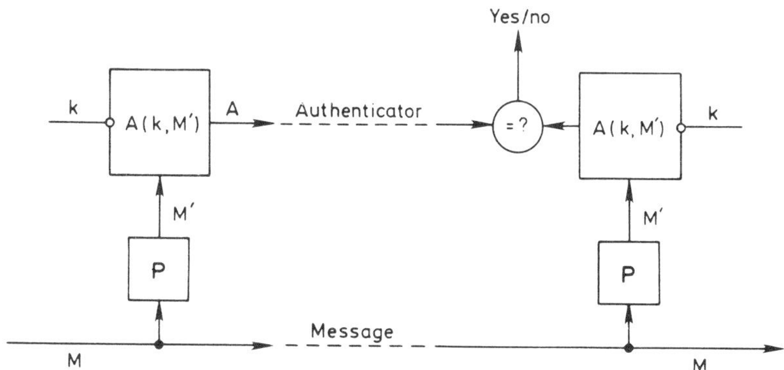

Fig. 5.1 Principle of data integrity checking

also computes A(k,M) and compares this with the received value A. If they are equal, the message is accepted. Sometimes this value A is known as a *MAC* or *message authentication code*. Another term is *cryptographic check value*, which is the OSI preference.

Authenticators have long been applied to financial messages sent by telex. With no processing equipment the computation must be done by hand and the nature of the algorithm, based on table lookup and addition, was matched to this requirement. In order to simplify the process, only selected fields are subjected to the algorithm. This is an example of *preprocessing* the message, shown in the figure by the function P. If a message is sent by a system that can change its coded form, the preprocessing must be done in such a way that only the essential and invariant features remain. For example a communication or storage system might treat 'new line' as equivalent to 'space' or it might remove or insert spaces or ignore the difference between upper and lower case letters. The preprocessing would then replace the 'new line' character by a space, reduce multiple spaces to one space and change lower-case letters to upper case. The message is transmitted without these changes; the altered form enters the authentication function. These operations carry a danger that significant distinctions may be obscured. Data communication networks of recent design are usually 'bit-sequence transparent', they preserve all details of the coded message — Teletex has this property for example. We recommend in these cases that there should be *no* preprocessing and that the unchanged message should enter the A(k,M) function.

The message conventions needed to complete the definition of authentication are, (a) the precise extent of M, where it begins and ends, and (b) the way in which the value A is formatted and located in the message as sent. A convention sometimes used is to reserve a place for A in the message format and replace it by a known constant value Q for the calculation of the authenticator. Figure 5.2 shows this method, where the fields F_1, F_2, . . . F_n comprise the message content. If Q, known to the sender and receiver, is otherwise kept secret, it becomes, in effect, part of the key.

The size of the authenticator field is fixed by the function A(k,M) and is typically between 16 and 32 bits, which is usually much shorter than the message. It follows that many messages have the same A value and this is a potential

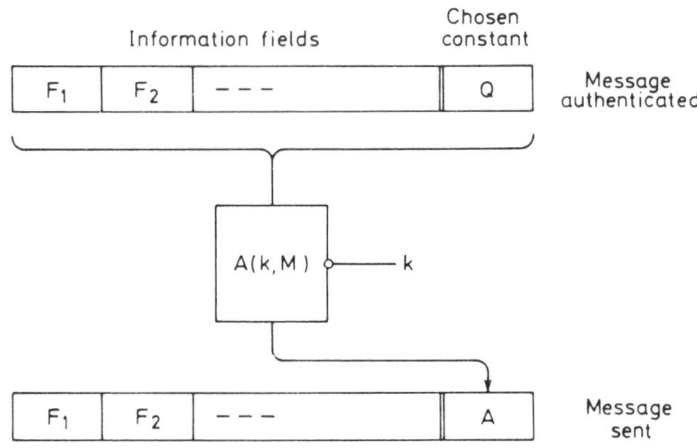

Fig. 5.2 Using a constant value, Q, in the authenticator field

weakness because the message could be changed. But without knowledge of the key the enemy does not know what other messages could be used. A good authenticator function produces pseudo-random output so the enemy is not helped by statistical knowledge, either. If the enemy relies on luck he can change the message and hope that the A value is still correct, but his probability of success is 2^{-n}, where n is the number of bits in the field. Even at 2^{-16} this is in many cases not a large enough probability to be useful. Trying enough times to have a useful chance of success would alert the receiver to a serious attack on the system. The customary size of the authenticator field is 32 bits.

An authenticator or cryptographic check value can be used on stored data in the same way as for communications. Anyone reading the data and knowing the key can check (with a very small chance of error) whether the stored data were changed by an enemy.

Like encipherment, the use of an authenticator value does not itself detect a replay attack, in which previously authenticated data are substituted, with the authenticator employing the same key. For example, during the currency of one key, data in a store may be changed by the legitimate user and the new authenticator stored. An enemy can restore the old values together with their authenticator and this interference is undetected by those legitimately reading and checking the data. The extra precautions needed to complete the defence against active attack are described later.

5.5. REQUIREMENTS OF AN AUTHENTICATOR ALGORITHM

The requirements for a good authenticator algorithm (integrity check) closely resemble those for a cipher but are somewhat easier to meet because there is no need for the inverse function to exist, as it must for decipherment. Like a cipher, the authenticator must protect its key from discovery, but this is not the full story because the function could be discovered, in principle, without knowing the key.

Cipher strength can be measured either by the 'known plaintext' or the 'chosen plaintext' criterion, and so can the strength of an authenticator. The most severe test would be to allow the enemy to submit a large set of messages and be given the corresponding A values. With this information, could he 'authenticate' a bogus message of his own? To make the criterion very strict indeed, he should not be able to find the A value for *any* message for which he has not been given it. In practice, authentication facilities are carefully guarded to prevent A values being given to unauthorized messages, so the 'chosen message' criterion is more severe than practice suggests. Also the enemy is interested in authenticator values for a specific class of bogus messages, not just any pattern of bits that suits his incomplete knowledge of the authenticator function. However, using the strict criterion is a good way to build in a large safety margin. The strict criterion may be the only one that can be defined easily, since it is impossible to characterize the set of 'good' messages for which A values will be 'supplied' to the enemy and the set of 'bogus' messages that he would like to introduce.

It is possible to describe how a good authenticator should behave, though this is not an effective criterion of strength. Changing any bit of message or key should alter about half of the bits of the A value. The full range of key values should be effective in the sense that there are no two key values for which the two authenticator functions $A(k_1,M)$ and $A(k_2,M)$ are closely related — whatever that means. For example these functions should exhibit no correlation as M is varied. These properties should be true for all keys and key-pairs respectively. As with a cipher, it is easy to state the properties a good function should have, but there is no set of tests which can be said to complete the testing of an authenticator.

As with a cipher, one of the possible methods of attack is an exhaustive search for the key. Whenever a message with its authenticator can be read, 'known plaintext' is presented to the enemy, but there are two reasons why key searching for authenticators is harder than for ciphers.

First the authenticator value is a small number, typically between 16 and 32 bits in length. If each key value is tested and there are, say, 56 bits in the key and 24 in the authenticator, the first message tested will pass about 2^{32} key values as candidates. A second message is needed and this reduces the number to 2^8 and a third message reduces these to the one actual key.

Secondly, each test of a key involves the whole of one message. Consider, for example, a 640-bit message using cipher block chaining (CBC) with the DES algorithm and compare this with the authentication based on the same method. There are ten blocks of message. Searching for the correct key out of 2^{56} possibilities requires on average 2^{55} tests. For the enciphered message, any block will serve for the tests except the first (because the initializing variable (IV) is not known). With one block alone, because it has 64 bits, there is a high probability that only one 56-bit key will be found to fit the ciphertext. In the case of authentication, 2^{55} tests on average must be made with the first message and each of these employs DES operations. The 10×2^{55} operations make a greater task for the searcher. After this, 10×2^{31} and 10×2^7 more operations are needed for the test with a second and a third message. Clearly all candidates passing the first test should be subjected at once to the other tests because the extra time involved is small and this enables the correct key to be identified before all 2^{56} key values have been tried. Key searching is only a 'last resort' method when other methods fail, nevertheless it is apparent that authentication is usually less vulnerable than encipherment.

The biggest practical use of authentication has been for financial messages, such as payments. For example the Society for Worldwide Interbank Financial Telecommunications (S.W.I.F.T.) system[4] (see page 289) uses an authenticator function but does not publish it. Another well-known authenticator is the 'Data Seal'[5] which is a proprietary system, also unpublished.

An ISO standard 8730[6] describes the way in which a Message Authentication Code is formed, in particular the format options for the text, field delimiters and character coding. There are optional rules for editing messages prior to authentication. ISO 8731[7] describes two approved algorithms. The first is the well-known application of DES in the cipher block chaining mode. The second is called MAA, for *Message authenticator algorithm*, described below.

The Decimal Shift and Add algorithm

A suggestion was made in 1980 by Sievi for a message authentication method. Sievi's method depends on the existence at sender and receiver of two secret ten digit decimal numbers, b1 and b2, which form the key that controls the operation of the algorithm. The two numbers provide the starting points for two parallel computations. The message to be authenticated is regarded as a string of decimal digits. This is immediately suitable for the numerical part of payment messages and for the alphabetic part it is proposed that a pair of digits is used to encode each alphanumeric character. The message is split up into blocks of ten decimal digits; any incomplete block at the end of the message may be padded with zeros on the right to bring it up to ten decimal digits. The algorithm works entirely in decimal arithmetic and is designed for use in a programmed decimal calculator having at least ten digit capacity. The method is called Decimal Shift and Add (DSA).

The blocks of the message to be authenticated are taken one at a time and enter two parallel computations until the message blocks are exhausted; then follow ten more cycles of the algorithm using zero blocks as input. The result of this is to produce, in the parallel computations, two ten decimal outputs, z1 and z2, and these are combined together to form the message authenticator.

The basic operation of the computations is a 'shift and add', defined as follows. Let D be a ten decimal digit number and x be a single decimal digit; $R(x)D$ is the result of a cyclic right shift of D by x places.

Thus:

$$R(3)\{1\ 2\ 3\ 4\ 5\ 6\ 7\ 8\ 9\ 0\} = 8\ 9\ 0\ 1\ 2\ 3\ 4\ 5\ 6\ 7.$$

Then the shift and add function is defined by

$$S(x)D = R(x)D + D \qquad (\text{modulo } 10^{10}).$$

Thus, in our example:

$$D = 1\ 2\ 3\ 4\ 5\ 6\ 7\ 8\ 9\ 0$$
$$R(3)D = 8\ 9\ 0\ 1\ 2\ 3\ 4\ 5\ 6\ 7.$$

Add, to produce

$$S(3)D = 0\ 1\ 3\ 5\ 8\ 0\ 2\ 4\ 5\ 7.$$

Since the arithmetic is modulo 10^{10}, carries beyond the tenth position are ignored.

Let us now consider the authentication of a message of two blocks, respectively m1, (1 5 8 3 4 9 2 6 3 7) and m2, (5 2 8 3 5 8 6 9 0 0). We select for b1 and b2 the values:

b1 = 5 2 3 6 1 7 9 9 0 2

and

b2 = 4 8 9 3 5 2 4 7 7 1.

The two parallel streams of computation begin by adding m1 to b1 and b2, respectively, modulo 10^{10}.

STREAM 1		STREAM 2	
m1	1 5 8 3 4 9 2 6 3 7	m1	1 5 8 3 4 9 2 6 3 7
+b1	5 2 3 6 1 7 9 9 0 2	+b2	4 8 9 3 5 2 4 7 7 1
result p	6 8 1 9 6 7 2 5 3 9	result q	6 4 7 7 0 1 7 4 0 8.

These results are the first intermediate values, p and q, now to be subjected to the first 'shift and add' operations. For stream 1 the number of places shifted is determined by the first digit of b2, (4), and for stream 2 by the first digit of b1, (5). So we have:

$$
\begin{aligned}
p &= 6\,8\,1\,9\,6\,7\,2\,5\,3\,9 & q &= 6\,4\,7\,7\,0\,1\,7\,4\,0\,8\\
R(4)p &= 2\,5\,3\,9\,6\,8\,1\,9\,6\,7 & R(5)q &= 1\,7\,4\,0\,8\,6\,4\,7\,7\,0\\
S(4)p &= 9\,3\,5\,9\,3\,5\,4\,5\,0\,6 & S(5)q &= 8\,2\,1\,7\,8\,8\,2\,1\,7\,8.
\end{aligned}
$$

Let us call these values r and s, and add the next message block, m2, in each stream.

$$
\begin{aligned}
r &= 9\,3\,5\,9\,3\,5\,4\,5\,0\,6 & s &= 8\,2\,1\,7\,8\,8\,2\,1\,7\,8\\
+\text{m2} &= 5\,2\,8\,3\,5\,8\,6\,9\,0\,0 & +\text{m2} &= 5\,2\,8\,3\,5\,8\,6\,9\,0\,0\\
\text{result } u &= 4\,6\,4\,2\,9\,4\,1\,4\,0\,6 & \text{result } v &= 3\,5\,0\,1\,4\,6\,9\,0\,7\,8.
\end{aligned}
$$

These are the second intermediate results, u and v; there follows the second shift and add operation. For stream 1 the number of places shifted this time is determined by the second digit of b2, (8), and for stream 2 by the second digit of b1, (2). So we have

$$
\begin{aligned}
u &= 4\,6\,4\,2\,9\,4\,1\,4\,0\,6 & v &= 3\,5\,0\,1\,4\,6\,9\,0\,7\,8\\
R(8)u &= 4\,2\,9\,4\,1\,4\,0\,6\,4\,6 & R(2)v &= 7\,8\,3\,5\,0\,1\,4\,6\,9\,0\\
S(8)u &= 8\,9\,3\,7\,0\,8\,2\,0\,5\,2 & S(2)v &= 1\,3\,3\,6\,4\,8\,3\,7\,6\,8.
\end{aligned}
$$

At this stage, the message data having been exhausted, the construction of the authenticator should properly involve ten more cycles with zero message blocks. We omit these to shorten our description and take the two last results to represent z1 and z2. The algorithm concludes by adding z1 and z2, modulo 10^{10}

$$
\begin{aligned}
\text{z1} &\quad 8\,9\,3\,7\,0\,3\,2\,0\,5\,2\\
+\text{z2} &\quad 1\,3\,3\,6\,4\,8\,3\,7\,6\,8\\
\text{result} &\quad 0\,2\,7\,3\,5\,6\,5\,8\,2\,0.
\end{aligned}
$$

This authenticator has ten decimals but evidently it was considered that a six decimal value is sufficient, giving a 10^{-6} probability of a false match. To reduce the size while using all the ten digits, the number is 'folded' by having the top four digits added modulo 10^6 to the other six, thus

$$
\begin{aligned}
&\quad 5\,6\,5\,8\,2\,0\\
&\quad 0\,2\,7\,3\\
\text{result} &\quad 5\,6\,6\,0\,9\,3.
\end{aligned}
$$

This final value is the message authenticator (except for the short-cut we made for ease of explanation).

Successive digits of b1 and b2 control respectively the amount of shift in streams 2 and 1 at each stage. The number of steps is always more than ten because of the added ten blocks of zeros. After each ten steps the digits of b1 and b2 are used again in succession. Thus each key value enters in two ways. It forms the starting value of one stream of calculation and controls the shift operations of the other.

The methods by which this algorithm defeats an enemy can be looked at in different ways. If all the calculation had been made with modulo 10^{10}, there might have been an arithmetic way to analyse the result, but the cyclic shift operations do not fall into this category. They are, in fact, calculations with modulus $10^{10}-1$, since each time a digit d is moved cyclically it changes the value by $(10^{10}-1)d$. It is this combination of calculation with two moduli that complicates analysis. Another source of strength is the combination of the result of two parallel calculations. If the ten digit result were known this would give no easy way to calculate backwards. The final set of ten zero blocks helps to reduce the vulnerability of the final message block. Otherwise a single digit change of the final message block would reveal the parameters in the last shift operation, which are two of the key digits. The use of each key number in two distinct roles complicates any attempt to separate out the influence of one key digit.

The DSA algorithm is well designed to exploit the abilities of a programmed decimal calculator. It was considered as a possible international standard, but, in the event not adopted. Small calculators employing the DSA algorithm are now used in home banking both to provide a message integrity check and also for the authentication of the user to the bank. This application is described in chapter 7.

Message Authenticator Algorithm (MAA)

The ISO working group recognized that more than one type of authenticator algorithm was needed and there seemed to be a need for an algorithm suitable for programmed computers of the binary type. Though it is known that the S.W.I.F.T. and Data Seal Algorithms meet this need, they are not available for use as a published standard.

This requirement and the need expressed by some UK banking organizations led one of us and a colleague to produce a proposal for an authenticator algorithm suited particularly to binary computation. We called it a 'main frame' algorithm because it is based on 32-bit arithmetic and logical operations which are traditionally found in a main frame computer. The advancing powers of microcomputers soon changed this situation. There are both hand-held calculators and microprocessors which can conveniently implement this algorithm, but it was primarily designed to give a combination of speed and security in any 32-bit computer which can rapidly produce the double-length result of multiplying together two 32-bit binary numbers. Its design was suggested by the DSA algorithm and it employs the same general principles. It is now incorporated in the international standard ISO 8731-2 and is known as *MAA*.[7]

The 'shift and add' operation of the DSA algorithm is similar to a multiplication with modulus $10^{10}-1$ in which the multiplier takes the binary form 10 . . . 01, having an adjustable number of zeros between the two 1 bits. Using an actual multiplication enables this same operation to be used for 'mixing' the digits between each introduction of a new block of message, with a less restricted multiplier value.

This multiplication scheme presents two dangers, the possibility that the result may become (and remain) zero and the property that, if the modulus has factors, any of these which appear in the intermediate result will remain there through the rest of the computation. To avoid the first of these problems the multipliers were deliberately constructed with certain bits fixed as zeros and ones, and to

```
                 V: = CYC(V);                  V and W derived from the key
                 E: = XOR(V,W);

X: = XOR(X,M_i):      Y: = XOR(Y,M_i);         M_i is a message block
F: = ADD(E,Y);        G: = ADD(E,X);
F: = OR(F,A);         G: = OR(G,B);            A,B,C,D, are constants
F: = AND(F,C);        G: = AND(G,D,);
X: = MUL1(X,F);       Y: = MUL2A(Y,G)          Modular multiplications, see text
```

Fig. 5.3 The basic loop of a 'main frame' authenticator calculation

avoid the second problem one of the parallel computations is done with modulus $2^{32}-1$ and the other with modulus $2^{32}-2$. The second of these is $2(2^{31}-1)$ and $2^{31}-1$ is a prime.

Figure 5.3 shows the basic loop of the authenticator calculation with the two streams of computation shown side by side. CYC(V) is a 1-bit cycle left shift of V, MUL1 is multiplication modulo $2^{32}-1$ and MUL2A is multiplication modulo $2^{32}-2$. A, B, C and D are constants which help to condition the multiplier values F and G. The ADD operation is modulo 2^{32} and the logical operations operate in parallel on the 32 bits. Each multiplier is derived from the other calculation stream using also the value E which is derived from V and W and changes at each step (repeating after 32 blocks).

As in the DSA algorithm, the final authenticator result is derived from both streams of computation, in this case by an XOR operation applied to the final results. A 32-bit authenticator is derived and it is left to the application to decide the size of authenticator field needed and the way in which the 32 bits will be used to generate this value.

The keys which control this authenticator are two 32-bit numbers J and K. These are used in a 'prelude' which is a complex calculation, employing multiplications with the same two moduli, and which generates six 32-bit numbers employed to control the computation. Though the prelude is a relatively complex calculation, it is used only once when new keys are introduced. The six values can then be stored and employed in any number of authenticator calculations. Two of these values are the starting values for X and Y in the two streams of computation, two of them are V and W used to derive the changing values E and two of them, S and T, are used as final blocks appended to the given message to round off the computation, this part of the procedure being known as the 'coda'.

The MAA algorithm derives its strength from the difficulty of analysis when calculations are carried out with a number of moduli. Not only are there multiplications with two different moduli, but also arithmetic operations modulo 2^{32} and the logical operation XOR. Also, the logical operations which generate the multipliers F and G are non-linear, so that any arithmetic analysis is exceptionally complex. By deriving the six quantities from the key in a complex manner any method of analysis which was capable of partly deriving one of these values would be unlikely to give a clue to the others. It is particularly important that the final blocks S and T which are appended to the message for authenticator purposes are secrets derived from the key values.

The computation time for authenticator values is proportional to the message size except for the small addition of blocks S and T. This makes it practicable to use the algorithm on short messages as well as on long files. For very long mesages the standard specifies a special mode of operation.

Authentication methods using the standard 'modes of operation'

Wherever DES hardware is available it is convenient to use it for authentication and for this purpose 'modes of operation' have been defined in Federal Information Processing Standard (FIPS) 81 and in ISO 8731-1.[7,8] Any block cipher algorithm could be used in these modes, but in practice the DES will normally be employed. This is convenient only if a hardware DES function is available because the algorithm is slow and inconvenient in software. There are two authentication methods based respectively on the cipher feedback (CFB) and cipher block chaining (CBC) methods used for enciphering messages (see pages 81–90). CBC operates always on 64-bit blocks, but a message can be padded out with zeros to fill the last block. CFB operates on appropriate-sized characters such as 8-bit bytes where ISO- or ASCII-coded character sets are used. The choice between the alternative methods will depend on whether the data are presented in characters or larger units. For use in financial messages a standard is available[9] giving more detail.

For the generation of an authenticator in CBC mode, the message to be authenticated is enciphered as a chain of 64-bit blocks, any incomplete final message block being padded out with zeros. The generated ciphertext is not transmitted. The authenticator is derived from the final output block, taking the most significant m bits. No additional DES operation is required at the end of the process because it is not possible for an enemy to make a change in the plaintext and a compensating change in the authenticator. Figure 5.4 shows how the process of CBC authentication operates. All the cipher blocks except the last are used only for feedback. The most significant m bits of the last block C_n is used to provide the authenticator. In US standards the authenticator is sometimes called the *message authentication code* (MAC).

In CFB mode the text to be authenticated is subjected to a CFB encipherment process. When all the data have been enciphered and the last unit of ciphertext fed into the DES input register (completing the CFB operation as normally understood), the DES device is operated once more, producing a new state of its output register. From the output register, the most significant m bits are extracted to form the authenticator. Figure 5.5 shows this method; note that if the m bits of MAC were taken from the DES input register at the end of the text

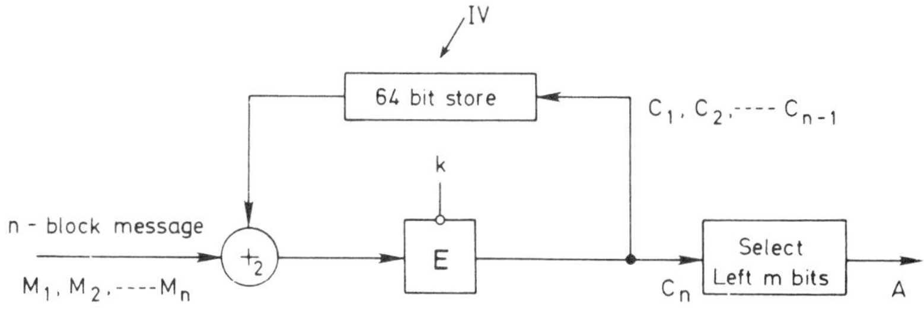

Fig. 5.4 Authentication by the cipher block chaining mode

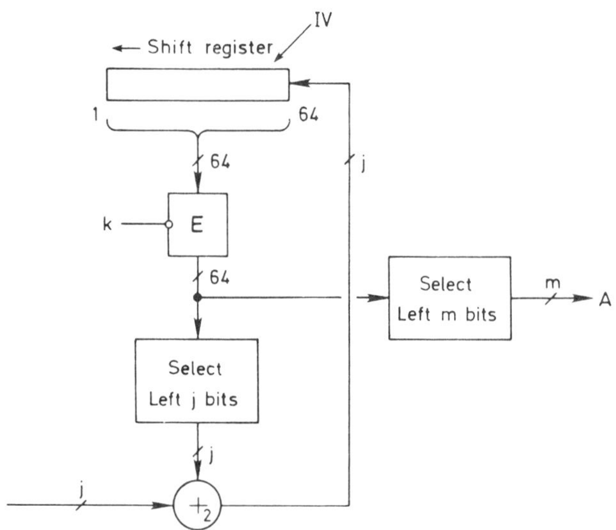

Fig. 5.5 Authentication by the augmented cipher feedback mode

encipherment, then for large enough m it would be possible to make compensating changes in the last plaintext character and the authenticator, hence the extra operation of the DES. This CFB method is not much used. The CBC mode is used in practice.

The number of bits used in the authenticator is subject to agreement between the communicating parties; it should be as long as is practicable. For US Federal telecommunications applications it is recommended[8] that the authenticator should be at least 24 bits in length. For financial purposes the US standard recommendation[9] is that m should be 32. m should be large enough that the risk of a matching value occurring by accident with a bogus message, which is 2^{-m}, is small enough. Making m longer than necessary would add to the message overhead due to the A value it carries.

The authenticator values obtained either by CBC or CFB mode depend on the entire message stream or chain. In any application the start and finish points of this stream or chain must be defined and also the location, size and format of the authenticator field. The authentication depends also on the IV which is loaded into the 64-bit register at the start of the operation. It is possible to use a secret IV, as is customary for CBC encipherment. However, the purpose of the IV is to prevent codebook attack on the first part of the ciphertext and this is not relevant to authentication. The simplest convention is to use a zero IV and this is adopted in several standards.

5.6. MESSAGE INTEGRITY BY ENCIPHERMENT

Encipherment of a message offers, in itself, a kind of integrity. If the sender and receiver of a message share the secret key, receipt of an enciphered message which deciphers into a sensible plaintext message gives the receiver some assurance that the message has arrived without unauthorized changes and came from the other owner of the key. 'Sensible' in this context implies that the message contains sufficient redundancy. When an authenticator is used, this provides the redundancy. Nearly all communications carry redundancy, but it is not always easy for a receiver's processor to test for correctness, so in practice redundancy is added to messages when encipherment is used for authentication. The extra redundancy is a new field in the message sometimes called a 'manipulation detection code' (MDC). The choice of a method for calculating MDC values, which is not a trivial matter, is considered in the next section.

An attempt by an intruder to alter selected bits of ciphertext generated in CBC mode leads to random changes in the deciphered message, which should be obvious to the recipient. A single bit alteration in one ciphertext block will result in random changes in the corresponding plaintext block; note, however, that the exactly corresponding bit will be affected in the next plaintext block. A garbled plaintext block in a CBC decipherment should be taken as a warning to mistrust the plaintext of the next block. It is safer to discard the whole message and request complete retransmission. If the transmission is in CFB mode, then the intruder can also select the plaintext bit he wishes to alter, but his action should be apparent to the receiver because of the error propagation property of CFB encipherment; at least sixty-four following bits will be subject to unpredictable change. The final character of a CFB transmission can be altered without detection since there are no following characters to betray the change. It should never carry vital data. In CFB the plaintext unit preceding a garbled section should be mistrusted and retransmission is again the safest course.

Choice of the plaintext sum check method for authentication

Drafts of the Proposed Federal Standard 1026[10] which contained a section on encipherment at the data link layer proposed a sum check on plaintext called the 'manipulation detection code' (MDC). This check field would be produced by splitting the message or 'chain' into consecutive 16-bit words (the last one padded with zeros if necesary) and adding these words together modulo 2. The result was a 16-bit field that was appended to the plaintext before encipherment; an alteration to the ciphertext was expected to be very likely to upset the relationship between text and MDC on decipherment, thus detecting an active attack. The proposed FIPS for 'Data Integrity' made a similar proposal. Unfortunately this choice of sum check method was bad and it will be replaced by a better method. The reason for this weakness is instructive and will be described.

Consider first the CBC method, because this is the easiest to analyse. Figure 4.2 of chapter 4 shows how a ciphertext chain $C_1, C_2, \ldots C_n$ is deciphered. If the IV is I, the resultant plaintext chain is

$$I +_2 \mathrm{D}k(\mathrm{C}_1),\ \mathrm{C}_1 +_2 \mathrm{D}k(\mathrm{C}_2),\ \ldots\ \mathrm{C}_{n-1} +_2 \mathrm{D}k(\mathrm{C}_n).$$

The modulo 2 sum of these 64-bit words is

$$I +_2 \sum_{i=1}^{n-1} \mathrm{C}_i +_2 \sum_{i=1}^{n} \mathrm{D}k(\mathrm{C}_i).$$

If a modification can be made to the ciphertext chain which leaves this 64-bit sumcheck unaltered, it will also leave the modulo 2 sumcheck of 16-bit words unaltered. Such modifications are easily made. Any pair of blocks can be interchanged since this only alters the sequence of blocks in both sums, except that the last block C_n cannot be used because C_n does not occur in the first sum. Any ciphertext block can be inserted into the sequence twice, adjacent or not, since equal blocks cancel in both sums, if the end of the sequence is avoided. Changes of this kind do not alter the 16-bit modulo 2 sum and will not be detected.

Cipher feedback is a little more complex, but an example of a modification which escapes detection can be given. Assume that 8-bit CFB is used, so that eight characters fill the shift register. Then the text consisting of the character sequence

A B C D E F G H

can be modified to

A B C D E F G X A B C D E F G X A B C D E F G H

without changing the 16-bit sumcheck. The proof of this is left as an exercise for the reader. The method of proof we used for CBC will serve, but the eight characters of IV must be assumed to precede the given character streams and these will enter into the calculation.

It is a simple matter to avoid this problem by addition modulo 2^{16} for the 16-bit word, that is to say ordinary addition with suppression of carries beyond the sixteenth bit. It is true that a ciphertext can be introduced if it is inserted 2^{16} times, but this is neither likely to be useful nor to go undetected.

It has not usually been suggested that a modulo 2 sumcheck can be applied to the output feedback mode of encipherment, since compensating modifications are so easy to produce. Nor will modulo 2^{16} arithmetic avoid the problem entirely because changes to the most significant bits of the 16-bit words add by modulo 2 addition without affecting other bits and these changes can be used freely if an even number of them are changed. This problem can be overcome by using modulo $2^{16}-1$ arithmetic, in which carries beyond the sixteenth bit are cycled round to add into the least significant bit. To add with modulus $2^{16}-1$ the 16-bit words are added into a register large enough to avoid losing carries. Assuming that a 32-bit register is long enough, the final step is to add the top 16 bits to the bottom 16 bits. If a carry is produced this carry is added into the least-significant bit (and no further carry is produced).

This example of the faulty MDC illustrates that it can be difficult to find a system's vulnerabilities. The methods using modulo 2^{16} and $2^{16}-1$ arithmetic have some weaknesses, but not, we hope, serious ones.

Encipherment or authentication?

We have seen that encipherment with a secret key provides one form of message integrity; it is now fair to ask: Why should it be necessary to consider any other method? Why not encipher with a secret key all messages that require authentication? The reason is that, whereas authentication may be required in a given context, encipherment may be an embarrassment.

Consider a system in which mesages are broadcast to many destinations, one of which has the responsibility of monitoring authentication. Here transmission must be made in plaintext, with an associated authentication field. One terminal checks the authentication field and holds the key for this purpose. The other terminals are simpler in design and possess no authentication capability. They rely upon a general alarm being given by the special designated terminal if an authentication failure is detected.

Another possible scenario involves a point to point communciation in which the receiving terminal has a heavy load of processing responsibilities, such that it cannot afford to decipher all received messages. Authentication is carried out on a selective basis, messages being chosen at random for checking.

Yet another very important application of authentication of a plaintext is that of a computer program text. Here it is essential that the text shall be capable of interpretation by a processor; decipherment each time the program was required to run would be wasteful of resource, therefore authentication by full encipherment is inappropriate. It is possible to authenticate the program text by appending an authentication field. This can be checked whenever assurance is required of the integrity of the program.

Authentication without a secret key

Neither encipherment nor authentication can be provided 'out of the blue'. The methods we have discussed so far have depended on a secret key, common to both sender and receiver. It is also possible to base authentication on an *authentic* number. This is no more paradoxical than basing the secrecy of enciphered data on the secrecy of a key because what has been achieved is that a large message is authenticated by a much smaller one.

In place of the authenticator A(k,M) let us use a 'one-way function' A(M) of the message alone. This must have the property that if a set of messages M is known together with their A(M) values it is nevertheless exceptionally difficult, given some other value A_0, to find *any* message M such that $A(M) = A_0$. A necessary condition is that the value A must have a large range, for example it could have 64 bits, large enough to prevent a table of messages being compiled and sorted on their A values so that for most A values a message could be found. Such a function is sometimes called a 'hash function' but these 'pseudo-random' functions are used for several purposes and some hash funcitons are not suitable for authentication. An example of a suitable function appears in chapter 9 (page 264).

To authenticate message M, the quantity A(M) is calculated and sent to the entity receiving or reading the message n such a way that A(M) is certainly authentic. With its aid M can be checked to see if A(M) agrees with the authentic value. If it does, M is pronounced authentic since it is assumed that no enemy could have found a bogus M that fitted with A(M). Applications for this kind of authentication are not widespread, but examples will appear later. Suppose, for example, that Ann has sent Bill a large file of data and Bill wants to ensure its authenticity. If both can agree on a one-way function A(M) then a telephone call will enable Bill's A(Mb) to be checked with Ann's A(Ma). If they are equal it is a good bet that Ma = Mb, so Bill can accept his version. The telephone call is assumed to confer authenticity because each knows the other's voice and can chat about mutual interests which would betray an impersonating enemy. Note that methods which do not use a secret key are vulnerable to the 'birthday attack' described in chapter 9, section 9.3.

5.7. THE PROBLEM OF REPLAY

Whether authentication is by encipherment or the use of an authenticator field, the defence is not complete because an authenticated or enciphered message can be read by an enemy and later used as part of a bogus transaction, provided this is done during the lifetime of the key. It is this possibility of 'message replay' which complicates authentication. Replay can take the form of changing the message sequence or replacing a stored data item by a value it held previously. In this section we shall consider the set of messages which make up a transaction between two communicating 'entities'.

The solution to the problem of replayed messages is quite simple in principle; it requires that each message should be different from all preceding messages using the same key and that the receiver should be able to test the 'newness' of each message. Furthermore, the sequence of messages which together form a transaction must be linked in some way so that their sequence can be checked. Protection against message replay would be complete if each communicating entity compared its received messages with all those it had previously received with the same key and rejected any messages which were copies, but this would be laborious and cannot be a general solution to the problem.

Authentication of transactions is basically a two-way requirement with each entity needing to authenticate the other. There are applications in which one-way authentication is sufficient, but the need for two-way authentication must always be considered. There is no longer any economic reason for transactions to do without two-way authentication.

In the conventional exchange of messages, each message is a response by an entity to the previous message which it received from the other. Typical short transactions employ one, two or three messages and there are circumstances in which a three-message protocol is essential for two-way authentication.

Use of a message sequence number

Two entities which frequently exchange transactions can maintain a sequence number n which increases by one for each message they exchange. Figure 5.6 shows how this number would be used in an exchange of messages M_1, M_2 . . ., with authenticators in each message. Each message contains a sequence number and each entity keeps a note of the next number it will send and the next number it expects to receive. The authenticator is computed over the whole of the remainder of the message, including the sequence number, so neither the message content nor the sequence number can be falsified without detection. Since each party knows which sequence number to expect next, deletion of messages, replay of previously transmitted messages or changing the sequence of messages is detectable by the receiver.

The requirement to link together the messages which make up a transaction can be met in other ways. Figure 5.7 shows a scheme in which the authenticator of each message is repeated in the following message. Message M_2 contains a copy of the authenticator A_1. Its own authenticator A_2 covers both the message content and A_1. For practical purposes, the authenticator is a random number which links together or 'chains' the sequence of messages so that it is extremely

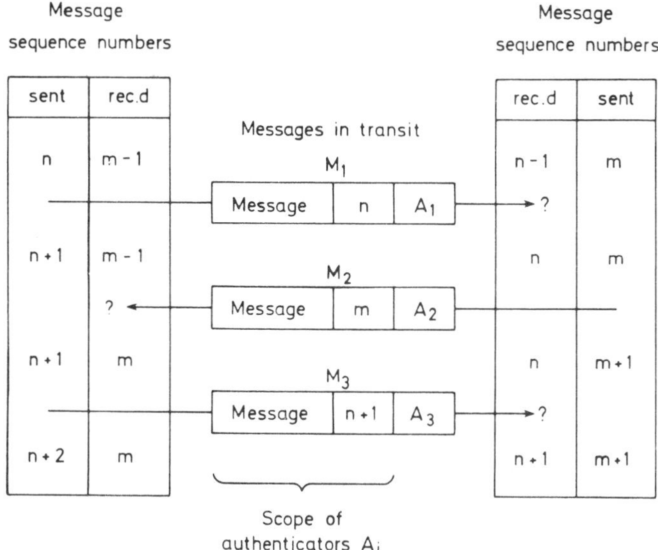

Fig. 5.6 Sequence numbering an exchange of messages

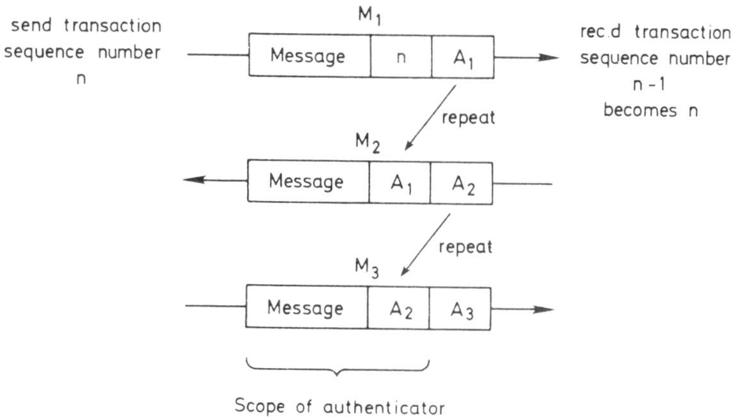

Fig. 5.7 Chaining messages by their authenticator values

unlikely that a false chain can be assembled by an enemy. The size for the linking authenticator is chosen with the criterion that the probability of an enemy finding a link by accident is too small to make it worthwhile to wait and search; typically from 16 to 32 bits are used.

Where messages are enciphered this provides a convenient way in which messages may be chained together by making the process of CBC run over from each message to its response. Thus the final 64 bits of each message become the initializing variable of the succeeding one. This form of chaining does not occupy space in the message. A similar method of chaining can be used with CFB.

When messages are chained in one of these ways (by repeating the authenticator or deriving the IV from the previous message) the sequence number n in the first message identifies the whole transaction. Supposing that transactions can be initiated by either entity it must be decided whether each maintains its own transaction sequence number or a common sequence is used for transactions initiated by either. If there are two separate transaction sequences, the origin of the transaction must be included in the first message, otherwise the enemy might record a transaction by A and later play this back to A as though it was a transaction initiated by B.

The sequence number can be initialized, for example at zero, whenever new keys are distributed. The field of n must be large enough to accommodate all the transactions which can take place before the next key distribution. In many systems this is an argument for changing the key frequently, to keep the size of the n field reasonably small. If, due to a communication or processing error, the sequence numbers held by the two entities get out of step, the resynchronization procedure must not allow the undetected insertion of stored messages by an enemy. This requires that the new value of n should be greater than any previous value held by either of the entities. Resynchronization according to this rule is simplified if the requirement of strictly incrementing n is dropped and the receiver merely checks that n has increased with each transaction he receives. This relaxed rule allows the sender, optionally, to hold a single sequence number for use in all the transactions originated (or messages sent), but this does not remove the need for each entity to keep a separate sequence number corresponding to each source for checking all received transactions (or all received messages).

The use of random numbers for entity authentication

In a large community, authentication can be based on session keys derived from a key distribution centre (see chapter 6), so that each entity does not need keys for all those which might communicate with it. Alternatively, public key ciphers (see chapter 8) can be used to simplify the key-distribution problem. In these circumstances, maintaining separate sequence numbers for each possible communicating entity is not usually practicable. The need to make each message different from others using the same key can be met (with high probability) by including a random number. Consider, for example, the three-message transaction shown in Figure 5.8. The originator of the transaction includes a random number R, which is returned as part of the response. Each message has an authenticator

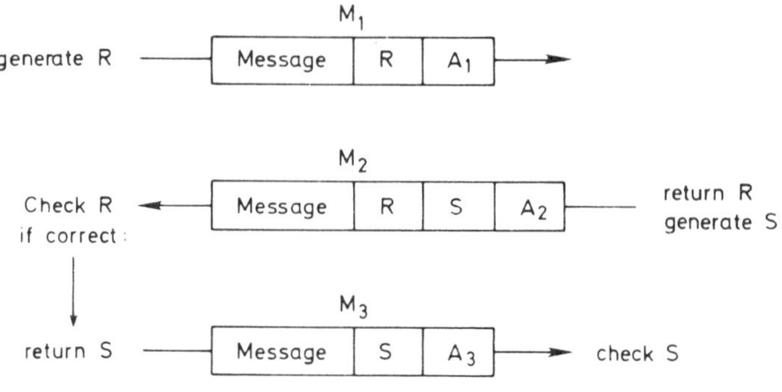

Fig. 5.8 Random numbers for entity authentication

covering all the content, including the random numbers, to enable tampering with the content in the communication network to be detected. The return of its random number assures the initiator that the response comes from a real entity and is not a recording of some earlier response. The size of the random number field is in this case governed by the Birthday paradox (see page 262). If there are likely to be N transactions using the same key, the random number field should be greater than N^2 in order to reduce to a low value the probability of a repeat.

For two-way authentication, the other entity includes in its response a random number S and this must be returned in the third message of the transaction as an assurance that the transaction is newly initiated and not a repeat. Each entity puts its own random number into the protocol in case the other is a masquerader.

Consider the possibility that M_1 is a replay of an earlier initial transaction message. The innocent receiver returns M_2, including the correct response to R and a new random number S. If M_3 is received in reply it must have come from an entity which possesses the secret key needed to generate the new authenticator value A_3. Could this be the genuine owner of the key if M_1 were a fake? This owner would not have accepted M_2 as valid unless it sent M_1. Therefore M_1 must have come from the genuine owner of the key and can be accepted as valid. But it is important that M_1 is not accepted as a valid message until M_3 is received and checked. For example, if M_1 were a request to update a data base, this action should be deferred until M_3 arrived.

If the transaction were to continue with further messages it would be necessary to chain them to the earlier ones by one of the methods we have described earlier.

Because of the Birthday paradox, the use of random numbers approximately doubles the size of field required for a given life of key. This presents no problem when a session key is used for just one transaction because the random numbers then have to be no bigger than normal authenticators.

The use of date and time stamps

Transactions employing a single message are vulnerable and should be avoided if possible. Where they are used, there is no possibility of the sender authenticating the receiver, but the receiver should be sure that the message came from the sender directly and was not a recording of an earlier message. A sequence number can achieve this but, in a large group, maintaining sequence numbers could present almost as much difficulty as storing all the received messages for comparison with new ones.

Where message transmission is reasonably rapid, an assurance of the 'timeliness' of the message is sufficient to prevent undetected message replay by an intruder. Store and forward systems delay messages, but do not prevent this method of checking if they can be relied upon to keep messages in sequence, since the time and date stamp provide a kind of sequence number which is universal for all messages received. Circuit-switched networks give delays less than 1 s and modern packet-switched systems should not exceed this figure very often; therefore, unless the rate of transactions is very high, a date and time stamp with a precision of 1 s is adequate.

If the date stamp is allowed to repeat, there is a risk of replayed messages. A very cautious view would be that no system is likely to be operating according to the same procedure for 100 years, so a date and time stamp containing a two-digit year, month, day, hour, minute and second is certainly adequate. Each of these fields has two decimal digits and when expressesd in hexadecimal characters this stamp occupies 48 bits. Employing a binary date would occupy a 16-bit field and a binary count of seconds in the day would use 17 bits. With these conventions a comprehensive date and time stamp would occupy 33 bits.

A simple quartz crystal clock provides a timing standard maintainable with 1 s accuracy, but, allowing for the need to correct it from time to time and the combined tolerances of two clocks, a margin of 5 s is likely to be the tightest criterion in general use. This can allow replays to be detected in all but the busiest systems. If public data networks provide a digital date and time source, this does not always mean that local clocks can be dispensed with, because of the risk that an enemy might be able to interfere with the time signal received. The public date and time source can safely be used to correct the local clock, with a warning if the correction is too large.

In theoretical treatments of the synchronization problem, much trouble has been taken to invent time criteria which do not require strict synchronization between the co-operating entities. With modern clocks and communication networks, 'real-time' can be re-instated.

In the very simplest terminals there may be cost criteria preventing the use of clock time, otherwise it presents no great problem. Active tokens, such as microcircuit cards, will usually have to accept a date and time value given by the terminals into which they are inserted and the security implications of this trust have to be examined.

Integrity of stored data

A file of data can be protected against illicit modification using authentication techniques closely parallel to those used for transmitted data. Authenticator values are calculated, using a secret key, and stored along with the parts of the file they protect. Users of the stored data also hold the key value and can check the authenticator.

The entire file could be protected by one authenticator, but any user would have to make the calculation over the whole file before using any of its data. Any change to the file, however small, would entail recalculating the authenticator over the whole file. It is better to divide the file into convenient 'units' which tend to be read or written at one time. Most files lend themselves to this kind of subdivision. But then a new threat appears — the units could be rearranged by an enemy if their authenticators are carried with them. If each unit has a built-in identifier, like the name and address in a file of data about people, this may not be a significant attack. If, however, a file is accessed by position (i.e. file address) the rearrangement threat is real. It can be met by including the address in the authenticated data. This does not require the address to appear in each stored unit, the address is inserted only for the purpose of calculating and checking the authenticator.

Messages and transactions are subject to the threat of a 'replay' during the currency of one key. This has a counterpart for stored data in the threat of 'restore'. After an item has been (legitimately) altered it can be restored by an enemy to its old value for which the authenticator is known. The attack could be an active attack on the stored data or it could be applied to a communication channel by which these data are accessed. The key used for authentication of stored data probably has a relatively long life because changing it requires a replacement of the authenticator values and that means processing the entire file. To protect against this threat we need a sequence number or a date and time stamp. Unless the units are large, adding date and time is an excessive overhead, but it might be a natural part of their contents. The sequence number becomes a *version number* which each unit carries in addition to its authenticator. If units are updated piecemeal all their version numbers will be different and there would be no way that a reader of the file would know that the latest version was being read. When the date and time or sequence number are not natural parts of the file, yet piecemeal updates are needed, there must be a separate 'update record'. This is a record which has its own authenticator and includes a date and time stamp to verify it has not been 'restored' by an enemy. It contains, for each unit in the file, a note of the current version number. It is not necessary to store the version number also in the item itself, but the number must be part of the authentication function for that unit. Therefore checking the authenticity of one unit of the file requires this procedure.

1. Check the authenticator and date/time stamp of the update record.
2. Read the version number of the required unit.
3. Format the version number (and the address, if used) with the contents of the unit.
4. Check the authenticator of the unit.

The use of a separately authenticated update record illustrates an hierarchical approach to authentication which we shall meet later in the application of digital signatures — an advanced form of authentication which has particular relevance to stored data.

5.8. THE PROBLEM OF DISPUTES

Authentication using a secret key is symmetrical — both parties know the key and either can generate an authenticator value. The secrecy of the key protects them both against third parties but not against each other. Disputes can arise between them.

Ann may send Bill an authenticated message, but if Bill is dishonest he can forge a different message and claim it came from Ann. The authenticator value provides no evidence to resolve a dispute, for it could have been given by Bill to a false message. Since it is feasible for Bill to falsify a message, Ann can deny sending a message she really did send. Bill cannot prove it is genuine by producing the authenticator value. Business is possible between honest people who trust each other, but there is no way to adjudicate disputes, even if the judge is given the value of the key. This problem — the inability to provide evidence to resolve disputes — is called the 'problem of disputes'. To avoid this is the purpose of the OSI service of 'non-repudiation'.

The most elegant solution to the problem of disputes is the 'digital signature' described in chapter 9, which continues this account of authentication.

REFERENCES

1. *Data processing — check characters.* Draft International Standard 7064, International Organisation for Standardization.
2. 'Data Transmission Over the Telephone Network: Series V Recommendations', *The Orange Book,* VIII.1, International Telecommunications Union, Geneva, 1977.
3. Davies, M.C., Barley, I.W. and Smith, P.E. 'The effect of line errors on the maximum throughput of packet type computer communication and a solution to the problem', *Proc. 6th International Conference on Computer Communication*, London, 943, September 1982.
4. Nacamuli. A.I. 'Integrating the processing of foreign exchange dealings', *Proc. International Conference on Computers in the City,* London, 317, May 1983.
5. Linden, C. and Block, H. 'Sealing electronic money in Sweden', *Computers and Security,* **1**, 3, 226, 1982.
6. *Banking — Requirements for Message authentication (wholesale),* International Standard ISO 8730, International Organization for Standardization, Geneva, 1987.
7. *Banking — Approved algorithms for message authentication Part 2: Message authenticator algorithm,* International Standard ISO 8731–2, International Organization for Standardization, Geneva, 1987.
8. National Bureau of Standards *DES Modes of Operation,* Federal Information Processing Standards Publication 81, December 1980.
9. *Financial Institution Message Authentication,* ANSI Standards Committee on Financial Services, ANSI Standard X9.9, August 1986.
10. National Communications System *Telecommunications: interoperability and security requirements for use of the Data Encryption Standard in the Physical and Data Link layers of data communications.* Federal Standard 1026, 3 August 1983.

Chapter 6 KEY MANAGEMENT

6.1. INTRODUCTION

The security of encipherment depends on protecting the key. In effect, encipherment concentrates the risk of discovery on the key in order that the data can be handled more easily. For this reason, the management of cryptographic keys is a vital factor in data security. Keys used for authentication and data integrity must be handled with equal care. In this chapter we describe some key management methods designed for the two main applications of encipherment, storage of files and communication of messages. In order to cover the subject in reasonable depth we have taken as examples two well-developed schemes, one coming from International Business Machines (IBM)[1] and the other an international standard.[2]

In a large network, messages are transferred between many terminals and hosts, on behalf of many users. The essence of end-to-end encipherment is to keep these users and these terminals separate so that they cannot read or interfere with each other's messages except where they are allowed to, by the sharing of keys. There will be many encipherment keys and their management is complex. Similar problems are presented by a host computer which stores files for different users and is asked to communicate some of these, under strict control to other users. A key management scheme should be able to handle a multi-user, multi-host and multi-terminal situation and protect data both in storage and in the communication network.

Key management includes every aspect of the handling of keys from their generation to their eventual destruction. The main complexities occur in the distribution of keys and their storage, so these will occupy most of our attention.

In order to move a key through a communication network, it must be enciphered with another key. For example, if we have a key *ks* which is used to encipher data and need to transport this key through the network, we use another key *kt* to encipher it as E*kt*(*ks*). There are advantages in using a data enciphering key like *ks* for only a short period and then establishing a new one through the network.

Since the key *ks* is typically used to encipher data for just one session it is called a *session key*. When the session key is being moved out to a terminal, the key *kt* used to encipher it for transit is called a *terminal key*. A terminal key is used for a longer period than a session key and it may have to be stored at a host computer with a number of similar keys for different terminals. In order to minimize the amount of secure storage needed, all these can be enciphered under yet another key *km* which is called a *master key*. Thus the storage of *kt* is in the form E*km*(*kt*).

The master key protects the secrets of the terminal keys and these protect the data-enciphering keys *ks*, which in turn protect the data, in a kind of hierarchy with the master key at the top. In their protected form, the lower keys (and the enciphered data) can be stored in ordinary memory or transmitted through a communication network. The whole system then depends on the master key which therefore needs the greatest attention to its security. Usually it is stored in a physically secure box. The key hierarchies of the IBM and ISO 8732 schemes will be described later.

6.2. KEY GENERATION

An ideal method of key generation would be one that chose the key at random — with an equal probability of choosing any of the possible key values. Unfortunately, this is difficult to achieve and we often have to make do with less. Any non-randomness which could give an enemy some way of predicting a key value or finding a value with higher than normal probability would reduce the task of searching keys and make cryptanalysis easier. Fortunately, a completely uniform distribution of key values is not essential; unpredictability is the essential requirement.

A classic method of generating random numbers is tossing a coin or throwing dice. It is perhaps surprising to find that methods like this, which are labour intensive, are suggested by IBM[3] for generating the top-level master keys. In fact, the labour involved is not significant because these keys are very rarely changed. The reason for choosing manual methods is to have them completely under the control of a trusted individual. Any mechanism or algorithm that is provided for generating keys carries with it a danger that someone has arranged the mechanism or algorithm to generate predictable, though apparently random, numbers. Tossing coins or throwing dice has the advantage that the operator has the whole procedure under his control. A small bias that might be introduced by wear on the coin or die would not significantly change the security. The IBM authors[1] publish a table for convenient conversion from the results of tossing a coin seven times to the two hexadecimal characters which represent one octet of a DES key. Fifty-six throws are needed for the whole key.

The top-level master key has such importance that the methods of double or treble encipherment described in chapter 3 are sometimes used. By having different people generate the various keys, the responsibility for safe generation of the master key can be spread.

Manual key generation is not suitable for the large number of keys in the levels below the master key. Some procedures need new keys very frequently. There are two possible sources of the random numbers required for numerous keys, a genuine physical source of random bits or pseudo-random number generators.

Random bit generators

Electrical noise is a problem in amplifier design but can be turned to good use. All resistors generate noise (though less at very low temperatures) and a wide-band amplifier will turn this noise into a signal which can switch a gate on and off. One method of generating random bits uses the zero crossings of this signal to produce pulses which change the value of a single-digit binary counter. The intervals between these pulses will vary in a random way and their probability distribution depends on the pass-band of the amplifier. At uniform time intervals the state of the binary counter is sampled to give the value 0 or 1. If the method works according to plan, this should be a random binary digit with its value uncorrelated with any other digit that has been produced. Careful design is needed to minimize bias or correlation. In order to remove any residual flaws of this kind, the random data can be used as key or data (or both) in a cipher algorithm to obtain unbiassed, uncorrelated results. This is the best source of random data for keys and the equipment cost is small.

Pseudo-random number generators

There are many examples of mathematically determined sequences which generate digits that are, to all appearances, random. Unfortunately, when a variety of statistical tests are tried, many of these 'pseudo-random number generators' can be shown to have a pattern in their output. Fortunately, a good cipher algorithm helps greatly in generating pseudo-random sequences. If the algorithm works well as a cipher then almost any method of using it to generate a sequence should generate random numbers, otherwise there is a pattern in the cipher output which would be a weakness in the cipher.

For example, if the key, k, is chosen then the result of enciphering the sequence of natural numbers 0, 1, 2, ... should generate a sequence of random numbers, each one unrelated to its predecessor. In other words, a random number sequence R_n can be generated as

$$R_n = \mathrm{E}k(n).$$

There is a danger that two sequences generated in this way might use the same key and therefore be identical. This illustrates the rule that random number generators should have random data to start them off. They use numbers called 'seeds' because a series of random numbers can be grown from them. A series grown from seed values cannot be infinite in length but it can be very long. The example we have just given does not stop generating random numbers until the value $n=2^{64}$ is reached.

The purpose of our random sequence is to generate key values (and perhaps initializing variable (IV) values) for a key management scheme. The essential requirement is that the keys it produces should be unpredictable, even if many of the earlier keys are known. The simple random number generator we described might not be able to conceal the value of n which entered into its calculation but, if the security of the DES cipher is good enough, this still does not allow the seed value k to be deduced and so none of the other random numbers can be found. It would not be wise to keep a key generator of this kind operating indefinitely with the same seed value. There must, therefore, be a source of seed values, obtained by a more reliable random process, such as tossing coins, throwing dice or the random noise source.

Most systems use random number generators that are a little more complex than our example, using many different sources of randomness or unpredictability, for extra safety. One source of unpredictability is a secret number which is already in the system such as a top-level master key. Though this master key is well protected against discovery, it is used in the cryptographic operations of the system. A pseudo-random number can employ these operations, which are not available to an enemy outside the computer system.

An an example, the seed values used in generating the random sequence can be obtained by cryptographic operations (using the master key) on a number representing the current date and time.

A useful source of randomness is the statistical nature of the traffic handled by a host computer. For example, the number of calls on the operating system can be counted. It would be difficult to predict this number exactly from outside the system. An unpredictable quantity can be obtained by timing an operator's reaction. When the system is started, it can ask the operator to pause a while and then press some convenient button. If the elapsed time before pressing this button is measured in microseconds, the range of possible values certainly extends over several thousand. By putting together a calculation employing as seeds a number of unpredictable or random quantities like these, it is possible to generate a number sequence which is for most practical purposes random.

An example of a pseudo-random number generator designed according to these principles is given in reference 3. The notation has been changed to simplify the appearance of the expressions. It has a starting value U_0 and generates a sequence U_i, making use of cryptographic functions which employ two secret master keys, shown here as k_1 and k_2. The generation of the sequence employs a further sequence of seed values Z_i each derived from some real-time clock readings with an input-output operation of unpredictable length between successive clock readings. The calculation of the sequence U_i employs these relationships

$$X_i = \mathrm{D}k_1(U_i)$$
$$U_{i+1} = \mathrm{E}k_2[\mathrm{D}X_i(Z_i)].$$

The random number sequence R_i is obtained from the U_i by the relation

$$R_i = \mathrm{E}k_2[\mathrm{D}X_i(U_i)]$$

The complexity of this method makes it extremely unlikely that anyone could predict its outcome or deduce anything useful from a sequence of R_i values that it creates. It is perhaps a matter of judgement how complex a random number generator is needed in practice.

Figure 6.1 shows, in block diagram form, the operation of the random-number generator we have described. The register is shown containing the value of U_i before returning it for the next iteration. The two compound operations shown in the broken rectangles are the cryptographic functions employing the secret master keys shown here as k_1 and k_2. This is the operation which we shall meet later in the IBM key management scheme and which is known as 're-encipher to master key'. It is provided for a specific purpose in the key management scheme and has been adopted in the random number generator simply because it employs the master keys and is available to the operating system but not to the users of the host computer.

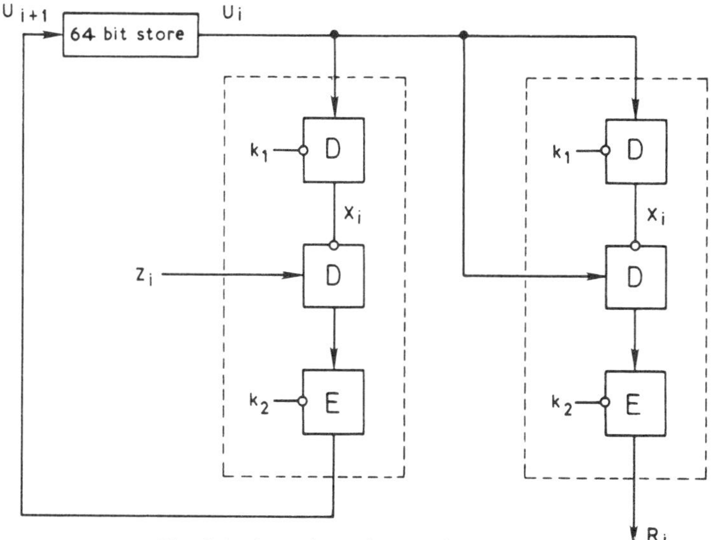

Fig. 6.1 A pseudo-random number generator

6.3. TERMINAL AND SESSION KEYS

Imagine a set of N terminals which we shall call T_i and that each terminal may wish to communicate in secret with any other. If all the keys to provide for the possible connections were produced in advance, each terminal would need to store $N-1$ keys and, supposing that the same key could be used for both directions of communication between a pair of terminals, there would be $\frac{1}{2}N(N-1)$ keys in total.

This method is clumsy and becomes more so as the population of terminals increases. In practice, few terminals are actually communicating at any time. The period of communication between two terminals is known as a *session*. It is better to provide keys only for those sessions which are actually in operation and these are the *session keys*. Then a procedure is needed to establish session keys whenever they are required.

The use of a session key has a worthwhile bonus of security. Because it exists only for the duration of one session, an enemy who could obtain one of these keys gets limited value from it. Of course, the establishment of session keys requires a key at a higher level in the hierarchy, in this case the terminal keys kt_i are employed. These keys are, in fact, better protected than the session keys because session keys can be attacked by a knowledge of the data that are being communicated, but terminal keys are used only to encipher session keys, so the information they encipher is both less in quantity and more random in nature than the data handled by the session keys. Effectively, an attack on the terminal keys would require a knowledge of the session key values so cryptanalysis must be successfully carried out twice to reach a terminal key.

When a session key, ks_{ij}, is needed for the communication between terminals T_i and T_j it can be transmitted through the network under two different encipherments, for these two terminals, $Ekt_i(ks_{ij})$ and $Ekt_j(ks_{ij})$ respectively.

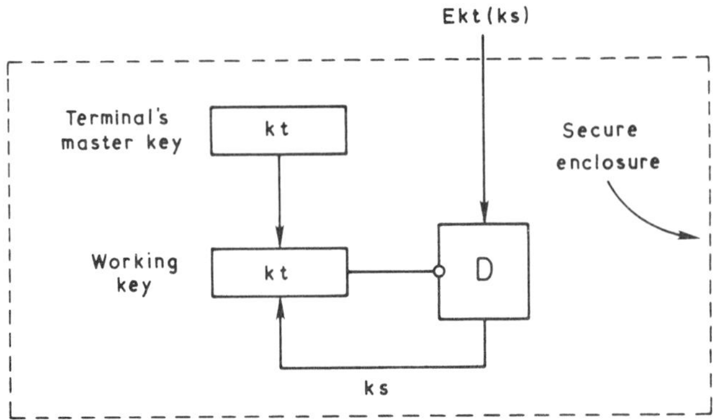

Fig. 6.2 Loading a session key at the terminal

These quantities are generated at some central point, which might be a host computer responsible for the operation of these terminals, or in a wider network there might be a specially appointed 'key distribution centre',[4] sometimes called a network security centre when it takes on additional responsibilities.

Figure 6.2 shows how the arriving session keys are handled at the terminal. The decipherment operation shown would typically employ a DES hardware device or chip which was capable of storing two keys. One of them, *kt*, is regarded in this context as the terminal's master key. It is held in a register where it can be made available each time a session key is to be loaded. Then its value is copied into the working-key register. This register supplies the key for the DES operation used to decipher the incoming session key and produce the clear value *ks*. In a well-designed chip, the value *ks* should not leave the device but be transferred after the decipherment into the working-key register, where it displaces the value *kt*. In this way, the physical security of the chip protects the session keys, since they do not come out of the chip at any time.

There are two other key management operations at the terminal, the loading of the terminal key and the destruction of the working key. Loading the terminal key is part of the process of transporting this key to the terminal and is discussed later. Destroying the session key could either be treated as a separate operation or regarded as a special case of loading a session key in which a dummy value such as 0 is loaded. It is useful for the destruction of the session key to be under the control of the host or key distribution centre but, on the other hand, any command from the centre can be stopped by an enemy, therefore, as a fallback, the session key should be cleared locally at the end of the session if the centre fails to do so.

Routes for distribution of session keys

Figure 6.3 shows a pair of terminals for which a session key *ks* is to be established. The key distribution centre (which might be a controlling host) must share a knowledge of the key kt_1 in common with terminal 1, and the key kt_2 in common with terminal 2. The most direct routes for the session keys are those shown on the left of the figure. These may be inconvenient in practice. One terminal takes the initiative in setting up the communication path, in this case we suppose it to be terminal 1. Terminal 1 must first communicate with the

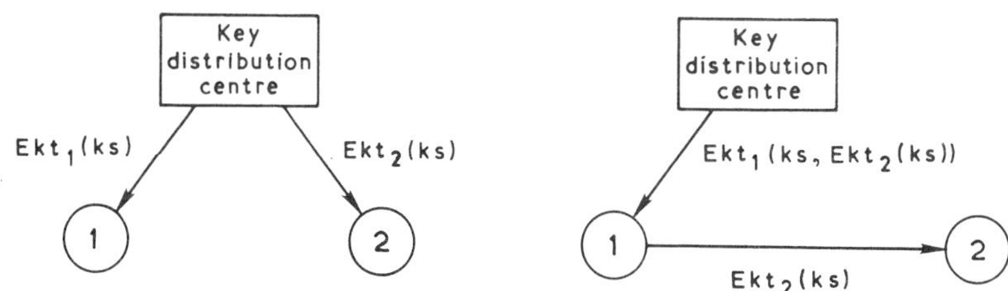

Fig. 6.3 Routes for session key distribution

distribution centre and request a session key, quoting the terminal 2 as the target for its future call. If the session keys are now distributed by the direct routes, terminal 2 will receive the session key before it receives the call from terminal 1. In fact it may be engaged in interaction with another terminal or unwilling to accept the call. At the time that it receives the enciphered session key it has no knowledge of which terminal is about to call it. In a busy situation, there will be a need for each terminal to be ready to receive several future session keys, storing them until a call arrives, then it must associate the call with one of these keys. Rules will be needed to decide the proper action in a number of error situations, such as a breakdown of terminal 1 before the call is completed, or a busy condition in the network which prevents the call. Time-outs could deal with these problems but time-outs do not produce neat and reliable solutions.

For these reasons it is usually preferred to transfer the session key to the target terminal (terminal 2) by way of terminal 1, as shown on the right of the figure. Terminal 1 must receive not only the enciphered form of *ks* for its own purpose but also the value of *ks* enciphered with kt_2, which it will pass on to terminal 2 when it is establishing the call. From the viewpoint of terminal 1, setting up this call has two phases, obtaining the key from the key distribution centre and then, after making the call to terminal 2, transferring an enciphered form of the session key to that terminal. In the figure we have shown that the key which is intended for transfer, namely $Ekt_2(ks)$, is included with the value of *ks* in the message sent to terminal 1 enciphered under kt_1. To do this, kt_1 should be used to encipher these data as a *chain*, to prevent an enemy separating the two parts for his own use.

This apparently simple key distribution function has been the subject of several security studies[5,6] which have considered a number of possible attacks and elaborated the protocol to frustrate them. There is a third possibility for routing the session keys, which is to send both keys from the distribution centre to terminal 2 and have terminal 2 relay the key required by terminal 1. Since the procedure begins with terminal 1 sending a request to the key distribution centre, this produces a complete cycle in which messages are passed successively from terminal 1 to the key distribution centre to terminal 2 and back to terminal 1. This circular route enables a random number chosen by terminal 1 at the time of making the request to be returned with the key he receives. By this means, a replay of old keys is prevented and the timeliness of the session key in use is ensured. This message sequence has not yet been elaborated and the examples we shall give below are of the kind shown on the right-hand side of Figure 6.3.

Session key distribution protocol

The two phases of key distribution are the interaction between terminal 1 and the distribution centre, called 'key acquisition' and the interaction between terminals 1 and 2 at the beginning of the call which we denote as the 'key transfer'. In both of these operations, the encipherment prevents the critical quantity *ks* being discovered by an enemy, but it does not complete the protection against an active attack. We have to guard against 'spoofing' which is one terminal pretending to be another. For example, an enemy might manipulate the communication links to take the place of the distribution centre and provide terminal 1 with a key that was distributed earlier. This could be done without knowing any of the keys, merely by repeating earlier messages. Another threat is that an enemy terminal might impersonate terminal 1 to the distribution centre and obtain a new key for calling terminal 2. A third threat, which is more serious, is that an enemy terminal might masquerade as terminal 1 in calling terminal 2, using a key that had been transferred earlier.

The principal danger lies in an enemy re-establishing a session key that has been used before, either by masquerading as the key distribution centre and persuading terminal 1 to establish this key, or by masquerading as terminal 1 and establishing the old key for a fraudulent call to terminal 2. The same messages would be used that passed between the terminals when the key was used earlier, so there is no need for the enemy to know kt_1 or kt_2.

The trick of establishing an old key can be exploited in one of two ways. Suppose that the enemy had obtained the value of *ks*, then at any time that he could establish this value again, he could either listen in on a call between terminals 1 and 2 or impersonate terminal 1 in a call to terminal 2. Finding the value of *ks* might be the result of cryptanalysis applied to an earlier session and the ability to re-establish *ks* defeats one of the virtues of the session key which is that it is used only for a short time. Another way of exploiting the masquerading trick, even when *ks* is not known is to use it to replay an earlier interaction between terminals 1 and 2. This might enable a transaction between these terminals to be repeated with an advantage to the enemy. Of course, transactions have their own protocol which usually employs checks to prevent transactions being replayed. Nevertheless, it is legitimate to expect some help from the key distribution protocol in preventing such abuses of the system.

Authentication at the key acquisition phase

When it calls the key distribution centre, terminal 1 should establish that it is conversing with an intelligent device that knows its key value is kt_1, and is not simply an enemy repeating the data from an earlier transaction. This *authentication* of the key distribution centre can be carried out by giving the centre a task depending on kt_1. The following interaction will do this:

Terminal 1 sends to key distribution centre in clear

$$a_1,\ r_1,\ a_2.$$

Key distribution centre to Terminal 1

$$\mathrm{E}kt_1\{r_1,\ a_2,\ ks,\ \mathrm{E}kt_2(ks,\ a_1)\}.$$

The commas denote that the terms are parts of a continuous message. Where the message is enciphered it is treated as a chain but the IV can safely be zero, because each chain begins with a random number: a_1 and a_2 are the addresses of the two terminals. They are sent with the request to the key distribution centre to identify the two terminals needing the session key. The quantity r_1 is a random number chosen for this interaction by terminal 1 and this same number appears in the enciphered reply, reassuring terminal 1 that it cannot be a recorded message, because it includes his random number. The other feature in the reply is the inclusion in the chain enciphered by kt_2 of the address a_1 of the calling terminal. When this reaches terminal 2, it helps to authenticate terminal 1 as the source of the message, because its encipherment with kt_2 shows that it must have come from the key distribution centre.

Does the key distribution centre need to authenticate terminal 1? Another terminal could make the call and present the same data a_1, r_1 and a_2. In fact, it would serve no useful purpose because in order to use the data from the centre it would have to decipher them with kt_1 which is a key known only to the key distribution centre and terminal 1. An intruder could change the value a_2 in the request and thus divert the subsequent transaction but the repetition of a_2 in the enciphered reply allows this to be detected. So authentication in this 'key acquisition' phase need only operate in one direction, in principle. But one factor might make it desirable for the key distribution centre to authenticate the calling terminals — the possibility that enemy terminals might overload it with bogus calls. The distribution centre can authenticate the calling terminal only by variable data, since fixed data can be replayed by masquerading terminals. The calling message could be replaced by

$$a_1,\ \mathrm{E}kt_1(n_1,\ r_1,\ a_2)$$

in this expression n_1 is a serial number for all the calls from terminal 1 to the key distribution centre since the key kt_1 was established. Its presence in the enciphered chain of this expression ensures that each call will be different and that the distribution centre can check the serial number, after decipherment, to verify that the caller possesses the key, kt_1. Alternatively a date and time can be used in place of n_1. The address a_1 must lie outside the encipherment so that the centre can identify the caller and use the correct key for decipherment.

There are other methods for authenticating the terminal to the key distribution centre but we question the need for further elaboration to solve this problem, since the only risk is the extra load on the key distribution centre.

Authentication at the key transfer phase

There is a clear need for two-way authentication between the terminals 1 and 2 enabling each to be sure of the other's identity and certain that it is not receiving a replay of earlier messages.

If terminal 1 is certain, because it authenticated the key distribution centre, that it received a new session key, *ks*, then the fact that the subsequent enciphered conversation with terminal 2 works will verify that terminal 2 is responding. No other terminal could respond sensibly because the establishment of the common session key *ks* requires it to decipher the key transfer message with its own key kt_2. If terminal 1 wants to get this reassurance early it can send another random number r_2 in the form $\mathrm{E}ks(r_2)$ and receive some function of r_2 in an enciphered message from terminal 2.

The authentication by terminal 2 of the calling terminal is more difficult. In order to detect a replay of a previous key transfer it can set terminal 1 a problem in the form of a random number r_3 transmitted as E$ks(r_3)$ and obtain a suitable function of r_3 in an enciphered reply. This does not guard against the masquerading calling terminal which has obtained the value of *ks* by cryptanalysis of an earlier exchange. Knowing the value of *ks* enables the bogus terminal to continue the masquerade.

Worse still is the case of the enemy discovering kt_1 because then he can obtain from the key distribution centre a stock of *ks* values for future masquerading. Even if terminal 1 changes its terminal key, the masquerade with old values of *ks* can continue, as long as terminal 2 keeps its master key unchanged.

The solution to this problem proposed by Denning and Sacco[7] is to time-stamp the session key. They propose the following squence.

Request from terminal 1 to key distribution centre

$$a_1, a_2.$$

Reply from key distribution centre to terminal 1

$$\mathrm{E}kt_1(ks, a_2, d/t, \mathrm{E}kt_2(ks, a_1, d/t)).$$

Key transfer from terminal 1 to terminal 2

$$\mathrm{E}kt_2(ks, a_1, d/t).$$

In these expressions, *d/t* is the date and time.

The value of *d/t* supplied to terminal 1 authenticates the key distribution centre. Any terminal which makes calls to the centre at shorter intervals than 1 min must keep a list of *ks* values for the last minute and check that these are not repeated. There will rarely be a need for such checking.

The value of *d/t* supplied with the key transfer to terminal 2 verifies the freshness of the session key. Keys obtained from the distribution centre must be used at once and cannot be stored for future use and the time required to establish calls must be at most a few seconds if the scheme is to work well. This immediacy should be easy with modern networks. Either terminal will reject a key that is too old, thus preventing replays.

Distribution of terminal keys

For the present, the possibility of non-secret methods of key distribution will be ignored (these are described in chapter 8) and we will describe more traditional methods using ciphers like the DES. Whatever key hierarchy is used, there must be at least one key at each terminal that has not been sent through the network. In the system we have just described this is the terminal key *kt*. Such keys must be loaded into the terminals manually and they may have to be changed from time to time for extra security. They must be physically transported to each terminal.

The older method of loading a key into a terminal was by rotary switches or a keyboard. This operation must be carried out only by trusted personnel, therefore a lock was usually provided into which a physical key was inserted and turned to allow the entry of a new key. Such key loading operations might be spied on by an enemy. If the key is written down on paper for its journey to the terminal, it is difficult to be sure that the key has not been copied en route. For these reasons it is now becoming customary to load the key from a transport module in which it has been electronically stored.

The key carrying module, sometimes called a 'key gun', can be of pocket-calculator size and its essential external feature is the means of coupling it to the terminal to deliver its key and to the source of keys when it is loaded. The means of coupling can be electrical, but optical coupling is popular since it seems more difficult to tap an optical coupling.

A key transport module can carry a number of keys for each destination and can be used to load keys at several places on the courier's route. The keys are selected by an address given to the module. All that is needed in such a module is a suitable store such as an 'electrically alterable read-only memory' or EAROM. But this provides no security against the reading of keys while the module is en route. For a more secure system the key-transport module is constructed with a microprocessor and a battery so that a random access memory (RAM) can be used for storage, which is cleared under the control of the microprocessor when security requires it.

Suppose that each stored key has one destination and can be cleared by the module as soon as it has been read. (Where two stations need the same key this value is stored separately for each destination.) Then an attempt to read a key illegally elsewhere than at its destination prevents that key being used and achieves nothing. It is essential to prevent the key being replaced by the enemy after it has been read. This can be done by giving each key transport module a secret which it shares with the key generator. It could be an encipherment key but, more simply, the loading of keys into the module can be controlled by a password specific to the module. However, 'one-shot' key transporters are not always convenient. Some users prefer to keep their transport modules at the destinations to act as back up stores for the keys, in case the equipment receiving the keys fails.

When reading a key does not clear it, password control can be used to prevent illegal reading. Each destination prepares its own module by giving it a password. When the module has been taken to the key generator and received its keys, it will not give them up until it receives its password. If it receives the wrong password too many times it destroys its keys, to prevent a password search.

Two kinds of attack remain. One is opening up of a module to discover all that it contains and then successfully closing it to make it acceptable for use (or replacement by an acceptable replica). The other is the tapping of the communication path when it delivers the key at the terminal. Those present at the time of loading the key into the terminal must be vigilant enough to detect these attacks.

Given that the module resists physical attack or reveals it when it arrives at the destination, the module could possibly be transported by a postal or other carrier service to save the cost of special couriers.

6.4. THE IBM KEY MANAGEMENT SCHEME

In the remaining part of this chapter we shall describe two different key management schemes, both of which have been worked out in some detail. They are not only valuable as examples of comprehensive schemes, they also demonstrate a range of techniques which can be used to develop key management methods for different applications. Like most working systems they contain a lot of detail and we shall not be able to cover every aspect.

The IBM key management system was described first in a group of papers in the *IBM Systems Journal*.[1,3] We have not used the notation of the IBM papers because we wanted to have a consistent notation throughout this book. In a few respects we have simplified the scheme to make it easier to explain without changing its essential features.

It is important to understand the environment for which the key management method was devised. Though it is like the key distribution scheme we described earlier, employing session keys and terminal keys, the environment is different because the terminals do not communicate directly with one another through a communication sub-network. Communication takes place through host computers, each host being responsible for a group of terminals. The host acts as a key distribution centre for its own group of terminals and encipherment is used for messages passing between the terminal and the host. There may be several hosts in a network and then the hosts organize the communication of messages or files between them. A key management system has two functions which the IBM scheme mostly keeps separate — the encipherment of messages to provide communication security and the encipherment of files to provide file security. Each of these functions has its own key hierarchy and the two key hierarchies have close parallels. We shall describe first the communication function and then the file security function, making comparisons between the two. The functions come together when files are transported from one host to another.

Physical security requirements

The handling of keys at the terminal is like that described in Figure 6.2. Each terminal contains a terminal key *kt* and at the start of a communication session, a session key *ks* is delivered to it, enciphered under the terminal key.

The physical security requirement at the terminal is the protection of the two keys. The loading of *kt* is a privileged operation. The key-management scheme must provide the value of *kt* for this purpose, whether or not a key carrying module is employed.

At the centre, the physical security requirements are much more severe. A number of terminal keys must be stored safely and at any one time many sessions may be in progress, so a number of session keys must be kept ready for use. The host computer is provided with a special, physically secure module containing all that secret information which must be protected. In the IBM papers it is called the 'Cryptographic Facility' and it is typified by the IBM 3848 cryptographic unit: we shall call it a 'Tamper Resistant Module' (TRM).

Both in the terminal and at the centre, the encipherment operation must be carried out in an appropriately secure place. At the host all encipherment operations occur inside the TRM. The TRM also contains the most important keys which, once they have been loaded into it, can never be read again. In the same way, none of the keys can be read from the hardware of the terminals.

The processing work of the host is carried out on behalf of many users, each with his own rights of access to certain terminals or files. As in all multi-user systems there is an elaborate access control mechanism administered by the operating system which decides the rights of any *subject* with respect to any *object*. Subjects may be users or processes and objects may be data, files, programs, terminals etc. The right to use a terminal with encipherment means giving a subject the right to employ a certain session key.

The prevention of illegal access by one subject to another's objects is the responsibility of the access control system. Encipherment cannot improve on the security of this access control, it merely extends the security to stored files and distant terminals. An enemy within the system who can undermine the access control mechanism can steal files or read messages illegally in spite of the use of encipherment. Therefore, the key management system must rely on the access control method to control access to keys.

There is always the possibility that one of the system's users could be an enemy and could steal data from the system and pass it to an enemy outside. If he could steal keys and pass these to an enemy outside, then various stolen storage media, or ciphertext recorded from a line, could be deciphered and all the security of encipherment with the stolen keys would be lost. It is, therefore, a cardinal principle that the key management scheme should prevent anyone, including legitimate users and system programmers, from obtaining any cryptographic key. Therefore, cryptographic keys (with a few exceptions) are allowed outside the TRM only when they have been enciphered by other keys. As a result of this principle, the damage that can be done by an enemy within the system is confined to the data he reads by misusing the access control system of the host. He cannot steal keys that can be used outside the system. To put it another way, the barriers inside the system between the various subjects are not strengthened by the use of encipherment. The encipherment of data, together with key management, places a firm barrier between what can be done inside the system and what can be done outside.

The main principle is that no keys can be taken outside the TRM in their clear form, the form in which they could be exploited by an enemy. The exceptions to this rule are the terminal keys which must be in clear outside the TRM, at least at the moment when they are loaded into the terminal. This operation of loading terminal keys is carried out under careful discipline by trusted people. The procedure for initializing the whole system by the creation of top-level master keys and loading them into the TRM is another exception to the general rule and it requires the greatest discipline and trust.

The key hierarchy

There is a three-level key hierarchy which will now be described as it applies to the communications security part of the scheme. This description is intended only to give a general picture and the details will follow later.

The three levels are shown in Figure 6.4. At the top level are two master keys, *km*0 and *km*1. In the middle level are the terminal keys, denoted *kt*, one for each terminal belonging to the host. The lowest level keys are the session keys, *ks*, which are associated with terminals on a more temporary basis. Only the session keys encipher data. The terminal keys, as we saw earlier, are used to encipher session keys for transmission to the terminals.

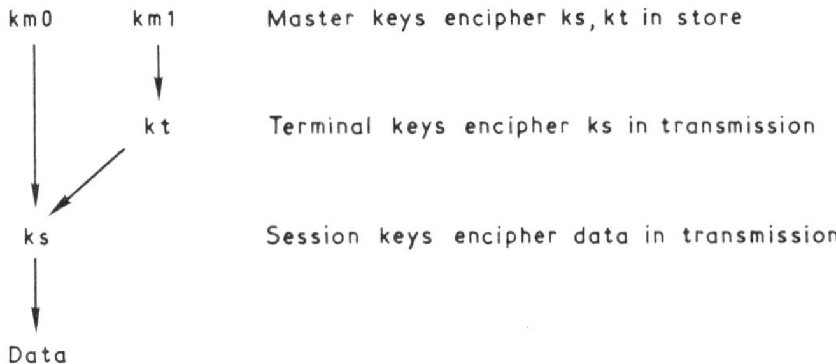

Fig. 6.4 The hierarchy of keys for communication security

The two master keys in the host, called *km*0 and *km*1, are used, respectively, to store all the session keys and all the terminal keys in the system. Because all the session keys are enciphered under the key *km*0 whenever they are outside the TRM, they can be stored in ordinary memory locations. The stealing from the host of a quantity such as E*km*0(*ks*) would not be disastrous because the master key *km*0 is special to this one host so the enciphered value of *ks* is useless outside it. Likewise, the terminal keys are stored as E*km*1(*kt*) in ordinary memory locations.

The purpose of storing the terminal and session keys outside the TRM is two-fold. First, it reduces the amount of store in the TRM and, secondly, it allows the control of access to these keys to be the normal access control provided by the operating system. There would be little point in having a duplicate or alternative access control system for keys.

The existence of the master keys inside the TRM makes all the operations of the TRM special to this one host. The various functions of the TRM are not all equally available. Some, such as the loading of the master keys, take place rarely and only with specially privileged people present. The operations of the TRM used for distributing new terminal keys are accessible only to the operating system. But at the lowest level, the operations of enciphering and deciphering data are available to ordinary user's programs, provided that they can gain access to the appropriate session key.

The encipherment and decipherment of data at the host

A program at the host computer can use the functions of the TRM to decipher data coming from the terminal or encipher data going to that terminal, if it can provide the TRM with the session key in use at that terminal. The operations carried out in the TRM are illustrated in Figure 6.5. They are not simple encipherment or decipherment because the key arrives in an enciphered form. The first action is to decipher the key using the master key *km*0 which is stored in the TRM. The resulting session key, *ks*, is then used for the encipherment or decipherment operation. This function of the TRM is given the name 'Host Encipherment' (HE) or 'Host Decipherment' (HD). In a sense, the host encipherment or host decipherment of each individual host is different because of the different master key contained in its TRM, so a session key taken from the store of one host cannot be used in another host.

The two inputs to the HE or HD operations must be distinguished because one is a key input and the other a data input. We have shown in our figure a small

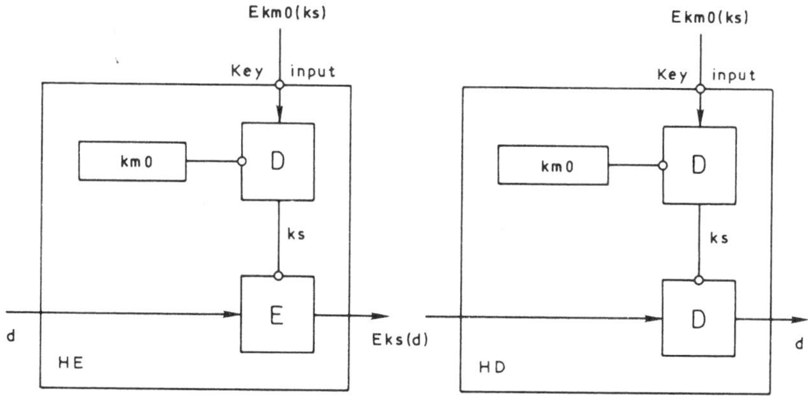

Fig. 6.5 Data encipherment and decipherment operations of the tamper-resistant module

circle where the key input enters the TRM by analogy with the notation used for the block cipher itself.

Generation and distribution of a session key

Figure 6.6 illustrates the function of the TRM which is used for distributing a session key. The new session key *ks* must be transmitted to the terminal in the form E*kt*(*ks*). Both *kt* and *ks* must be protected by encipherment whenever they are outside the TRM. The terminal key *kt* is stored under encipherment by the master key *km*1 and the first operation in this key distribution function is to decipher it and produce the plaintext value, *kt*. Then the stored session key E*km*0(*ks*) can be deciphered inside the TRM by *km*0 and re-enciphered by *kt* to produce the required quantity E*kt*(*ks*). This operation is known as 'Re-encipher from master key' (RFM) since it changes the session key from encipherment by the master key to encipherment by the terminal key. This function of key distribution is carried out by the host's operating system, responding to a request from the

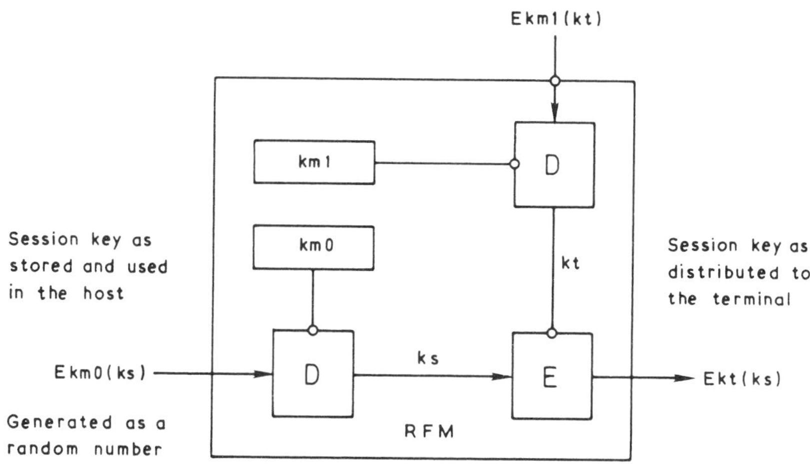

Fig. 6.6 Session key distribution

present 'owner' of the terminal. The enciphered values of the terminal keys, kt_i, can be stored in ordinary, unprotected storage but their use is controlled by the operating system.

It is not allowable to generate a new session key in its plaintext form because that form may only appear inside the TRM. The pseudo-random-number generator is employed to generate a new value of E*km*0(*ks*), the stored form of the session key. The resulting value of *ks* will then be a random number, perfectly suitable for its purpose. Generating the enciphered key E*kt*(*ks*) at random is useless because the RFM function works only one way, there is no function available which will take E*kt*(*ks*) and re-encipher it under the master key. It is important that there should not be such a function, otherwise an enemy within the system could obtain the session key by tapping the line while it was being transmitted to the terminal and from it obtain E*km*0(*ks*) which he could then exploit in the HE or HD functions to encipher or decipher users' data taken from the line. So the one-way property of the RFM function is essential.

In this operation there are two master keys, *km*0 and *km*1. If only one key were used it would be possible, by misuse of the TRM operations, to expose a key in its unenciphered form as shown in Figure 6.7. Suppose that the same master key, *km*0, was used to encipher both *ks* and *kt*. Then the quantity E*km*0(*kt*) could be used as the key for the HD operation and applied to the value E*kt*(*ks*) obtained by tapping the line to the terminal during session key distribution. This trick would produce the session key *ks* in plaintext form and this could be employed anywhere to decipher users' data obtained from the line. The use of two different master keys prevents this trick.

Generation and loading of a top-level master key is time-consuming, so the need for loading two different master keys must be avoided (and as we shall see later, a third is needed). In the IBM scheme this is achieved by deriving *km*1 from *km*0 simply by inverting a specific set of bits in the value of the key. Although these keys are then closely related, their encipherment properties are completely different and, since both are kept secret, their relationship is not of great importance to the strength of the system.

Since the random value used for starting the key distribution is E*km*0(*ks*) the value of *ks* produced is a random block of 64 bits and does not have the correct

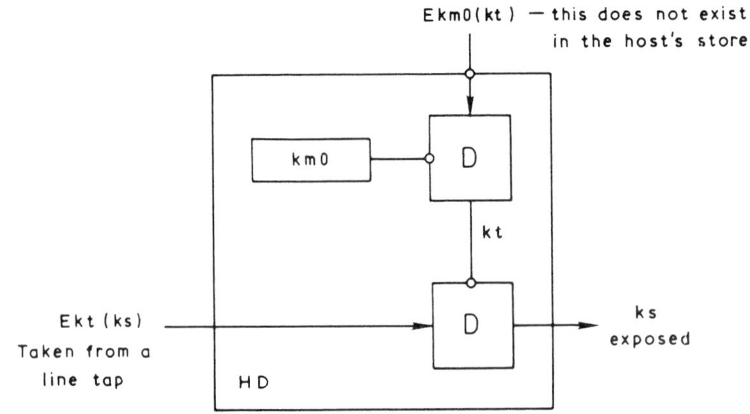

Fig. 6.7 How ks *could be exposed if* km1 = km0

parity check properties of a normal DES key. It would be possible to arrange that, when the value *ks* is produced in the TRM, its parity is corrected and thus when it arrives in the terminal and is loaded into the working register it has the correct parity. It seems that IBM did not do this, but allowed the value of *ks* to reach the terminal without its correct parity checks. The DES chip ignores the eight spare bits instead of treating them as parity digits.

Generation and distribution of the terminal key

New session keys are produced frequently but the terminal key is changed less often. When it is changed, the new key value has to be taken to the terminal and loaded into it and for this purpose it must be available in its cleartext form, *kt*. It is therefore an exception to the normal rule that cleartext keys are never exposed. Since it is important that there should be no way of regenerating *kt* at the host after it has been distributed, the only way to organize its distribution is to generate the key in the cleartext form and this presents a new problem, how to form the quantity E*km*1(*kt*) which is the stored value of the terminal key. This operation is, of course, carried out by the operating system and it uses the special quantity E*km*1(*km*1).

A function is provided in the TRM for enciphering under the master key *km*0 and this is known as 'Encipher under Master Key' (EMK). It is a privileged operation which is given the quantity *x* and returns the quantity E*km*0(*x*).

The quantity E*km*1(*km*1) is produced when the system is initialized, then it is stored away and kept only for use in the *kt* distribution process. To produce it, *km*0 is used to produce *km*1, a trivial operation, and this is used with the EMK function to generate E*km*0(*km*1). This quantity is used as the key in the HE operation with *km*1 as data, as shown in Figure 6.8, to generate the special quantity needed. After these processes, all stored values of the master keys are destroyed except those in the TRM. A written copy of *km*0 is kept in a safe for disaster recovery. If the system is destroyed, but enciphered files are available in a back-up store, these would be useless unless the master keys could be retrieved.

Using the special quantity E*km*1(*km*1), it is possible to complete the distribution of the terminal key in two stages. First this key is enciphered under *km*0 using

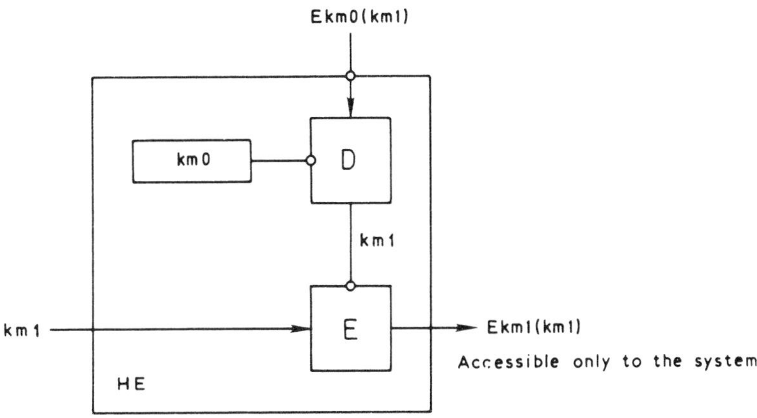

Fig. 6.8 An operation used at system initiation

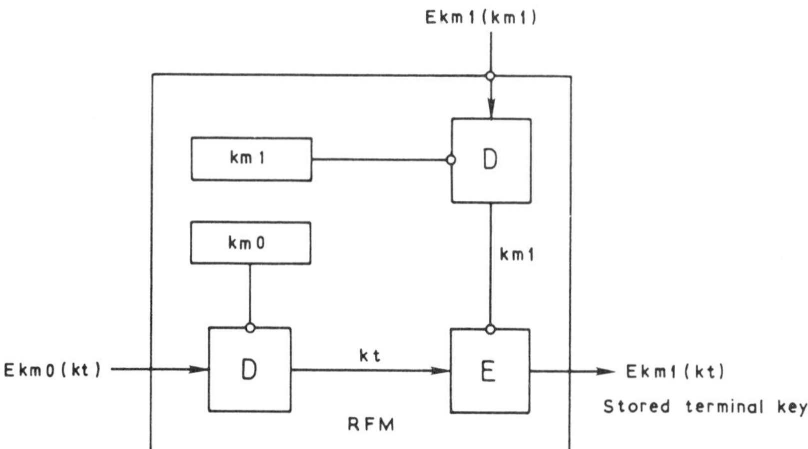

Fig. 6.9 An operation used to prepare a new terminal key, kt

the EMK function, then the RFM function is applied to this quantity, using as key the special quantity E*km*1(*km*1), as shown in Figure 6.9. The result is the quantity E*km*1(*kt*) which can then be kept in store and used for distributing session keys as shown in Figure 6.6.

Notice that the intermediate quantity E*km*0(*kt*) is one which earlier we said should never exist, because it could be misused to expose session keys stolen from the enciphered line. This means that terminal key distribution must be carried out with extreme care, using trusted software and making sure that the intermediate values cannot be accessed by another program.

The principles of file security in the IBM key management scheme

The hierarchy of keys used for file encipherment is very similar to the one used for communication security and is shown in Figure 6.10. The data in the files are enciphered by file keys, *kf*, and file keys are stored under secondary file keys, denoted *kg*, which are held by the owners of the files. A single value of *kg* creates a 'protection domain' which may contain a number of files each having its own data enciphering key. The quantities E*kg*(*kf*) need not be concealed because they cannot be used unless *kg* is available. For example it may be convenient to store

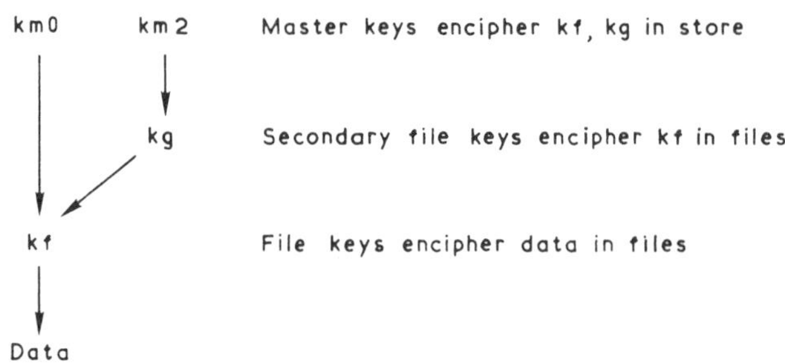

Fig. 6.10 The hierarchy of keys for communication security

the value of E*kg*(*kf*) in the header of the file itself, to make sure that it is always available when the file has to be deciphered. By having these individual file keys, the benefit to an enemy of breaking the cipher to obtain one key is very limited unless the knowledge of the file keys could be used in a further stage of cryptanalysis to obtain the secondary file key *kg*. Handling groups of files together under one key, *kg*, can also enable a group of files to be communicated to another user by giving him the *kg* value. These appear to be the main reasons why the intermediate or secondary file key is included in the hierarchy.

At the top level of the hierarchy are the master keys. The key *km*0 is used for enciphering file keys (as well as session keys) in the host. The key *km*2, derived like *km*1 by a trivial operation on *km*0, enciphers secondary file keys for storage in the host.

The encipherment or decipherment of a file is allowed for any user or process which has access to the secondary file key in its stored form E*km*2(*kg*). This can be used, as we shall show later, to generate the file data encipherment key *kf* in the form E*km*0(*kf*) required for employment in the HE and HD functions. These quantities E*km*0(*kf*) are not stored in the system, they are simply recreated whenever a file is to be deciphered. Thus the control for deciphering a file rests in controlling access to the secondary file key *kg*.

Generating and retrieving a file key

Figure 6.11 shows the TRM function which is used when preparing to encipher a file or retrieving a file key for decipherment. It is called 'Re-encipher to Master Key' (RTM) and is almost a mirror image of the RFM function but using the master key *km*2 instead of *km*1. At the top of the figure, the 'key' input is the secondary file key in the enciphered form E*km*2(*kg*) in which it is stored in the host. The data quantity entering the RTM function is a newly generated pseudo-random number when the file is first enciphered. This gives rise to a random number *kf* inside the TRM which is the file data encipherment key and is never exposed outside. The result of the RTM function is the file key enciphered under the master

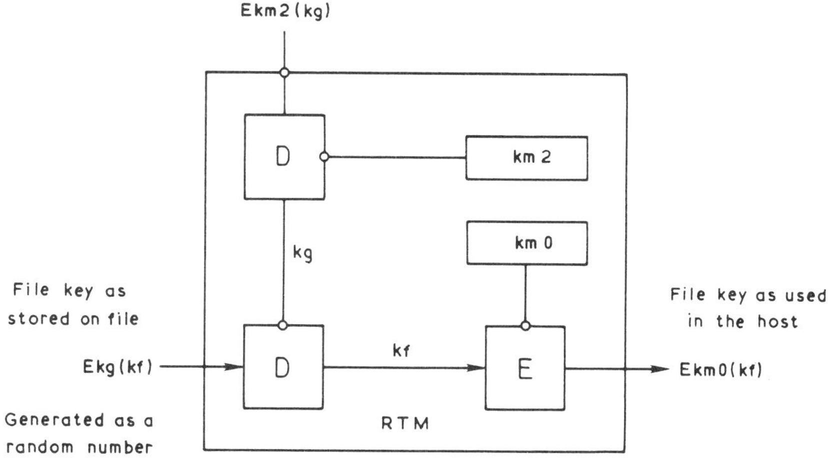

Fig. 6.11 Preparation to encipher or decipher a file

key *km*0 and this is a quantity needed for encipherment or decipherment using the HE or HD function. The quantity E*kg*(*kf*) can be stored along with the file in order to be ready for later encipherment or it can be held by the owner of the file. When decipherment is required, this quantity is passed again through the RTM operation to generate the value E*km*0(*kf*) used as the 'key' for the HD function.

The clever feature of the design of this system is that it makes the RFM and RTM functions almost the inverse of each other but employs different master keys, *km*1 and *km*2. By employing one of these functions for communication security and the other for file security, it is made impossible to misuse these functions in the wrong context. If *km*1 and *km*2 were identical, RTM could be used to reverse the effect of RFM and thus values of E*kt*(*ks*) stolen from the line would be usable to generate E*km*0(*ks*) for decipherment.

We have described both the communication security and the file security functions inside a single host. When two hosts are connected together in a communication network, two further functions are required, the transfer of messages under encipherment between hosts and the movement of files from one host to another in an enciphered form. These are provided by extensions to the key management scheme which will now be described.

Transfer of enciphered data between hosts

All we have described so far about the IBM key management scheme concerns a single host. In order to extend this to a number of hosts which are nodes of a network the keys in the system must have an additional index which will be a superscript *i* or *j*. For example, the master keys are $km0^i$, $km1^i$ and $km2^i$ for the *i*th node. A method of transferring data from node *i* to node *j* will be described. For this purpose, all that is needed is to be able to establish as many session keys as we need, which are known to both nodes. A session key would be created in node *i* as $Ekm0^i(ks)$ and then, in the ordinary way, the operation RFM would be used (see Figure 6.6) to encipher this session key under a terminal key. The session key can be sent to another node if there is a suitable 'terminal key' for the purpose. The proper name for this key would be 'secondary communiction key' but to avoid the change of notation we will call it kt^{ij} signifying that it serves for moving session keys from node *i* to node *j*. It is important that this 'inter-node terminal key' is directional; it can only function from node *i* to node *j* not the other way round.

Figure 6.12 shows how the session key moves from its normal place of storage

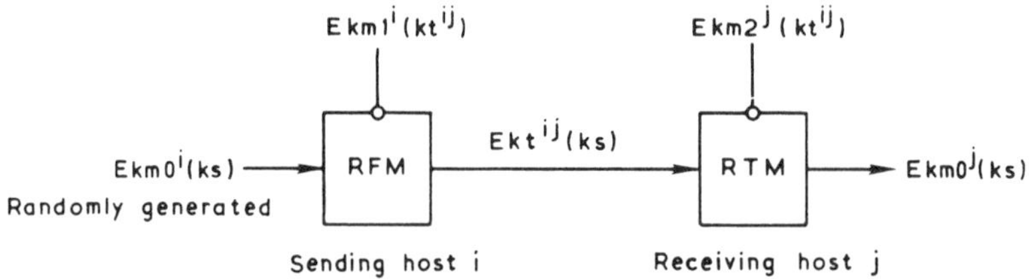

Fig. 6.12 Preparing a session key for inter-host communication

in node i through the intermediate form E$kt^{ij}(ks)$ to the form in which it is stored in node j. Different users or processes in node i will establish their own session keys in this way, all using the same inter-node key kt^{ij} for the purpose. Then the session keys can be used with the HE and HD functions to enable data to be transferred from one node to the other. In fact the session key can be used to transfer data in either direction. The directionality of the inter-node key relates only to the movement of the session key. For example, kt^{ij} is used when a user or process in node i takes the initiative in transferring the session key to node j.

The operation shown in Figure 6.12 is made possible because the inter-node key is available in each node in a suitable form, using the $km1$ master key in the second node and $km2$ master key in the receiving node. In this respect, the receiving node is anomalous because a terminal key is not usually stored in this fashion. Its existence in node j requires the use of the function RTM to be controlled very carefully. The inter-node key kt^{ij} must be transported between the two nodes outside the communication system, probably in a key carrier, and then established in the appropriate enciphered form for storage at each node. We have already seen how E$km1(kt)$ can be established at a node in a privileged operation using the secret number E$km1(km1)$. A similar procedure can establish E$km2(kt)$ and for this purpose the secret number, E$km1(km2)$, is needed. This is generated when the system is initialized, immediately after the master key $km0$ has been loaded, using a procedure like that shown in Figure 6.8. Obviously these functions which operate on clear values of kt^{ij} and establish their appropriate enciphered form at each host must be very carefully controlled, and they are not used very often. When an inter-node key is established between a pair of nodes, it is available for a relatively long period to transmit session keys. A different key, known as kt^{ji}, is established for sending session keys in the opposite direction.

Transfer of enciphered files between hosts

To complete the description of the IBM key management operations, it remains to show how a file can be transferred from node i to node j. This is illustrated in Figure 6.13. The middle part of this figure is very similar to the operation shown in Figure 6.12 but with the session keys replaced by file security keys. The secondary file key kg^{ij} can be regarded as an 'inter-node file transfer key' for sending files from node i to node j. Its establishment uses the procedure described earlier.

The procedure in Figure 6.13 is more elaborate than the communications procedure because the quantities E$km0(kf)$ do not normally exist in the system, but are called into being when an operation is to be carried out on a file. The

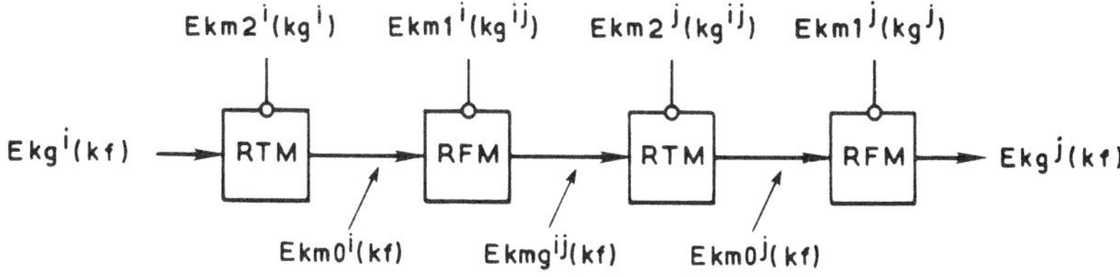

Fig. 6.13 Moving an enciphered file from host i *to host* j

notation here shows secondary file keys such as kg^i, which are the usual secondary file keys belonging to a file-owner, with a designation added to show which node they belong to.

The starting point of any file operation is the enciphered file key such $Ekg^i(kf)$, usually stored along with the file to which it refers. When this quantity goes through the procedure shown in Figure 6.13 it emerges enciphered under a new secondary file key in node *j*, namely the secondary file key of the user or process which is to receive the file. There is no change to the enciphered data since the file key *kf* remains unchanged. All that is necessary at the receiving end is to place the new enciphered file key $Ekg^j(kf)$ in the appropriate place in the header of the file.

In this procedure the RFM operation which prepares the file key for leaving node *i* is unusual, in that secondary file keys are not normally held under encipherment by *km*1 and stored file keys, like E*km*0(*kf*), cannot normally be transferred to a new secondary file key by the RFM operation. Therefore, it is important to control carefully the use of the RFM function in this context.

The security given by the IBM key management functions comes from limiting the cryptographic operations provided by the TRM to a specific, small set of operations each working with the appropriate set of keys. This restriction prevents the exposure of keys in their clear form or the use of any stolen keys in a different host from the one that they belong to. But the functions used for transferring data or files from one node to another cannot not be made available to every user and these are among the functions which must be controlled by the system in order to maintain security. Other things that must be controlled are the key loading functions and the use of the special numbers E*km*1(*km*1) and E*km*1(*km*2). Ultimately, security depends on the control of access to storage and processing functions within the host computers.

6.5. KEY MANAGEMENT WITH TAGGED KEYS

The IBM key management scheme uses encipherment by variants of the master key in order to separate different kinds of keys and ensure that they are used in the correct way. Three variants of the master key are needed so that functions provided for data encipherment and decipherment, session key distribution and file key retrieval cannot be misused for other purposes.

Jones[8] has described a method of handling keys which achieves the same separation of functions with only one master key. The separation of functions in this new method can be applied to an extended key hierarchy without any further complication. The principle employed is to attach a *tag* to each key in its clear text form which identifies that key's function. The idea of tagging words in a computer has often been used, for example to distinguish between capabilities and other words in a capability-based computer. The tag is typically a short extra field attached to the word and preferably one which cannot be separated from the word or altered by non-privileged users of the system. In this section we shall describe the use of the eight 'parity bits', which are attached to the 56-bit DES key, as a tag field. Later, we shall see that the number of tag bits needed is quite small, but the availability of eight bits helps to simplify the description of the method.

These eight bits were intended to be parity bits and some DES hardware implementations refuse to accept a key unless the parity is correct. There seems no special reason for sum-checking the key, because it is easy to verify that keys have been installed correctly by giving them test patterns. The proposed international standard for DEA1 in its draft form regarded these simply as eight undefined bits without specifying how they were to be used. The ingenious aspect of Jones's scheme is that, since keys are not exposed outside a TRM in their clear form, it is not possible for an enemy, even one with access to the computer, to modify these tags.

To begin our description, assume that each type of key employed in the key management system has its own tag value. The top level master keys are an exception because they never leave the TRM so their method of use is entirely controlled. The other key types are the communication keys, *ks* and *kt*, and the file security keys, *kf* and *kg*. We denote the tags attached to these keys as *Ts*, *Tt*, *Tf* and *Tg* respectively. The key values are concatenated with their tags to form plaintext values which are then stored in enciphered form. A typical example of a key in this system would be E*km*(*ks*; *Ts*). The semicolon denotes concatenation where the tag is actually interleaved with the key, occupying the spare bits. Since there is only one master key, we call it *km*.

In order to use the tags as part of the control of key management, each of the operations of the TRM checks the tags of incoming keys and, where appropriate, places the correct tags on outgoing keys. We shall now consider the action of these various functions in turn, beginning with the encipherment and decipherment of data.

There is little change to the functions depicted in Figure 6.5. The incoming session key is tagged in the form (*kg*; *Ts*) and the HE and HD functions will only accept a session key if it has the correct tag. With any other value of the tag, the TRM must return a 'reject' signal.

The HE and HD functions are also used, with file keys *kf*, to encipher and decipher files. For this purpose, the HE and HD functions should be prepared to accept keys with either of the tags *Ts* or *Tf*. The essential restriction on these operations is that they should only be used on data, not on keys. So if an enciphered key-enciphering key were presented in place of the enciphered session key at the 'key' input of one of these functions, it would be rejected. Essentially, the *Ts* or *Tf* tags verify that a *data* enciphering key is being used, making it safe to offer the encipherment or decipherment function to a user or process.

Next, consider the RFM and RTM operations. These are operations entirely with keys but the 'key' input of the function takes in an enciphered key-enciphering key whereas the 'data' input takes an enciphered data-enciphering key. Figure 6.14 shows the RFM operation in its new form. It checks the types of the two keys it is handling when their plaintext is exposed inside the TRM. The key which enters at the data input leaves the module (under different encipherment) carrying the value of tag with which it entered. Because of this checking of key types, there is no need for the master key variants.

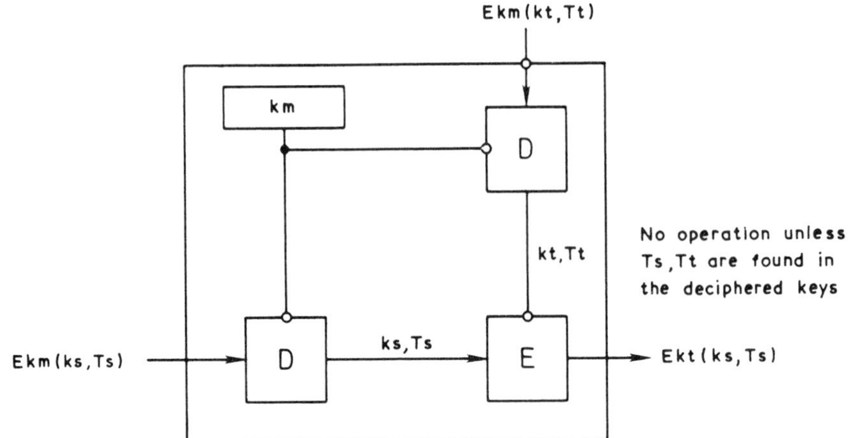

Fig. 6.14 The operation RFM with tagged keys

Generation of new tagged keys

New keys are produced by a random number generator but before they can be used they must be correctly tagged and enciphered. The key enciphering keys, *kt* and *kg*, present no problem because the processes which are allowed to generate these keys have access to the 'encipher with master key' operation which, in this key management scheme, could attach either the *Tt* or *Tg* tag, whichever is asked for. But although a terminal key must be available to the system in clear form for distribution to the terminal, the same is not true of a secondary file key, which is only ever used in the enciphered form outside the TRM. It is, therefore, more secure to insist that a new secondary file key be generated entirely within the TRM, each such key generation starting with a new random number which is not revealed outside the TRM. This prevents an enemy who has penetrated this part of the system from generating identical keys with two different tags so that a key value could be used in both the RFM and RTM operations, something which the system is intended to prevent.

The generation of new session and file keys in the IBM key management scheme is performed in such a way that different operations are used, RFM for the session key and RTM for the file key. The starting point in each case is a random number which represents an enciphered key. With the tagged-key method, a random number will not create a key with the correct tag, so a different method of key generation is needed. One possibility is to include in the operations of TRM, two key generation functions designed for these types of keys. A session key is generated in the form E*km*(*ks; Ts*) and can be used for any terminal. In order to generate a file key, the secondary file key *kg* must be provided to the TRM as E*km*(*kg; Tg*) and with this input a random key *kf* is chosen and its enciphered form E*kg*(*kf; Tf*) is supplied from the TRM. Thus each type of key requires a function for generating it in the TRM, designed in such a way that the same key cannot be generated with more than one tag value. Only *kt* keys begin their lives in plaintext form.

Extending the key hierarchy

We have been using a key hierarchy with three levels, at the top the master keys with no tag, at the next level two kinds of 'secondary' keys with different tags and at the lowest level the data enciphering keys (session and file) with their distinctive tags. If we wanted to extend the key hierarchy to four, five or more levels we would need to introduce new tag values and require the various operations of the TRM, in particular the RFM and RTM operations, to recognize new combinations of tags which they will allow.

The proposal by Jones has a different and simpler scheme which allows the key hierarchy to be extended without introducing new tag values. The main distinction which it makes is between data enciphering and key enciphering keys. Only data enciphering keys are allowed to enter the operations HE and HD. The categories *ks* and *kf* that we have distinguished are both included in the data enciphering category. All key enciphering keys belong to a second category, which includes *kt* and *kg*, but these two types of key must be distinguished because we only allow *kt* as a key input to RFM and *kg* as a key input to RTM. This distinction is made by providing two bits in the tag field which determine whether the key can be used for encipherment or decipherment or both. From Figure 6.6 it can be seen that *kt* is an enciphering key and from Figure 6.11 that *kg* is a deciphering key. Therefore three bits form the complete tag; one distinguishes data and key enciphering keys, the others allow encipherment and decipherment respectively, or both. The operations RFM and RTM can be carried out at several levels of an extensive key hierarchy because there are key encipherment keys of several levels. Only data enciphering keys can be used in the operations HE and HD.

The distinction between enciphering and deciphering keys has so far been applied only to key-enciphering keys, leaving the data-enciphering keys such as *ks* and *kf* marked for both encipherment and decipherment. But data-enciphering keys can also be produced for a single role, either encipherment or decipherment. Within a system controlled in this way it should be possible to generate corresponding pairs of enciphering and deciphering keys. This makes possible an asymmetric form of cryptography.

A particular session or file key, with its tag, can be transmitted throughout a network of inter-connected nodes which employ the same tagged key management system. Their security depends on the physical security of all the TRMs in the system and in particular on the safe distribution of the 'inter-node keys' such kg^{ij} and kt^{ij}. The inter-node keys create a problem for the key generation method which we described earlier because the same key is tagged in two different ways in the two nodes it interconnects. We saw earlier that the handling of inter-node keys requires privileged operations. In the tagged scheme it seems to require that the inter-node key, generated as a terminal key for use in RFM at one node, can be converted into a secondary file key for use in RTM operations at the other node. The establishment of inter-node keys uses operations that must be very carefully guarded, whether the IBM method or Jones's variant is used.

6.6. KEY MANAGEMENT STANDARD FOR WHOLESALE BANKING

The international standard ISO 8732 entitled *Banking–Key Management (Wholesale)*[2] describes key management procedures for safeguarding the secret cryptographic keys used to protect banking messages. The scope known as wholesale banking includes messages between banks, between banks and their corporate customers or between banks and a government. This makes it clear that messages of some consequence are being protected and that such applications as home banking and point of sale transactions are not within its scope. The standard is complex and occupies more than one hundred pages. It describes the procedures for manually exchanging keys in order to initiate the system and then, with the aid of these manually exchanged keys, the automatic exchange of the keys which are used to protect the messages and, in particular, to provide message authentication. The major part of the standard is concerned with the automatic methods, giving their procedures, the contents of the messages and the detailed coding.

For someone coming to this standard for the first time, the main problem is the quantity of detail. Our account cannot be a substitute for the standard itself and we therefore do not attempt to cover the detail but describe instead the main principles, leaving out features where this does not prevent its principles being understood. For example, we give little attention here to error messages or error recovery procedures.

The standard is very significant because it is the first internationally agreed principle for key management. Consequently, some of its notation and concepts have already been adopted in other schemes, such as proprietary key management procedures which do not follow ISO 8732 in detail but claim some similarity. Departing from our usual terminology, we shall adopt the word *encryption* in place of *encipherment* to align with the standard, where appropriate.

We begin our account by describing the key hierarchy and the way it uses keys of both double length and single length. The standard was derived from the US standard ANSI X9.17 where the intention, at the time, was that the DES algorithm would be used. For generality, the international standard does not specify that DES shall be used, though the standard ANSI X3.92 (which describes the DES algorithm) is quoted.

Each participating bank or other (corporate or government) organization is expected to have a *key management facility* which includes the cryptographic equipment and provides a secure environment in which the cryptographic functions can be performed and the keys held. The standard does not describe how secure arrangements can be made for the storage of the keys, though their encipherment under some unspecified master key is implied. These are details which are the concern of each user and they are not specified because they do not affect the exchange of messages between the participants.

A short section of the standard, little more than one page together with an example in an annex, deals with the manual distribution of keying material which sets up the basic keys on which the key hierarchy depends.

We continue with a description of the three different environments under which key distribution can be carried out, according to this standard. For each of these, the messages which are exchanged in normal operation will be described, together with a partial account of the content of these messages.

The standard does not claim to cover all aspects of key management. The banks themselves can decide on the lifetime of keys, the number of keys they use and the provisions they will make to have reserve keys available in case one is compromised. The standard provides for communicating the date and time at which the key becomes effective and for exchanging messages which withdraw a key from service. How these tools will be used by individual banks is left to their judgement. The user can also choose the type of key hierarchy (within limits) and whether to use double or single length keys for some of the functions described. Thus a decision to use the ISO 8732 standard for key management does not complete the agreement between users, who would then have to detail the chosen options and the schedule of keys and key changes.

There are two technical features which are central to this standard, known as *notarization* and *key offset*. Notarization employs the modification of a key using the names of the source and destination of a communication as the modifying data. This key is then used to encipher a subordinate key in a key management message and the receiver of that message can only interpret it by using the correct key modifications by the names employed by the sender. This prevents any intruder diverting a key intended for one purpose into another purpose. There are other ways to do this, but the *notarization* principle is used here. A similar modification of the key encrypting key uses a counter which is incremented throughout the whole life of that key. In this way, the subordinate keys sent with this key encrypting key carry a built-in note of their sequence number and an attempt to reuse an old key would be detected. Notarization and key offset are generally used together.

These two features, and the general complexity of the standard make it questionable whether it should have a wider validity than the scope for which it was intended, but it is invaluable as an example of a fully developed key management standard.

The key hierarchy

Figure 6.15 shows the two alternative key hierarchies employing two or three layers of keys respectively. It also shows the notation for the three types of keys employed.

A key used for enciphering or authenticating data is known as a *data key* (DK) and is the lowest level, layer 1, of each hierarchy. These keys are used for either encipherment or authentication but should not be used for both except when the standard specifies this. Wherever data keys are distributed for use in protecting banking messages, either one or two keys can be sent so that one can be used for encipherment and the other for authentication. The encipherment function of the data keys also includes the encipherment of initializing values for use in the standard modes of operation described in chapter 4. Data keys are always single length keys.

For the distribution of data keys, a *key encrypting key* (KK or KKM) is used. In

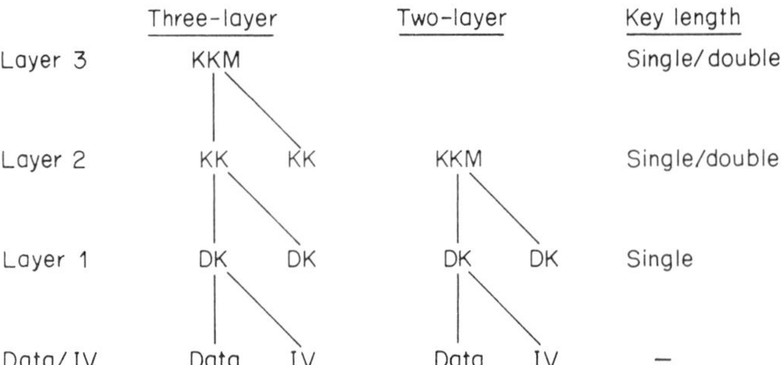

Fig. 6.15 Alternative key hierarchies of ISO 8732, KKM, master (key encrypting) key: KK, key encrypting key, to encipher DKs; DK, data key, to encipher data or IVs or authenticate messages

the two-layer hierarchy a master key is used, which has been established by manual key distribution for this purpose. In the three-layer hierarchy any KK which is used to encipher a data key for transmission can itself be distributed using a master key encrypting key (KKM).

Since it is widely accepted that the 56-bit key size of the DES algorithm weakens it, under certain circumstances, there is sometimes a need for multiple encryption to obtain an effectively longer cryptographic key. Where keys are in use for a long time, the opportunity for an attacker to perform a key search is greater, therefore this standard envisages the possible use of double-length keys at the higher levels of the hierarchy. The method by which encipherment is performed with these double-length keys is described in the next section.

The use of double-length keys is mandatory only for master keys which are used for key management between a participant and one of the two kinds of key management centre (either a key distribution centre or a key translation centre). Apart from this requirement, key encrypting keys (KK or KKM) can be either single or double length but a single length key must not be used to encipher a double length key. The standard does not give any criterion for the use of double length keys but we would expect a double length key to be used wherever the life of the key exceeds about one month. The presence of a double length key is signified by the notation ∗KK or ∗KKM.

Encipherment and decipherment with double length keys

Two single length keys each of 64-bit size can be concatenated to form a double length key. Thus a double length key ∗KK is considered as being made up from two single length keys KKl and KKr where the notation signifies the left and right-hand halves of the concatenated double length key. The encipherment and decipherment processes using these keys are shown in Figure 6.16. In each case three cryptographic operations are employed. For double-length encipherment these are encipherment, followed by decipherment, then by encipherment. Both encipherments use the left-hand part of the key and the intervening decipherment uses the right-hand part. This triple processing was chosen because the simpler double process could be weakened by a 'meet-in-the-middle' attack, employing a large amount of storage in order to reduce the computational task of key searching.

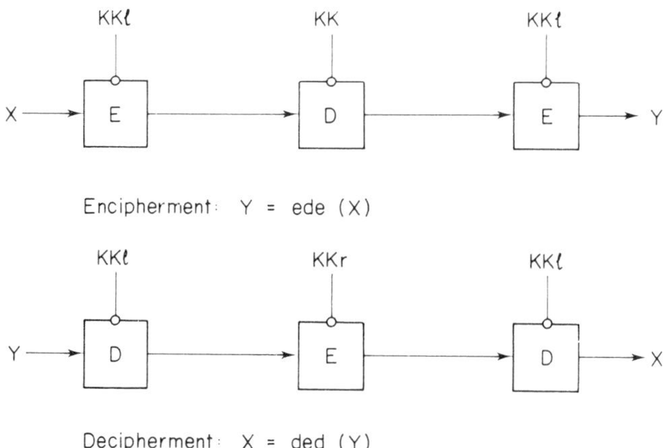

Fig. 6.16 Triple encipherment with a double length key

The corresponding decipherment function shown in the figure employs decipherment followed by encipherment then decipherment with the same key sequence. The notation for encipherment using this process is *ede* and correspondingly for decipherment *ded*.

It is not entirely certain that the use of double length keys provides an effective key length which is double that of a single length key, but the structure of DES and the experimental evidence that it does not form a group supports the view that multiple encipherment is more secure. There is, however, no need to use multiple encipherment unless the threat of a key-search attack is significant.

Key distribution environments and messages

The simplest environment is that two banks wish to establish keys for the exchange of messages and they already have in common a key encrypting key, either a master key (KKM) or a key at layer two of a three-layer hierarchy (KK). Their aim may be to distribute a KK for the three-layer hierarchy or to distribute data keys or perhaps to do both. No other party is involved. This is the *point-to-point* environment.

To avoid the need for banks to have master keys or key encrypting keys in common with all their correspondents, it is possible to use a key management centre which has an established cryptographic relationship with both parties to the communication. Then, using the key already established they can obtain the necessary material from the key management centre to establish contact. There are two kinds of key management centre, one of which generates keys, called a *key distribution centre* and the other which accepts a key from one party, re-enciphers it and returns it in a form suitable for communication with the other, called a *key translation centre*. Thus there are three environments in the standard, known as *point-to-point, key distribution centre* and *key translation centre*. For each of these there is a procedure for carying out the key exchange which we shall describe with reference to Figures 6.17, 6.18 and 6.19 respectively.

In describing these procedures we do not detail all the messages which can be used. In particular, any message may be in error and receive a response which is an *error service message* (ESM). In the context of the key translation centre there is an *error recovery service message* (ERS) which reports errors to the centre and requests resynchronization and reinitiation. In this account, we shall confine ourselves to the messages used in normal operation, when no errors occur.

The coding of these messages employs the character set of International Standard ISO 646, often known as the ASCII code. The messages are coded in a way which makes them readable when printed on the page. There is an intention to introduce another version of the standard with binary coding of messages. In the coding, both the message type and the tag for each field of the message is a set of three capital letters. Within the message, the various fields are tagged to indicate their significance and separated by space characters. Additional format characters can be inserted to improve the ease of reading the message. For our purpose, we shall not define the format and coding in detail.

Point-to-point distribution of keys

When two banks A and B already have a key encrypting key in common they can use it to exchange other keys of the form KK or KD. The messages they use are shown in Figure 6.17. For example, bank A can send a *key service message* (KSM) to B containing a key or keys enciphered under the already established key. Bank B will return a *response service message* (RSM) and this completes the exchange. But suppose that bank B wishes to establish keys in this way and does not have the ability to generate keys. It is then possible for bank B to send the optional *request service initiation* message (RSI) shown by a broken line in the figure, which stimulates bank A to begin the procedure already described.

All messages in this and the other protocols we shall describe contain three particular fields, the first defines the message type, for example a key service message would have a field MCL/KSM where MCL is the tag for the message type. The other fields which all messages contain are the identities of the originator and receiver of the message. We shall not repeat these fields because they occur in all messages.

All messages also contain either an authenticator (MAC) or else an error detection code (EDC). Whenever a message is of cryptographic importance and a suitable authenticating key is available, a MAC is included. In the point-to-point procedure, the KSM and its response RSM have MAC values but the RSI message does not. This is because authentication always uses a data key and the purpose of this procedure is to establish a data key (or keys) so the existence of such a key cannot be assumed. The KSM and RSM messages both include a MAC using the data key which has been transmitted in the key service message.

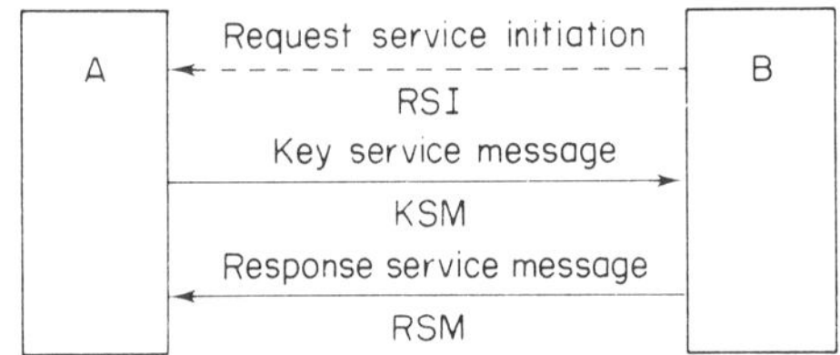

Fig. 6.17 Basic messages in point-to-point key distribution

The error detection code (EDC) is formed in the same way as a MAC but uses a fixed cryptographic key value 0 1 2 3 4 5 6 7 8 9 A B C D E F in hex notation.

The key service message can contain a key encrypting key, enciphered under a master key already in existence but this is optional and will not occur in a two-layer hierarchy. It always contains at least one data key and can contain two if it does not also have a key encrypting key. If there are two data keys present, the key used for authentication of the key service message is formed from their exclusive OR. The KSM may also contain an initializing value (IV) enciphered with a data key, the second one if there are two present. The RSM also contains the date and time at which the data keys become effective.

For each key encrypting key shared by the two banks, a counter is maintained which starts at the value 1 when this key is first used to encipher KD values. It is incremented for each subsequent use and the counter value is sent in the message. The receiver checks the counter and responds with an error service message if it is wrong.

Notarization is an optional feature in the procedure for point-to-point key distribution. If it is used the field NOS/ with no parameters is included in the message. In this case the KK is subject to a transformation by notarization and key offset before it is used to encipher the data key.

The response message contains no more than the fields contained in all messages, mentioned above and the MAC. The MAC is formed with the same key that was used for the key service message, therefore the receiving bank must decipher the KD value or values before using them to check the received authenticator and generate an authenticator on the response. Reception of a correct RSM by bank A, when its MAC value has been checked, verifies that the correct KD was received and hence a correct KK, if one was sent. There is no corresponding check on the correct receipt of the IV.

In addition to the messages shown in Figure 6.17 either bank may send the other a disconnect service message (DSM) which, in addition to its standard fields includes the identity of the key to be discontinued. The DSM also includes the identity of a key used for authenticating it and a MAC value. The response to the disconnect service message is an RSM containing the identity of the discontinued key and a MAC using the same key as for the DSM. Note that the content of a response service message varies according to the type of message to which it responds.

Key distribution centre

If bank A does not have a data key in common with bank B it may be able to establish such a key through the agency of a key distribution centre (CKD). For this purpose, bank A first communicates with the distribution centre and obtains from it a new data key in two forms. One of these it can interpret using a KK which it holds in common with the distribution centre. The other it passes on to the bank B which itself can interpret this using another KK which it has in common with the same distribution centre. The scheme of messages used is shown in Figure 6.18.

There is an optional request service initiation (RSI) from bank B to bank A, in

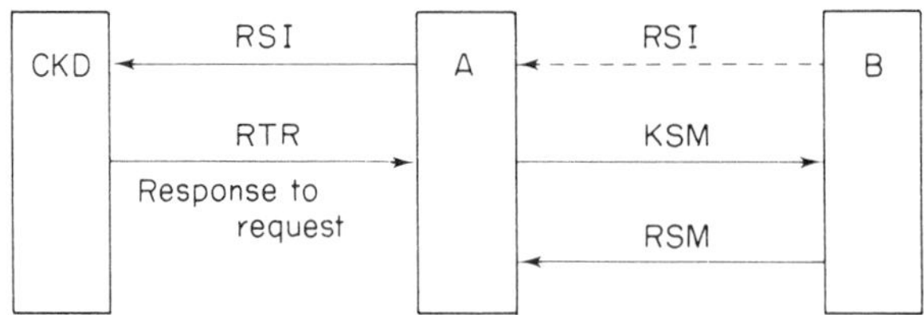

Fig. 6.18 Basic messages using a key distribution centre (CKD)

case bank B wishes to stimulate bank A into carrying out this protocol. Normally, bank A would initiate the protocol with an RSI sent to the distribution centre. In this procedure, only the RSI messages use an error detection code, all the others employ a MAC. The RSI from bank A indicates the service required and the identity of the 'ultimate recipient' which is bank B.

The service provided by the key distribution centre is the supply of one or two data keys and optionally an IV enciphered under a data key, the second one if there are two. Note that a key distribution centre does not supply key encryption keys. This contrasts with the key translation centre which is able to translate key encryption keys if a superior master key encryption key exists.

The response to request (RTR) is returned by the distribution centre and contains the usual three fields identifying the message type, the originator and receiver and also a MAC using as key a data key or, if there are two, a key produced by their exclusive OR. In addition, the RTR includes the identity of bank B, the data key or keys in two forms, an optional IV and the date and time at which the data keys become effective.

The key distribution centre maintains a counter which is incremented each time it uses the key encrypting key, either the one for bank A or the one for bank B. The two values of these counters (CTA and CTB) are sent in the message. Concerning the data key or keys, these are provided in two forms called KD and KSU, the first intended for bank A and the second for bank B. Each is notarized, that is to say the KK used is modified according to the source and destination of the key and it is also offset with the value of the counter CTA or CTB.

Bank A receives the RTR, extracts the KD value or values from their notarized encipherment and verifies the MAC. It can then generate a MAC for the key service message which it sends to bank B containing the keys in the form KTU, together with the IV, date and time of effectiveness of the keys and the counter for bank B (CTB). Bank B can then extract the KD value and verify the authenticator, after which it returns the RSM. Both the KSM and the RSM contain the identity of the key distribution centre. Finally, bank A verifies the MAC received in the RSM.

Key translation centre

A more powerful facility is provided by the key translation service offered by a key translation centre (CKT). In this scheme, bank A generates the keys it needs and passes them to the translation centre for conversion to encipherment under the keys held in common between that centre and bank B. The message which initiates this procedure, request for service (RFS) contains the key values that will

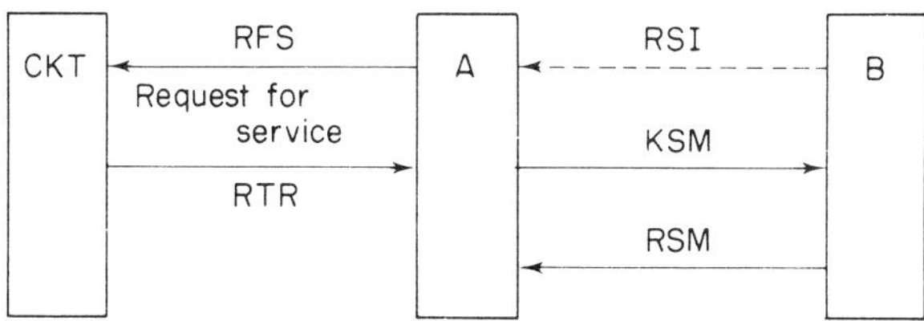

Fig. 6.19 Basic messages using a key translation centre (CKT)

eventually arrive at bank B, see Figure 6.19. As before, the whole procedure may optionally be initiated by bank B using an RSI message to prompt bank A into starting the procedure. Apart from the RSI, all the other four messages in this procedure are protected by authentication using a MAC based on a data key which is transmitted (in enciphered form) with the message for that purpose. Though the names of the messages are similar in these various procedures, the content is different according to the procedure in which they take part. The handling of the counters CTA and CTB is similar to that of the key distribution procedure, so also are the identifying data, such as the identity of bank B in RFS and RTR and the identity of the key translation centre in KSM and RSM. Effectively, each important message holds the identity of all three parties.

The key translation centre translates either one KK or one or two data keys. It never translates keys of both types. Keys to be translated are sent enciphered under existing superior keys but not notarized. If a KK is sent for translation a data key is also sent, enciphered under this KK, but is not intended for translation, it is merely for calculation of the MAC values.

The response, RTR, contains the translated keys, either a KK or a KD or a pair of KDs. They are notarized for transmission between the translation centre and bank B, and offset using the counter CTB. The value of CTA is sent in the RFS and returned in the RTR. Bank A receives the RTR and verifies its authenticator using the data key it originally sent. It then transmits the notarized keys to bank B, together with the CTB value received from the translation service.

If a key encrypting key has been translated, new data keys are generated by bank A to include in the KSM, enciphered under the new KK. At least one such data key is needed to provide the means for authentication of the KSM and its response. Two data keys may be sent in the KSM and may subsequently be used between the two banks as working data keys. In addition, bank A may generate an IV and send this enciphered under a data key, the second key if there are two of them. Bank A also decides the effective date and time (EDK) for the new keys and includes that in the message. Bank B deciphers first the KK, if any, then the data key or keys before verifying the authenticator and returning the RSM, which contains only identifying data and a MAC. Finally bank A verifies the authenticator of the RSM.

The full procedure for operation of the key translation centre includes error service messages (ESM) and error recovery service messages (ERS) which help to report and recover from loss of synchronism of the counters. As with the other two environments any one of the parties can send a disconnect service message (DSM) to terminate the operation of a specified key and receive a response to verify that this has happened.

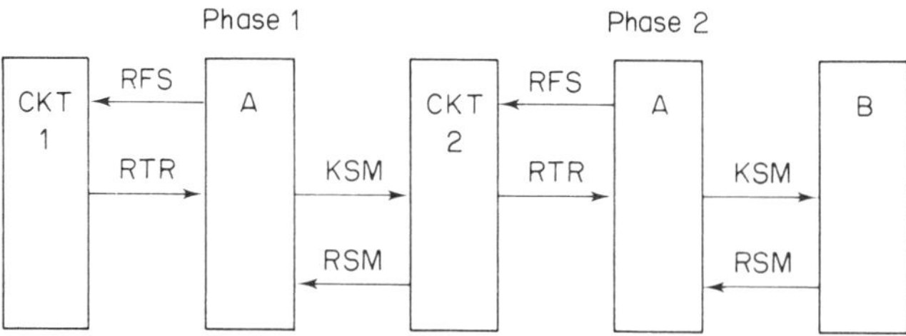

Fig. 6.20 Using two key translation centres in turn

Consecutive use of two key translation centres

Suppose that bank A has a key relationship with translation centre CKT1. By key relationship we mean that the two parties share a key encryption key. Suppose also that CKT1 has a keying relationship with another key translation centre CKT2. Bank A wishes to communicate with bank B for which the normal key translation centre is CKT2. There is a scheme which enables bank A to communicate with bank B by using the procedure shown in Figure 6.19 twice over. The method is shown in Figure 6.20 in which, for convenience, bank A is shown twice.

In the first stage, bank A uses the normal procedure with key translation centre 1 to establish a key encrypting key in common with CKT2. Having done this, bank A then interacts with CKT2, and since CKT2 has a key in common with bank B, it is able to translate bank A's key for delivery in a KSM message to bank B. At the end of these two stages, banks A and B have a key encrypting key in common. They can then exchange data keys by the point-to-point procedure and use these data keys to protect their messages.

It is clear that this principle could be extended to translate keys through a number of translation centres which are linked by mutual KK values. A group of translation centres could organize themselves to provide a comprehensive service in a number of ways, for example appointing one of them to translate keys between any of the others. These possibilities are not described in the published standard, which quotes the arrangement of Figure 6.20 as an application of the key translation procedure.

This extension to more than one centre is not possible with the key distribution centres because they can distribute only data keys and therefore cannot establish a new KK relationship.

Key notarization and offset

Key notarization and offset are applied to any keys supplied by a key distribution centre and to the translated keys from a translation centre. These keys are supplied under encipherment by a superior key and the notarization and offset procedures consist of modifying the superior key before using it for encipherment. Notarization means that the key is modified according to the source and destination of the message, offset means that the key is modified according to a counter value, which provides a sequence number for the successive keys enciphered with this superior key. The purpose of notarization and offset is to prevent the replay of a message carrying encrypted keys and to prevent the substitution of one message for another. This same purpose could have been achieved in other ways but in this standard the notarization and offset techniques are mandatory in the key distribution and key translation functions. In the case of point-to-point key distribution, use of notarization is optional.

Figure 6.21 shows the calculation involved in key notarization and offset. We begin with a key encrypting key of double length which has two parts, KKl and KKr. The application to single-length KK will be described later. The identities of the source and destination of the key are given by double-length fields entitled *TO* and *FM*. The two halves of TO are called TO1 and TO2 and similarly the two halves of FM are called FM1 and FM2. The figure shows how a combination of encipherment and exclusive OR is used to combine these values to form a double-length value called the notary seal (NS). This is the quantity which is used to modify the key *KK before it is used for encipherment.

The figure also shows the use of key offset, in which the counter value CT is combined by exclusive OR with each part of the notary seal before this is used to modify the double-length KK which is to be used for enciphering the key. The result of these operations is the double-length value *KN which is used in the triple encipherment process *ede* to encipher the key for transmission. Normally this transmitted key will be a data key, but it could be a key encrypting key enciphered by a master key encrypting key. A similar procedure can be used with single-length KK values employing the convention that KKr equals KKl. A single-length notary seal value is generated by combining the 32-bit halves of the double-length value as shown at the right of the figure. The subsequent operations are applied to a single-length value to generate a single-length modified key KN used for the encipherment of a transmitted key.

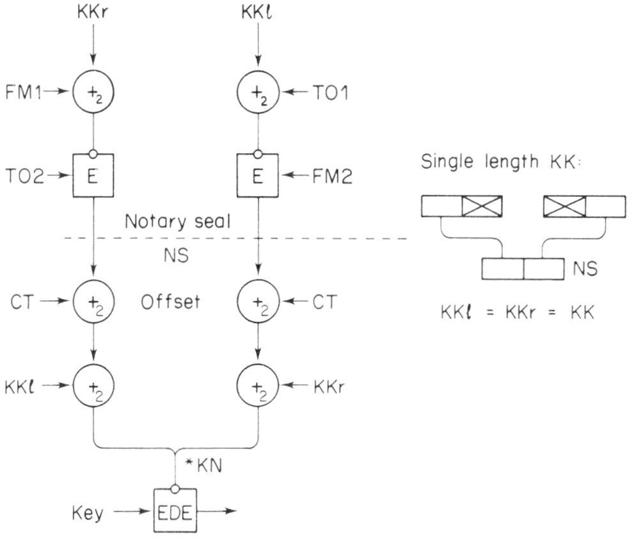

Fig. 6.21 Notarization and key offset

6.7. ALTERNATIVES FOR KEY MANAGEMENT

We have described (but not fully) two elaborate key management schemes designed with special circumstances in mind. In each case it is possible to see how alternative and sometimes simpler schemes could have been developed with much the same properties. It seems that the science of key management is still developing.

General purpose cipher algorithms and modes of operation are now an established fact but key management is tailored to individual applications. The need for this adaption will become even clearer when we describe the special circumstances of point-of-sale EFT in chapter 10. For all kinds of applications, key management standards will become necessary and they will have some features in common and many points of divergence. Progress in this field still seems unclear. The international standard ISO 8732 includes some valuable concepts and notation but should not be taken as a point of departure for all future key management standards.

There is great potential for the use of public key cryptography in key management. This technology is introduced in chapter 8 and the application of these asymmetric algorithms to digital signature is described in chapter 9. Both are valuable techniques for improving the security of key management. Standardization in their key management role has not begun but ISO 8732 envisages that there will be, in due course, provision for the use of asymmetric algorithms for key distribution. The most immediate use is to employ public key cryptography to distribute keys for symmetric algorithms as typified by DES. The security requirements for the safe use of public key cryptography are described in chapter 8 and digital signatures, as described in chapter 9, provide a superior method of checking the integrity of key distribution messages and resolving disputes concerning messages sent and received.

Key management in electronic funds transfer systems (EFT) has been developed both with symmetric algorithms and public key cryptography. Chapter 10 contains descriptions of key management of both kinds.

REFERENCES

1. Ehrsam, W.F., Matyas, S.M., Meyer, C.H. and Tuchman, W.L. 'A cryptographic key management scheme for implementing the Data Encryption Standard', *IBM Systems J.*, **17**, No. 2, 106–125, May 1978.
2. *Banking–Key Management (Wholesale)*, International Standard ISO 8732, International Organization for Standardization, Geneva, 1988.
3. Matyas, S.M. and Meyer, C.H. 'Generation, distribution and installation of cryptographic keys', *IBM Systems J.*, **17**, No. 2, 126–137, May 1978.
4. Heinrich, F. *The network security center: a system level approach to computer network security*, NBS Special Publication 500–21, January 1978.
5. Needham, R.M. and Schroeder, M.D. 'Using encryption for authentication in large networks of computers', *Comm. ACM*, **21**, No. 12, 993–999, December 1978.
6. Popek, G.J. and Kline, C.S. 'Encryption and secure computer networks', *Computing Surveys*, **11**, No. 4, 331–356, December 1979.
7. Denning, D.E. and Sacco, G.M. 'Timestamps in key distribution protocols', *Comm. ACM*, **24**, No. 8, 533–536, August 1981.
8. Jones, R.W., 'Some techniques for handling encipherment keys', *ICL Technical J.*, **3**, No. 2, 175–188, November 1982.

Chapter 7 IDENTITY VERIFICATION

7.1. INTRODUCTION

The security of a system often depends on identifying correctly the person at a terminal. A familiar example is a bank's 'Automatic Teller Machine' which gives cash to account holders that it can identify. Direct person-computer interaction is good for efficiency, but it opens possibilities for fraud by a *masquerade* — the adoption of a false identity. Access to computers and entry to buildings should be controlled according to the identity of the person, but like many human skills that we take for granted (such as language and vision) recognizing a person is surprisingly hard for a computer.

Suppose that a bank has a million customers. If it is to recognize them by a personal identification number (PIN), that number must have at least six decimals and if instead it recognizes by a signature, the measurement of the signature must separate each person's signature from all the others. The bigger the population, the harder it is to make a positive identification. Fortunately for us, this is not the problem usually presented. We assume that the terminal user wants to be known. We can ask him his identity and it can be entered on a keyboard or from a card he carries. We can use this claimed identity to access a file giving the reference 'measures' of his signature or his reference PIN. If the given data match the reference, we accept him. Suppose he is an impostor who does not have the correct signature or password, then there is a small probability that he may be lucky, for example a four-digit password, well-chosen, makes this probability 10^{-4}. With the right precautions, this can be adequate because nobody would risk being caught for such a small chance of success. Nevertheless, the probability of a successful masquerade must be chosen according to the amount at stake — an instant pay-off of £10 million is worth a 10^{-4} chance try if the risk of detection is small enough. The important characteristic of identity verification is that the precision of the method need not be increased as the population increases. It depends rather on the ratio of the potential illegal gain to the 'loss' value of being caught. Practical systems do not *identify* the individual but do verify the identity he claims to have.

Methods of personal identity verification can be divided into four broad categories, though particular systems can contain elements of more than one category. These categories employ, (a) something known by the person, (b) something possessed by the person, (c) a physical characteristic of the person, (d) the result of an involuntary action of the person.

A password and a passport are examples of categories (a) and (b). Categories (c) and (d) are not always distinguished — we can quote a fingerprint as an example of a physical characteristic and a signature as the result of involuntary action, since each movement is not individually controlled.

Any successful identification system must, in addition to being economically viable and adequately secure, be acceptable to the system users. An inconvenient or objectionable method is likely to provoke users into finding ways to circumvent the procedures, making the task of any potential intruder much easier.

7.2. IDENTITY VERIFICATION BY SOMETHING KNOWN

Passwords

The best-known example of identity verification by something known is the password. Passwords have been used very widely in identity verification in an enormous range of contexts, from Ali Baba's opening of the magic cave, through military applications, to protocols for accessing computer systems. Much attention has been given to methods of generating and managing passwords. A standard for PIN management was published by ANSI in 1982.[1]

We can divide passwords into several categories, (a) group passwords which are common to all users on a system, (b) passwords which are unique to individual users, (c) passwords which are not unique to individual users, but serve to confirm a claimed identity, (d) passwords which change every time a system is accessed. Each of these types of password has seen some use in our context.

Group passwords are sometimes used as part of the log-on procedure for a service being accessed from a terminal. However, being common to all users, they are likely to become known outside the circle for which they are intended. Disclosure usually happens through a breach of security such as writing the password on the wall by a terminal, a regrettably widespread practice. In any system where more than the minimal security is needed, group passwords should be ruled out.

Passwords unique to each user potentially provide a much higher level of security. If illegal access takes place, then a system log should be able to show under what password the access took place. Control of this kind can detect the use of a loaned password, disclosed by the user to some acquaintance. On the other hand, careless security in storing or handling a password may lead to its unintentional disclosure; this could also be said to be the fault of the person to whom the password has been allocated. Good security practice entails memorizing the password and not keeping a printed copy. If unique passwords are implemented, then a separate user identifier need not necessarily be used, though an added confirmation of identity can be generated. In a user population which is very large, as, for example, the users of a bank's automatic teller terminals, use of unique passwords is impracticable — such a password would have to be too long for easy memorizing.

This implies a third kind of password which is becoming increasingly common — the non-unique password. Here a relatively short password, of, say, four decimal digits, is allocated to each user. Identification of the user depends on a much longer number which the user is not required to memorize — it may be held on the magnetic stripe of a card, for example. The fact that the same password is allocated to a number of different users presents no problem if the password allocation bears either no relationship or no easily determined relationship to the user identity. In an on-line system using non-unique passwords, a list can be held centrally which gives user identities and their corresponding passwords; identity verification depends on looking up the presented identity and password on this list. In an off-line system, the password must be related to the identity token, for example the magnetic-striped card, in such a way that the off-line terminal can interpret it whilst interpretation by the user (or intruder) is too difficult; this implies some kind of encipherment.

There is a fourth kind of password — one which changes every time the system is accessed. To operate such a system it is necessary to prepare a list of passwords and give a copy of the list to the user. At each access one password is used and then becomes invalid for future access. This prevents an intruder from acquiring a current password by tapping a communication line because the traffic on the line gives no indication whatsoever of the next acceptable password. Clearly the system is stronger than those systems which use the same password(s) repeatedly and is useful where the highest security is required. However, there are certain disadvantages; the user must protect his printed list of passwords and the password list requires storage and protection at the central installation. The amount of password material depends on the frequency of access. One-time passwords of this type are in use in the Society for Worldwide Interbank Financial Telecommunications (S.W.I.F.T.) network of the international banking community described in chapter 10. For greater security the S.W.I.F.T. password tabulations are prepared in two halves, so that each list contains only half passwords. The two halves are sent separately to users, reducing the chance of complete passwords being intercepted.

It is of course necessary to design the password handling by the system in such a way that it is protected against disclosure. For this reason, any password that is transmitted from a terminal to a central installation should be protected during transit by encipherment, otherwise a passive line tap will suffice to betray it to an intruder. Equally, the response from the central installation to the terminal accepting the validity of the identity and password must be protected by encipherment. If this is not done, then the accepting response can be reorganized and reproduced on the line by an intruder with an active line tap; if the terminal were an ATM this would allow an intruder to defraud the system without actually knowing any password. Furthermore, some variability must be included in both request and reply (for example, serial number or time/date), otherwise a standard request or reply would lead to a standard enciphered format. All the intruder need then do would be to record the standard message and replay it whenever he chose; a replayed request would deceive the central system or a replayed reply would deceive the ATM.

We have referred to lists of identities and passwords held at central installations. Such lists could easily become available to operational staff or systems programmers; on this account it is important not to store passwords in clear form. The usual solution to this problem is to encipher the passwords in some way; the original idea for using enciphered password lists is due to Needham, cited by Wilkes.[2] If the encipherment is carried out with an algorithm such as the Data Encryption Standard (DES), then the risk is transferred to disclosure of the encipherment key which must be given suitable protection. At the time when the DES algorithm had been specified, but not implemented in hardware, the algorithm was used by the Bell Laboratories for protection of passwords in their Unix system.[3]

The original suggestion by Needham made use of a one-way cipher function; this is a function that is easy to compute in one direction, but very difficult to invert. This type of function is considered in detail in chapter 8. The Bell password protection system made a novel use of the DES algorithm as a one-way function in a programmed implementation. Instead of using a secret key to encipher the password list, the first eight characters of each individual password were used as a DES key to encipher a constant; the DES algorithm was iterated twenty-five times and the resulting 64-bit field repacked to become a string of eleven printable characters. The result of this operation was stored in the password list. The strength of this system was considered adequate until fast hardware implementations of the DES became available; it was feared that fast DES operation would allow exhaustive search for passwords, using the enciphered forms from the password list. For this reason the programmed algorithm was altered so that the internal E permutation (see chapter 3) became dependent on a random number chosen (and entered into the password file) at the time of registration of each password. Thus standard DES devices could no longer be used for a systematic attack on the password security.

Figure 7.1 shows the flow of variables in password checking with a one-way function.

The Bell study considers the time taken to search for a password. On the assumption that an intruder has the opportunity to program a system to try a sequence of passwords without being detected and thrown off the system, it is possible to predict how long it would take to test all possible passwords against a given identity. On a PDP11/70 it took 1.25 ms to encrypt and test each potential password and compare the result with the password file. For a sequence of four lower case letters the time taken to check all possible passwords was 10 min; for a four-character sequence selected from ninety-five printable characters the time was 28 h; for a sequence of five characters chosen from the set of sixty-two alphanumeric characters the time was 318 h. Evidently a password consisting of

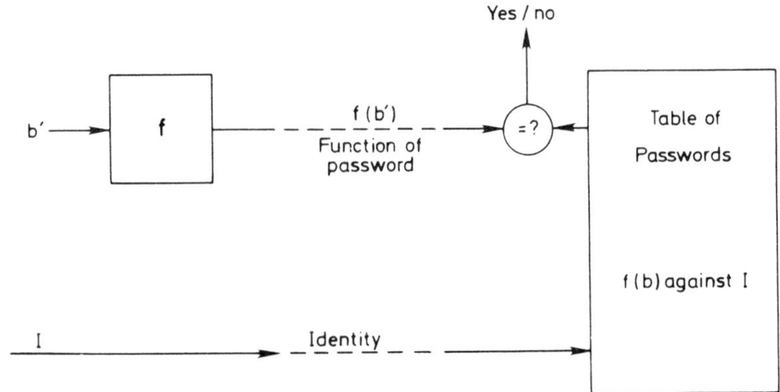

Fig. 7.1 Password checking with a one-way function

four lower case letters would be pitifully inadequate. A judgement must be made that weighs password length (with problems of memorizing longer sequences) against the value of the protected entity.

Choice of password presents some interesting problems. It has been shown that users asked to choose passwords without any advice or constraints are prone to select character sequences that are easily predictable; this is because they choose sequences that they consider to be easily memorable. The Bell report casts a fascinating light upon user habits when no constraints were placed on password choice. Of 3 289 freely chosen passwords examined the character strings used were as follows: 15 were a single American Standard Code for Information Interchange (ASCII) character, 72 were two ASCII characters, 464 were three ASCII characters, 477 were four alphanumeric characters, 706 were five letters, all upper case or all lower case, and 605 were six letters, all lower case.

They also discovered that favourite choices of passwords were dictionary words spelt backwards, first names, surnames, street names, city names, car registration numbers, room numbers, social security numbers, telephone numbers, etc.; to an intruder these are all worth testing before considering less-likely character sequences. Of the 3 289 passwords the Bell authors collected, 492 fell into one or other of these categories. Added to the list of short sequences already given, this meant that 2 831 passwords, or 86 per cent of the sample, were either too easily predictable or too easily searched for.

It seems from this evidence that the practice of leaving choice of password to the user can result in the use of unsatisfactory passwords. An alternative is to have the choice of password made by the system. For this purpose a series of random characters can be generated, but the user will find it difficult to memorize totally random character streams, even if the string length is limited to eight characters, say.

Memorability is known to be enhanced if the character stream can be chosen so as to be pronounceable, because users find syllables more easily memorable than characters. Therefore a generator that is constrained to obey certain built-in pronunciation rules may be able to produce passwords which are difficult to predict but not impossibly difficult to remember. For example, 'SCRAMBOO', 'BRIGMERL', 'FLIGMATH' and 'LIDFRANG' are all nonsensical, but easily pronounceable and therefore memorable. The total number of different eight-letter character strings is about 2.1×10^{11}. One particular password generation scheme[4] produced a measure of the fraction of all possible eight-character strings that were pronounceable; this was found to be about 0.027, giving a total population of pronounceable eight-letter strings as 5.54×10^{9}. A large commercial English dictionary would contain as many as 2.5×10^{5} different words of all lengths, so the random number generator used in this mode produces vastly more potential passwords than are found in a dictionary; hence such passwords are far less predictable.

The task of secure transmission of passwords from a central system to the users deserves attention. Unless the passwords are transported securely, their value is nullified. Encipherment is no solution because verifying the correct identity of the receiver presents the same problem that the password is intended to solve. It seems advisable to use some other means of transport such as the letter post for password transmission. The banks use this technique when sending PINs to their customers. In order to prevent interception of the password during transit, it is recorded inside a sandwich envelope at the time of generation by a computer. A character printer imprints the password on the sandwich; as there is no ribbon in the printer the password does not appear on the outside of the envelope; inside there is a sheet of carbonless copy paper, so the password gets printed internally, but is only readable when the envelope is torn open. If the envelope is interfered with in transit, then this is obvious to the intended recipient, should he eventually receive it; any apparent interference should lead to the rejection of the password. In banking practice the PIN is used in conjunction with a plastic card which is sent separately from the PIN; in order to receive a PIN the authorized user must acknowledge safe receipt of the card to the bank. An intruder must obtain both card and PIN before being able to secure any profit from his action. A modest degree of security is obtained by this technique, but it could not be applied where high security is called for. In the latter case it may be necessary to use a trusted courier service to transport passwords or hand them over personally to the user.

There is a school of thought that believes that the password should not be known to the issuing organization. One system[5] that allows a personal choice of password unknown to the system makes use of a one-way function of a combination of personal account number and password; the password is chosen by the user on the occasion of a visit to the bank and keyed on a keyboard unseen by any bank official, combined with the personal account number and then stored in the bank system. To identify the user at a subsequent visit to the bank, the same one-way function is used. The password is not known by the bank at any time; if the user forgets the password, then another personal visit to the bank is necessary in order to choose and activate a new password. Since the password is not known to bank employees, manipulation of the account cannot take place by normal system channels, though this does not rule out improper access by privileged system operators, for example using the command to change to a new password.

When the screen of a VDU requests 'Enter your password' and the user taps his password on the keys, he is taking a risk. The system has not yet authenticated itself to the user so he might be seeing a masquerade of the actual system, designed only for the purpose of capturing his password. This trick has been widely used as a student prank. When a guard holding a gun asks for your password you have no choice, it is an unequal situation. Perhaps this led to the habit of using one-way passwords in computer systems. But if each party needs to authenticate the other, there should be an exchange of passwords.

When system and user exchange passwords there is a dilemma. One party gives his password before the other and he has no guarantee at that moment that the other party is not a masquerader. This problem can be overcome in a way that is illustrated in Figure 7.2. Here there are two parties, Ann and Bill, holding passwords P and Q respectively. Each knows the other's password for checking purposes. They do not directly exchange passwords but use a one-way function to reply to a challenge from the other party. For example Ann challenges Bill by sending the random variable x_1 and Bill replies with a one-way function y_1 of this value, namely

$$y_1 = O(Q, x_1)$$

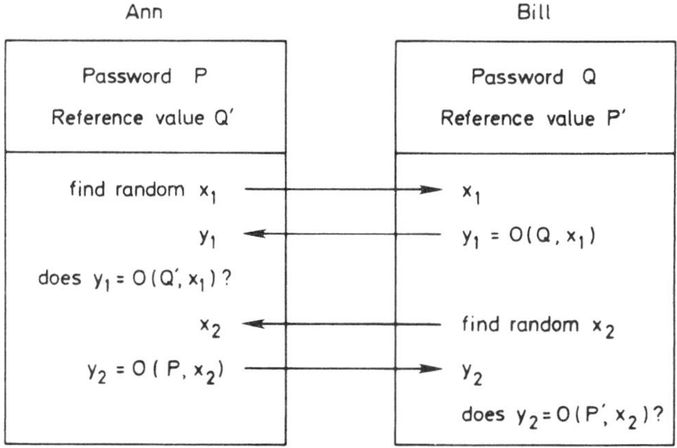

Fig. 7.2 Two-way password checking

The one-way function must be such that even when x_1 and y_1 are both known it is not possible in practice to determine Q. Although Ann cannot determine Q from this reply she can check her own value of Q against the value used by Bill in applying the same one-way function to the value x_1 which she created. If this check is successful she accepts Bill as the genuine owner of the password Q.

In a similar way Bill issues the challenge x_2 and receives the response y_2 in the form

$$y_2 = \mathrm{O}(\mathrm{P}, x_2).$$

This can also be checked so that Bill now has determined that Ann knows the password P.

This method fails to be secure if the range of the passwords is insufficient. For example, in banking practice the personal identification number is often only four decimal digits long. The system described above using such a short password is very weak because the 10 000 possible values can be checked and, for one of them, the response will match the response received. The challenger can determine the password of the other party.

It might seem that this problem can be overcome by padding the password with a random number and revealing the padding only after the challenges and responses have been exchanged. For example, let the 16-bit password Q be concatenated with a 40-bit random number R_1 to form a 56-bit variable and then let Bill respond to the challenge x_1 from Ann with

$$y_1 = \mathrm{O}(\mathrm{Q}\|R_1, x_1).$$

In this expression, $\mathrm{Q}\|R_1$ is the 56-bit concatenation of these two variables.

Suppose that each party has carried out this challenge and response, then it remains for them to exchange the values of R_1 and R_2 so that they can verify the responses and thus authenticate each other. But the problem has not been solved because the dilemma 'Who should give the password first?' has been converted into the dilemma 'Who should first reveal the random padding?' As soon as one party reveals his random number, the other party can carry out a password search to find his password.

Variable passwords based on a one-way function

The variable password gives improved security against line-tapping but requires the distribution of new sets of passwords to replace those used up. An ingenious method of using a one-way function has been proposed which provides a sequence of passwords by distributing only one number. It uses a one-way function $y = O(x)$ and iterates this function many times. We use the notation that $O^n(x)$ is the result of operating the one-way function n times, beginning with the variable x. For example

$$O^3(x) = O(O(O(x))).$$

To establish the password sequence, Ann takes a random variable x and forms the value y_0 to send to Bill:

$$y_0 = O^n(x).$$

This transfer need not be made in secret since the seed value x cannot be discovered.

For the first password Ann sends the value

$$y_1 = O^{n-1}(x).$$

Bill can easily check this since $O(y_1) = y_0$ and he has already received y_0 as the initializing value.

The sequence of passwords continues in this way. The ith password is

$$y_i = O^{n-i}(x).$$

Each password can be checked by its relation with the previous password, which has itself been checked. The sequence terminates at the nth password which is x itself.

If passwords are lost or an error occurs, provided that Bill has a recent authentic password value P, a new offered password can be checked against this value by repeated operation with O to cover the gap between the offered password and P. In this way the system can re-synchronize after a failure such as might be caused by a line error.

Questionnaires

Another method of identity verification makes use of information known to the authorized user but unlikely to be known to others. On first introduction to the system the user is asked a series of questions relating to such apparently irrelevant things as name of school headmaster, colour of grandmother's eyes, name of favourite author or football team, etc., rather than to the more usual components of a curriculum vitae, such as place and date of birth, maiden name of wife, etc. Not all the questions need be answered, provided that enough are answered. The user might add his own questions and answers, if the system allows this feature. The selected items are such that the user is likely to find them easy to remember, but an intruder is likely to find them difficult to discover. If desired, the answers may be totally fictional, in which case an intruder may find them even more difficult to discover. On later visits to the system for access, the user is asked a selection of questions to which the system has stored the answers.

Correct replies allow the desired access. Greater security can be obtained if, during an access session, the system asks further questions from time to time; this would detect a session taken over in some way by an intruder.

The questionnaire system of access control has the advantage of using easily memorable information, but it does involve a rather lengthy exchange between the user and the system, which may be found tedious and inconvenient by some, especially if repeated. It is hardly likely to be acceptable as the means of controlling access to a bank automatic teller terminal, for example. Furthermore, if the value protected in the computer by the access control system is great enough, the determined intruder can take the trouble to research the background of the authorized user sufficiently to provide the questionnaire answers. Therefore this technique is not likely to be used in systems requiring a high level of security.

7.3. IDENTITY VERIFICATION BY A TOKEN

In the widest sense, access control by means of something possessed by the authorized user is very familiar. Nearly all of us possess keys to operate locks in cars, filing cabinets, house doors etc. Computer system access may be made dependent upon possession of a key which fits a lock installed in the computer terminal. The first action of a user is to enable the terminal using his key, after which normal log-on procedure may follow. For adequate security the lock must be of a complicated (and therefore expensive) type, preferably 'unpickable'; there must be a wide range of possible key variations (known in the locksmith trade as 'differs') to fit locks of the particular class in use, so that the intruder cannot make use of a full set in his possession, trying each until one is found that will fit. Unfortunately, if a key is lost or loaned to someone, it is usually fairly easy to copy it, even though the lock mechanism may be complicated. Locksmiths have agreements that keys of particular types are not copied without proper authorization, but even this will not deter the determined intruder. The copied key may be restored to its owner without the latter even being aware of its being missing. A lock and key is a useful way to restrict access to one or two selected people, such as those authorized to load cryptographic keys, but it is not of general use for access control. The problem of 'copyability' is typical of all token-based identity verification.

Magnetic stripe cards

A token which is becoming as common as the physical key is the plastic card with a magnetic stripe. Such cards are increasingly being used for identification purposes, for automatic teller machine (ATM) and point-of-sale operations, for credit validation, for access control to secure sites etc. An International Organization for Standardization (ISO) standard[6] gives dimensions for card and stripe; other proposed standards define the use to which each of the recording tracks on the stripe shall be put, including the relevant data formats. The general card dimensions and layout are indicated in Figure 7.3. Amongst other data the user identity is stored on the magnetic stripe. Generally speaking the magnetic stripe card is used in conjunction with a type of password called a personal identification number (PIN). In off-line systems the PIN must also be stored on the stripe

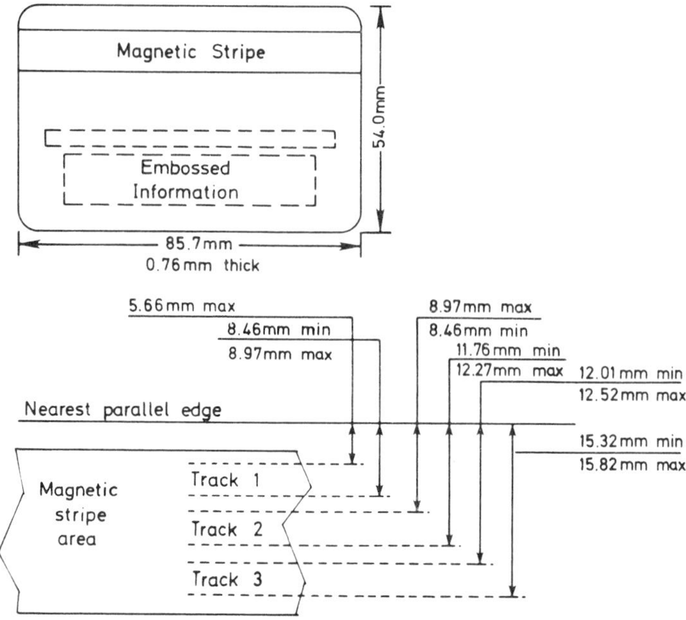

Fig. 7.3 Layout of magnetic-stripe cards

in enciphered form. The recognition equipment reads the identity, deciphers the PIN and compares the latter with the PIN keyed in by the user. In on-line systems the reference PIN need not be carried on the card, but can be stored in the central system; the claimed identity is notified to the central system together with the offered PIN; if the latter agrees with the reference PIN, then access is permitted.

As we have remarked already in our discussion of passwords, the user must be educated to take proper care of his PIN. It is not unknown for users to write their PIN on the card itself in order not to forget it; this is clearly asking for trouble. There are also slightly more subtle forms of attack. Cards often carry the user name imprinted so that an intruder can easily discover the user identity corresponding to a stolen card. If the card is an ATM card, the enemy telephones the user, claims to be representing his bank and asks for the PIN 'in order that we may restore your card to our system'. If the card does not carry the user identity, the latter may still be discovered because the bank national number and user account numbers appear; these identifiers appear together with users' identity on cheques.

Ideally, PINs should be memorized by their users and not written down in diary or notebook. However, this may be asking too much of the average user, especially if a large number of cards is carried. It is said that the average number of credit cards carried by American businessmen is eleven, each of which may have its distinct PIN.

Unfortunately it is very easy to copy the usual kind of magnetic stripe card. Cards can be manufactured, without excessive difficulty, to resemble true cards sufficiently well as to deceive most people having to deal with them. One widely used method of hindering the production of forged embossed cards is the hologram covering one or two digits of the embossed number; any attempt to change the embossing destroys the hologram. The contents of the magnetic stripe can be transferred to another stripe without expensive equipment. Therefore it is important to consider means of making cards much more difficult to counterfeit. Methods devised for this purpose tend to concentrate on hindering the reproduction of the data on the magnetic stripe. Many 'security features' have been invented to improve the security of magnetic recordings on cards; two will be described here.

Watermark tape

The Emidata/Malco system[7] establishes a permanent, non-erasable, magnetic encoding in the structure of the tape. The permanent recording is made during the manufacture of the tape by exposing it to a sequence of magnetic fields whilst the magnetic particles (held in suspension in a resinous lacquer) are still wet. The particles in most tapes (for example gamma ferric oxide) are in the form of long thin needles (acicular); to produce the watermark pattern the particles are first aligned by application of a steady magnetic field at 45° to the longitudinal axis of the tape. A pattern of current pulses is then applied to a special recording head as the tape passes beneath it; when the current is present the particle orientation is changed to the other 45° orientation. The tape then passes through a drier and the orientation of the acicular particles is fixed permanently. Figure 7.4 shows the arrangement of particles. Emidata use the trade name 'Watermark' for tape prepared in this way.

To check the permanent record on a stripe with this structure a special reader is required; in this reader the tape is exposed first to a constant magnetic field and then the fixed recording is read by a head suitably oriented. Because the tape is first exposed to the constant field before reading, this operation is restricted to track 0 of the stripe, a track which is not present on most current magnetic stripes. On tracks 1 to 3 recording and reading is carried out by heads with normal orientation; the normal recording and reading is not affected by the presence of the underlying structure in the magnetic tape. It is, therefore, possible to read a permanent pattern of ones and zeros from track 0 which is not alterable by a forger; this recording pattern can be used to provide valid identification of the stripe and therefore of the card. It would be virtually impossible for a forger to reconstruct the underlying recording in such a way as to deceive a Watermark reader; any attempt to record the 45° pattern on normal tape would be frustrated by the initial exposure of track 0 of the tape to the constant field in the Watermark reader. The layout of the various tracks is shown in Figure 7.5.

The watermark contains 50 to 100 bits of data which are fixed and it is not easy to code these to suit the user of each card. In many cases, random numbers are

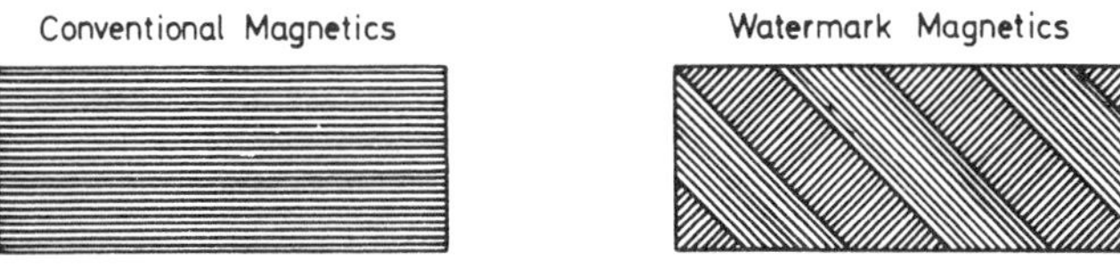

Fig. 7.4 Arrangement of 'Watermark' particles

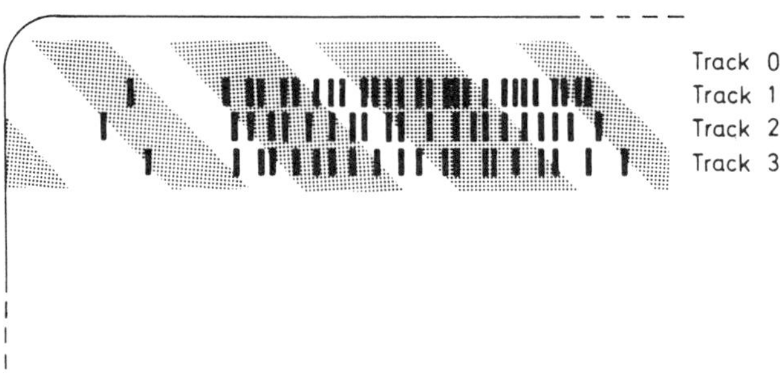

Fig. 7.5 Layout of 'Watermark' tape

used. If W is the watermark number and D the data on tracks 1 and 3 to be protected, an enciphered check field, $C = Ek(W, D)$ can be stored on the magnetic stripe as well as D. Thus the reader can verify this relationship if the secret key, k, is used in its checking device. An enemy cannot compute C and therefore cannot use one card to hold another's data, since their W values are different. The data D can be changed in a device where k is available by altering C to match the new data, but generally it is fixed data that are protected, such as the account number and PIN validation data. It is a weakness that k must be available at all places where the watermark is checked (the checking need not be done at the terminal, however, if a link to a central place is available). The use of public key ciphers (see chapter 8) makes it possible to check using only non-secret data, but increases the data storage requirement a great deal.

The whole strength of the system depends upon the difficulty faced by a forger in counterfeiting watermarked tape. The manufacturing process is such that a forger would find it very difficult to do this, since the process requires significant capital investment and unique technical skills.

Sandwich tape

An alternative to Watermark tape is the tape sandwich in which low- and high-coercivity recording media are bonded together with the low-coercivity tape nearest the recording heads. In order to write on such a tape so that the record is made in both layers a head is required which will produce a high-intensity field. In the tape reader, the tape first passes through an erase field such that any recording in the low-coercivity layer is removed, but the recording in the high-coercivity layer is not disturbed. The 'permanent' recording is then read in the usual way. Figure 7.6 illustrates the structure of the tape and the nature of the recording.

Security of a system using sandwich tape is based on two premises; first, if a forger uses ordinary tape in the manufacture of a forged card, this will be frustrated by exposure of the card to the erase field in the reader; second, if a forger seeks to use the sandwich tape on a stolen card and make a recording of his own, he will find it difficult to create the high-intensity field required for writing into the high-coercivity layer. There is no doubt that use of ordinary magnetic tape is frustrated by this technique, but it seems that high-intensity recording heads might not be too difficult to obtain. There is, therefore, some doubt whether the sandwich tape provides an adequate solution to the problem of making the recordings on plastic cards unforgeable.

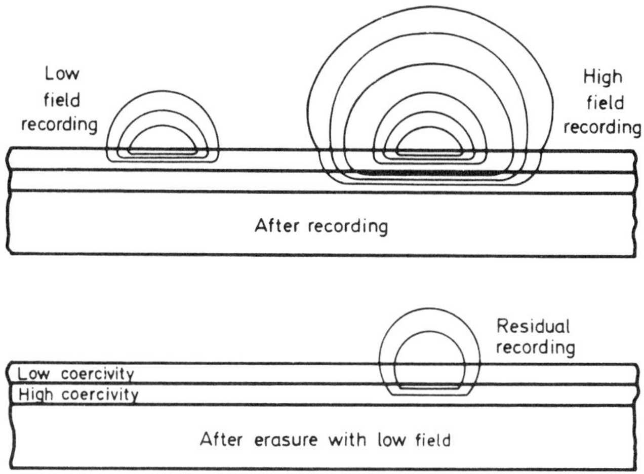

Fig. 7.6 Structure of 'sandwich' tape

Active cards

A recent development is the 'active' or 'smart' card, in which microprocessor or store chips are embedded in the plastic card. This enables the card to extend its storage capacity beyond what is possible on a magnetic stripe (about 250 bytes) and to engage in some data processing. The problem until recently has been the difficulty of introduction of the chip into the card. It has been seen as a requirement that cards of this type shall be of the same dimensions as the standard card with magnetic stripe (8.57 cm × 5.40 cm × 0.76 mm); it is also desirable that the chip card shall be as durable as the standard card. Developments in chip technology have now advanced to the point where the chip is small enough to fit into the required space, thickness being the critical dimension; new methods of mounting and connecting to the chip allow sufficient flexibility. Contacts to the circuitry are via plated areas on the card surface; power supply will almost certainly be applied via this route so that the card is active when it is in a terminal device.

At present the type of chip inserted in a card can be a processor with about 8 Kbytes of store in electrically alterable ROM; on the other hand, other cards contain only memory. This of necessity limits the kind of operation that may be carried out.

Information stored on the card may be semi-permanent, stored in ROM, in which case it may have the same sort of function as that usually carried at present on magnetic stripes — user identification, etc. Illegal alteration of the ROM contents would almost certainly be much more difficult than alteration of a magnetic stripe; therefore, security of such a card is enhanced. Such a card can play a role in identity verification similar to that which we have already outlined, though the additional security and amount of storage might allow more-sophisticated systems.

One application of a card with stored data is for viewing pay-television. If the television transmission is sent out enciphered, such that a decipherment key is required in order to make it usable, then the decipherment key, changed every

20 s, say, can be broadcast with the television signal but encoded under a key (a key enciphering key (KEK)) which is held on the plastic card of the viewer. Decipherment of the transmitted key takes place in the card, so that the KEK is not taken out of the card. Possession of a valid card enables the successive transmission encipherment keys to be decoded and the television signal to be made intelligible and used in the receiver. From time to time the KEK is changed, so that a card becomes out of date and a new card must be obtained.

If the card has encipherment capability, this can be invoked in an identity verification procedure. The secret key is held on the card and each card issued has a different key according to the user to whom it has been issued. Identity verification can then be made to depend on the card making the correct response to a challenge issued by the verification terminal; the terminal issues a test pattern which the card encrypts and returns to the terminal; the latter checks the response against the encipherment key which should be contained in the card, correct response implying presence of the right key. This system has the advantage that the response depends on the challenge and, therefore, an attempt by a potential intruder to tap the channel between card and verification terminal would yield little of use in making an attack upon the system.

If a system can be devised in which sufficient rewritable store can be established on a plastic card, then many applications become possible. Bank customers may be supplied with cards which allow them to make purchases in shops at point-of-sale terminals; the card is initially 'charged' at the bank with value (debited to the customer's account) which is decremented at the point-of-sale terminal, the amount of the decrement being transferred as a credit to a corresponding data store held by the retailer. When the customer's card is exhausted, it is returned to the bank for 're-charging'. The retailer periodically takes his stored data to the bank for credit to his account. This type of system is described in chapter 10.

The physical security of data on a 'smart card' is improved by storing all the sensitive data on the same chip that holds the processing and encipherment facilities. Active or passive attack can be made difficult, but there is not yet available an independent assessment of the level of physical security obtained. The thin plastic card is not the only format for an 'intelligent token' containing storage or processing chips. Tokens could be physically similar to a small calculator or like a physical key that turns in a 'lock' which is really a reading terminal. It is possible that greater physical security can be provided in enclosures a little larger than a card.

Another development uses plastic cards which contain stores which are read once and then erased. One such storage medium is the hologram. The kind of application where a system of this type may be used is the purchase of a card with a set initial value, to be used against the provision of a service, for example, transport, telephone or pay-television; a recent example is the 'Cardphone' service of British Telecom, which uses a holographic medium. As the service is used, so the storage on the card is progressively read and erased, bit by bit; finally the card has no more value. When the card is exhausted it is thrown away and another card purchased. To the supplier of the service it is important that the erasure of the card value is permanent, otherwise illegal 'refreshment' of the card may take place with loss of revenue. It is also important that it is excessively difficult to manufacture false cards or to make copies of genuine cards.

Authentication by calculator

With a smart card as a token it is possible to use the challenge and response principle described on page 174. This is suitable for new systems where the terminal can be fitted with a smart card reader but in existing systems this may be too expensive or not feasible at all. Some of the advantages can be obtained by employing, in place of the smart card, a device in the shape of a small hand-held calculator which has a key pad and a display and can perform the necessary cryptographic functions. A number of these calculator-type tokens are now on the market and as examples we shall describe two which illustrate useful principles.

The first of these is called Watchword and is supplied in UK by Racal–Guardata. The hand-held device is called the *watchword generator* and presents to the user a typical calculator keyboard and LCD display. It can function as a calculator but when a red button is pressed becomes a challenge–response device using the principle shown in Figure 7.7.

The device is initialized by loading a cryptographic key and a PIN value. At the host, which is seeking to authenticate the user, the functions carried out in the watchword generator can be duplicated using the PIN and key values stored there. The authentication procedure starts with a random number of seven decimal digits generated by the host and sent to the user as the *challenge*. The user reads these from his terminal's display and enters them into the token where they are converted by a one-way function into a seven-digit response, which it displays. The user reads this display and enters it on his terminal's keyboard, returning it to the host. The same calculation is made at the host and the results compared to see if the authentication of the user is valid.

To protect the watchword generator against misuse by someone who finds or steals it, it will not function correctly unless it is given a correct PIN value by its user. The secret key stored in the device ensures that each watchword generator has its own transformation of the challenge into the response. When the device is used, it first requests, on the display, that the PIN be entered, then requests the challenge. If the wrong PIN was given, a different response is produced but

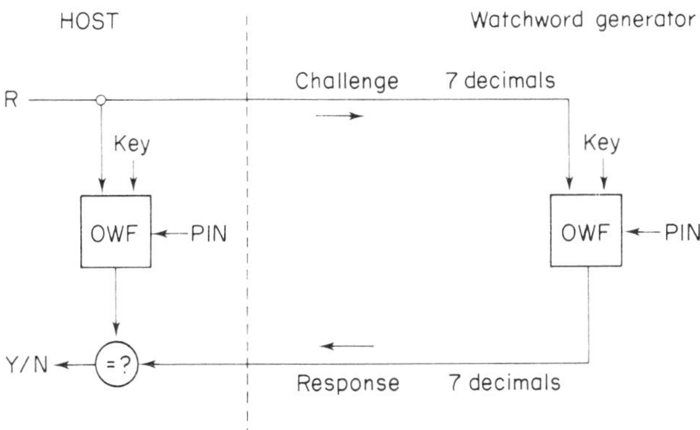

Fig. 7.7 Principle of the watchword authentication system

this response is also a function of the PIN and the challenge otherwise it would be possible, by experiment, to find the correct PIN.

To initialize the device it must be cleared of all data, then the cryptographic key is entered. If desired, it can then be handed to the user who will choose his own PIN and complete the initialization of the device.

The implementation of the host end offers a number of possibilities such as a program in a mainframe, a dedicated microcomputer or, for the greatest security, a security module with tamper-resistance features.

The watchword system has other features not shown in the diagram. It can employ two keys, the selection being made by the user. Using the same PIN, two different hosts can be accessed, each having its own secret key. There is also provision for a second value of PIN and the host can detect that the second PIN has been used when it receives the response. The main purpose is to indicate that the user is operating the watchword device under threat.

The second device we shall describe is the *Safe 200* supplied in UK by Computer Security Ltd. This is a slightly smaller calculator than the watchword generator and the principle it uses is shown in Figure 7.8. There is no challenge, but the S200 device provides a new 5-digit number each time it is used which, in effect, is a variable password that the host system can recognize. This has been called a *session PIN* or SPIN. The generation of new SPIN values is shown in the figure as a one-way function of register value S of 20 decimal digits which enters a one-way function with a secret key and generates the new S value. Each time the S200 is used, S is updated in this way.

The generation of a new SPIN value is subject to the entry by the user of the correct PIN which is checked by the S200 against a stored value. If more than three incorrect PIN values are entered, the value of S is changed in a characteristic way which can be recognized at the host in order that searching for a correct PIN value will be detected. Note that here the criterion is applied in the token whereas the watchword system allowed the host to apply its own criterion for detecting PIN searching.

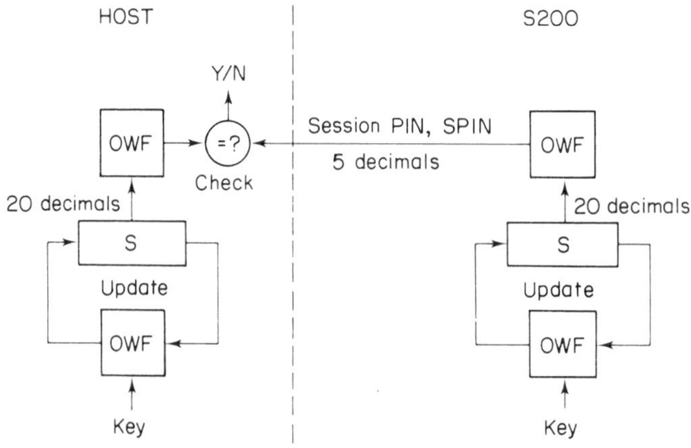

Fig. 7.8 Principle of the S200 authentication system

In order to verify the SPIN value, the host performs the same calculations and maintains a value of S which should be synchronized with the S200 device. Loss of synchronism is most likely to result from operation of the S200 without completing the procedure at the host. Therefore, if the host does not obtain a match it tries again with the next value of SPIN in the sequence and again with the next after that, accepting a match with any of these values.

The S200 system has its own special features. The user is allowed to change his PIN value by a procedure involving the host and the new value is enciphered in the S200, the result being presented on its display for the user to transfer to the host.

The SPIN is used to authenticate the user to the host. In a long series of transactions contained in one session there may be a requirement to verify that communications are still coming from the user holding the authorized S200 device. This is achieved without re-entry of the PIN using a sequence of transaction numbers obtained by pressing a button marked TAN. These are cryptographically derived from the S register but do not modify it so the synchronism between host and S200 is not disturbed. This feature is also useful if the user's terminal requires a secret key for message authentication, since this can be obtained from successive TAN values and the same data can be obtained by the host. In its original application for home banking, the prototype of the S200 provided an authenticator calculated from the account number and the amount to be transferred. These features are also available in the S200 device and may find some application.

Because of the need to maintain synchronism, each S200 works with a single host, and operation with more than one host is not provided.

The principal advantage of the S200 is that it does not require a challenge value to be entered into the token. The watchword system requires the user to transcribe both the challenge and the response but the S200 requires only the transcription of a single number, the SPIN of 5 decimal digits.

The host implementation can be a program in the host computer, a dedicated microcomputer or a security module. The latter gives greater security.

There is yet another way to employ this principle of synchronism between the host and the user's token. Instead of updating the stored variable (S in the S200) each time the device is used, it can be updated at regular intervals as determined by an accurate clock. Thus the value used as password depends on the secret key and the precise time. This is the principle employed in the *Safe 100* device. The clocks in the token and at the host may gradually get out of synchronism but, provided the device is used often enough, the host can verify the received value against a number of values in the sequence and correct its clock to restore synchronism.

The principles described above and embodied in calculator-like devices have the special virtue that they can be applied easily to existing systems without any hardware changes. When doing this it is important to study the security of the implementation at the host and in particular the security of the secret keys which must be stored. Improved security without special hardware is difficult to obtain. Compared with the use of a password alone, one of these calculations can greatly improve user authentication.

7.4. IDENTITY VERIFICATION BY PERSONAL CHARACTERISTICS

Since passwords may be disclosed by carelessness and tokens may be lost or forged, the degree of security which is afforded by the systems which we have described may in some circumstances be less than satisfactory. For this reason much attention has been given to methods of identity verification which measure the personal characteristics of the authorized user, in the belief that these are much more difficult to forge and are not susceptible to careless use.

Measurement of physical characteristics has a long history, particularly in applications connected with crime detection. The Bertillon system, used, mainly in France, for about 40 years from 1870 onwards, depended on taking body measurements such as length of forearm and fingers, height, width of head, length of foot etc. This method, known as anthropometry, depended on keeping structured records of known criminals and seeking to prove the identity of suspects by measurements taken on arrest. During use of this system it was not unknown for pairs of people to be found with the same measurements within the tolerances allowed. The Bertillon system was largely superseded by the introduction of identification by fingerprinting. This seems to have been suggested as a technique in the mid-1880s and it was adopted by the London Metropolitan Police at about 1900.

Recognition of known individuals is a frequent experience for all of us. This we do on the basis of many different physical characteristics, facial appearance, colour and length of hair, stature and stance, sound of footfall, voice, even smell etc. We also recognize written signatures when we receive letters or documents; a signature is written as a habit and does not involve the normal channels of muscle control; it can, therefore, be regarded as almost a physical attribute. Some physical attributes lend themselves conveniently to a machine devised to carry out recognition; others are not easily adapted to machine measurement and verification.

Machine recognition

It is astonishing how many different recognition techniques have been tried in all seriousness by researchers. Some of the more realistic schemes include recognition of handwritten signatures, fingerprints and voice; we shall consider each of these techniques in some detail. Some of the less obvious schemes have tried to use head bumps (machine phrenology), lip prints, prints from the soles of feet, vein patterns in the hand or wrist and even the response of the skeleton to a physical stimulus. Weight of the individual is another parameter which has been suggested; high discrimination cannot be expected if verification depends only on measuring weight, though it might be used as confirmatory evidence.

We have already stressed the need for an identity verification system to be acceptable to those required to use it. Some of the ideas just mentioned fail this requirement, though they might otherwise provide a perfectly good means of identity verification. To measure lip prints the user would be required to kiss some interface on the machine; this is hardly likely to be acceptable on the grounds of hygiene; it is even aesthetically unpleasing. Sole prints can only be obtained if users are prepared to take off their shoes and socks — possibly acceptable for access control to a swimming pool, but hardly appropriate for an office block. Head bumps can be measured by allowing one's head to be sensed by probes contained in a helmet attached to the machine; the system could be devised to remove the helmet if the identity were verified, but take a firm grip on the head of any impostor! If such a system made occasional errors, user confidence is unlikely to be high. The mechanical response of the skeleton to a calibrated stimulus implies that a controlled kick must be applied to some part of the body — again not likely to achieve easy acceptance.

One very important issue is the degree of risk to which an intruder is exposed if his attempt to intrude fails. If it is very likely that he will be apprehended, then it is less likely that he will make his attempt unless the probability of successful deception is high and the reward for intrusion also high. At a site access point a failure to verify identity can be arranged to set off an alarm to draw the attention of a guard. If this is combined with an 'entrapment module' which the subject has to enter before identification, then intrusion is made even more risky. Whether such a daunting physical arrangement would be acceptable to authorized users, with the possibility of occasional false alarms, is a matter which should be considered carefully.

Such strong physical measures are not possible in many (or most) environments where verification of personal identity is required. At the unattended bank ATM the potential intruder can go completely unobserved to attempt his deception. In these circumstances it is important not to allow unlimited repeated attempts to gain recognition. At an ATM it is possible to retain the user's card after three failures to present the correct PIN. This measure has not proved acceptable everywhere. In the USA and elsewhere some credit cards are also used in ATMs but deprivation of a credit card is a great inconvenience. In such circumstances the system must return the card to the user even though security checks suggest some irregularity. An on-line system could note the fact of the multiple failed attempts and black-list the card until the authorized user was contacted.

System tolerance

Measurements of the physical characteristics of any individual are likely to show appreciable variation from time to time. Therefore, a realistic identity verification system must take this into account and allow a measure of variation among the measurements of the individual. The problem created by allowing this tolerance is that confusion between individuals can be expected to increase as the tolerance increases. If a system fails to recognize a properly authorized individual, this is known as a false alarm or a Type I error. On the other hand, if a system accepts a false individual, then this has been called a false acceptance or an impostor pass; impostor passes are called Type II errors. We shall see that there is generally a trade-off between the incidence of Type I and Type II errors.

Because of the degree of variation between measurements taken with the same individual, it is possible to say that if the measure taken on a particular visit to the system matches exactly in every particular the profile of the individual claimed identity, then it is very likely that an attempt at intrusion is taking place using some illicitly obtained copy of the user profile. Precise correspondence is sufficiently unlikely to warrant investigation.

We now consider several identity verification systems which have been the subject of much research and development, leading in some cases to a marketable product. We will reserve comments on comparative performance until later in the chapter.

7.5. HANDWRITTEN SIGNATURE VERIFICATION

Handwritten signatures have long been accepted as a means of indicating assent to a document, for example, a contract. More recently they have also come to be used as a means of personal identity verification, for example, in the use of credit cards. Traditionally signature verification has depended on visual inspection of the signature after it has been written and has involved the exercise of critical judgement on the part of the verifier, who is usually not particularly well qualified and in some cases not motivated to carry out this task. Where the validity of a signature is in dispute, as in courts of law, expert witnesses are often called to testify to the authorship of a particular piece of handwriting.

Signatures are written so frequently that they become a reflex action, almost a physical attribute of the signer, not subject to deliberate muscular control. The imitation of a signature is a very difficult process, especially if an attempt is made to carry it out at the normal speed of writing. Much attention has been paid to finding methods of processing handwriting by machine, both with the aim of interpreting the sense of that which has been written and the aim of recognizing the writer from the style of the handwriting. In signature verification we are not concerned to interpret the writing, only to establish the identity of the author. Verification can be carried out on signatures which have already been written or on signatures observed during the process of writing. Most research projects on automated verification have concentrated on the latter, but the former has an important function also, for example in verifying signatures on cheques as these are processed by banks.

Forged signatures on documents such as cheques may be of three types, freehand forgeries, simulated forgeries and traced forgeries. Freehand forgery occurs when a cheque book is lost by its owner and found by someone who has no idea how the owner normally writes his signature. Since account holders' names are now printed on cheque blanks, the finder makes up a signature to correspond to the name on the cheque. In simulated or traced forgery the forger has an example of the true signature to work from. Machine detection of freehand forgery can be based on an analysis of the true signature in terms of dimension ratios and slant angles, measured for the signature as a whole and also for specific letters in it. These measures are stored in a computer file against the identity of the account holder. On presentation of a cheque for verification, the cheque signature is compared against the profile of the true signature as stored, allowing rejection of freehand forgeries. Very promising results have been obtained with a system designed on this basis by Nagel and Rosenfeld.[8] These techniques are not applicable if the forged signature has been simulated or traced.

Another significant development[9] in the field of static signature verification is due to the programmable vision systems group at NPL. The NPL device measures empirically chosen 'features' of signatures which are consistent for individuals, but different for different individuals; only 'global' measures involving the whole signature are employed. Seven features are measured, each on a scale from 0 to 9. The degree of separation possible with these parameters is sufficient for the bank and credit card applications which were the original design aim.

If the dynamics of signature writing can be brought in as an additional set of parameters in the verification process, then this can make the verification task much easier and the difficulty facing the forger is made correspondingly greater. Furthermore, it is with signatures written in real-time that we are concerned when using signature verification as a means of checking personal identity in access control. Verification equipment using such a technique is designed to measure the rhythm of the writing, the way the writing implement successively touches the writing surface and is raised from it etc. A number of measurements which apply to the whole of the signature can be used. For example the time taken by a person to write his signature varies very little, so that this measure, applied as a first test, can often detect a forgery. A significant research project at the National Physical Laboratory (NPL) produced a dynamic signature verification system called VERISIGN.[10] The VERISIGN system used ten measures quantifying distinct characteristics of the signatures, including number of contacts, turning points, loops, slopes, velocity and acceleration.

Techniques for recording pen movement

Techniques for measuring pen movements have developed rapidly over the hundred or so years that the subject has been studied. The earliest technique appears to have been that of the Teleautograph, which used a system of levers to record the movements of the stylus; the movements were then digitized for transmission over a communication line. The Teleautograph Corporation has since developed more sophisticated ways of sensing pen movements and still markets devices with the same basic purpose. Other manufacturers have developed inventions with the same object; more recently pens have been fitted with orthogonal accelerometers, recording acceleration in two directions; other devices have used strain gauges which measure writing pressure. All devices which measure some parameter on the stylus itself and signal it to the analysis equipment suffer from the problem of connecting a wire or cable to the pen in order to carry the signal. Such an addition may be found inconvenient by the writer and this may cause the writer to change the style of his signature and even to fail to achieve the kind of consistency in writing which he normally produces. Therefore, it seems desirable to find some method of recording the pen movement that does not impede the writer. Techniques which have been tried include a writing pad which is able to plot pen movement in terms of two-dimensional space and also to detect the making and breaking of contact with the surface. A writing pad of this type was developed at the National Physical Laboratory under the name of CHIT. This used two conducting membranes, normally just out of contact with each other, in which application of the writing stylus caused the two membranes to touch. It was then possible to measure the X and Y co-ordinates of the point of contact by treating each membrane successively as a potentiometer. The pad was arranged to sample the co-ordinates fifty times each second. The CHIT pad was used in the NPL VERISIGN system. Some other designers recording signatures with special pads have used only the time dimension of contact with the writing surface; this provides much less information about the signature and is unlikely to discriminate as well between true and false signatures.

Use of signature verification

To introduce a signature verification system into access control each user is first asked to supply a number of sample signatures. The verification system analyses these and enters their characteristics into its records. Several signatures are required from each subject and the system can be programmed to demand more signatures until it is satisfied with the consistency of the parameters. For individuals with highly inconsistent signing habits, the system may not be able to achieve enough consistency. For such individuals special measures may have to be taken. These can include relaxing the tolerances or requiring these users to employ a different means of access altogether. If the tolerances are relaxed, then this may mean that it is easier for an impostor to masquerade as such a person.

Once accepted by the system, users present themselves for signature verification on subsequent occasions and will be allowed a limited number of attempts to get a signature accepted. Failure to achieve acceptance should lead to the user being required to gain access by another channel or, in an unattended situation, access is denied. As users' writing habits may be expected to vary slowly with time, the system may have a degree of adaptivity built in; the degree to which this is allowed must be chosen with care, since this may make successful forgery easier.

7.6. FINGERPRINT VERIFICATION

It has long been recognized that individuals differ in the ridge patterns on the skin of the fingers and other parts of the body. The study of these patterns is known as dermatoglyphics; it is said that the presence of inherited physical disorders can be detected from the patterns. Furthermore the variation between the ridge patterns of the fingerprints of individuals is so great that it is generally believed that no two people can be found who have identical fingerprints. Even though the overall dimensions of the fingers change as the body grows, the geometry of the fingerprints does not alter; unless plastic surgery is undertaken, we have our fingerprint patterns for life. Because a latent fingerprint is easily left on a surface touched, fingerprints have long been used in criminal investigation. Various techniques are used to 'develop' latent prints, making them visible for photography and permanent record.

In order to match suspect individuals against records, it is necessary to be able to classify fingerprints according to their characteristic features. The system used by most police forces is based on that devised by Sir Edward Henry, the Metropolitan Police Commissioner, in 1897. The major features are arch, loop and whorl, illustrated in Figure 7.9. According to the Henry system each type may be subdivided into two or more subtypes. In addition we have composite

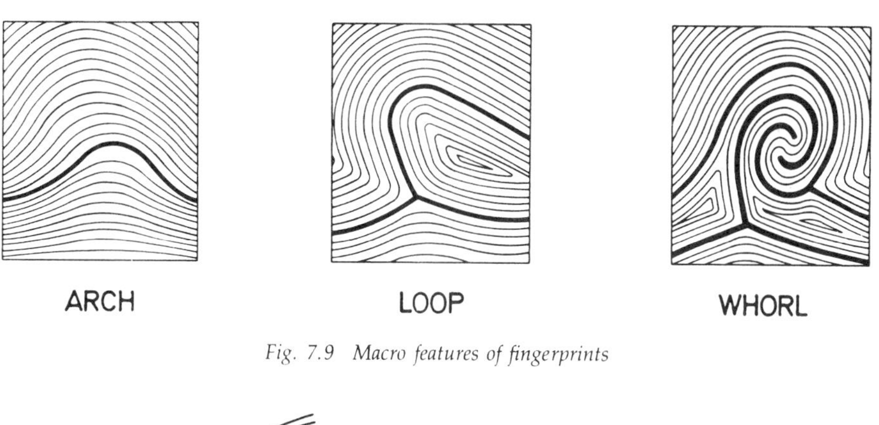

Fig. 7.9 Macro features of fingerprints

Fig. 7.10 Fingerprint minutiae

types made up of components of the other types. Each fingerprint has one or more of these characteristics. However, the smaller features, or minutiae, play an even more important part in classification of fingerprints. These small features are such as ridge end and ridge bifurcation, see Figure 7.10; typically there are between 50 and 200 such features on one finger. It is on the location and orientation of minutiae in fingerprints that positive identification really depends. It is said that detection of twenty minutiae can lead to positive identification of an unknown with a known print.

Inspection of scene-of-crime fingerprints to identify the characteristic minutiae is normally the function of very highly skilled officers. They have access to libraries of classified fingerprints of known criminals. Systems have been designed to computerize the storage and access of fingerprint records. Indeed attempts have been made to scan scene-of-crime fingerprints by machine and thus automate the whole process, but this is not easy if the prints are smudged and incomplete. It can be difficult even for a skilled operator.

Machine recognition of fingerprints

When reading the prints of willing individuals who present themselves for identity verification, it is generally possible to obtain a more consistent and reliable impression of the print because the conditions under which the print is read can be carefully controlled. No ink pad is needed in a print reader used for personal identity verification; instead the subject places his finger on a designated area. One reading method involves the use of total internal reflection on a glass plate. Because the precise orientation of the print may vary between different occasions, it is necessary that the design of the reader allows the image to be rotated over a small angle. The reader looks for the minutiae and identifies their position and orientation. This information is then compared with the fingerprint data obtained when the person was introduced to the system. The data file is accessed according to the claimed identity of the subject.

Difficulty might be encountered by somebody who has suffered an injury on the fingertip. This might temporarily prevent verification by the system. After healing has taken place the print can be expected to regain its normal pattern, unless the degree of tissue damage was great.

As regards acceptance of a fingerprint verification system by its 'customers', there is, unfortunately, a psychological drawback because fingerprints are associated with criminals. Many law-abiding citizens are very reluctant to have their prints on record in a computer system. Acceptance must relate to the success of the system in identifying people and to the convenience of use. If it could be shown that the system was very reliable, making very few mistakes, and was very easy to use, it is possible that it might gain acceptance, but the natural bias against it must be overcome. A fingerprint system is expensive because the reader is complex and the verification process difficult. In this respect it compares badly with dynamic signature verification.

Various possibilities of 'fingerprint forgery' can be imagined but no evidence seems to be on record that there are practicable ways to deceive these machines.

7.7. VOICE VERIFICATION

Recognition of people by their voice is a human skill which is very highly developed, exercised in all sorts of circumstances, even with limited bandwidth and in the presence of noise. This ability is often used in the transaction of business, where parties to a deal recognize each other in a telephone conversation. Many attempts have been made to produce a reliable method of identifying people by their voice in such a way as to provide evidence for a court of law; the production of voice prints and their analysis by skilled individuals is a subject which has generated a good deal of controversy. Expert witnesses can disagree over the interpretation of voice prints.

In voice verification what is required is not that the machine shall recognize what is said, but that the machine shall compare the characteristics of the sounds against a reference set held for the individual. It is important to create the right conditions for attempting to verify the identity of a speaker, avoiding distortion and high ambient noise. An enclosure with reverberation might cause problems. Ideally the speaker should be in a similar acoustic environment on all occasions when verification is attempted, which could require the provision of a special booth. Furthermore, there should be some control over the words to be spoken. It is not generally acceptable to allow the subject just to pronounce his name. Some names, for example Kim King, convey very little acoustic information, whilst others, for example Paddington Bear, offer quite a lot. Ideally, the system should present the subject with a short phrase, selected to maximize the number of elements which allow distinctions to be made between individuals. The words chosen should be ones that the individual often uses.

In the voice verification technique developed by Texas Instruments each subject on introduction to the system speaks a standard set of sixteen words. As the recognition process uses only the vowel sounds, the words are chosen so as to contain different vowel or diphthong sounds. From the vocabulary of sixteen words, thirty-two sentences are generated. By using Fourier analysis, with sampling at 10 ms intervals, the sentences are scanned for regions where the amplitude is large and significant phonetic information is available, such as the rapid transition of spectra with time. For each identified transition the energy in various bands between 300 and 2 500 Hz is measured for a period of 100 ms surrounding the transition time, giving a reference template for the speaker. On a verification visit, the subject speaks a sentence presented to him and this is Fourier transformed every 10 ms. When the sentence is complete each relevant template is scanned along the sentence to see if there is agreement within the required limits.

Voice verification has the disadvantage that it takes some time to speak the words (possibly twice if the system uses an aural prompt). The analysis process then takes more time. Therefore the throughput may be low and the delay time considerable. Furthermore, the voice is affected by the physical state of the person; respiratory infections cause the voice to change its quality and the emotional state also has an effect on the voice. A request to repeat the identification procedure might cause stress which makes agreement worse.

7.8. RECOGNITION OF RETINAL PATTERNS

The retinal blood-vessel pattern of a human eye — retinal vasculature — is highly characteristic of the individual. An identification system making use of this property has been developed by Eyedentify Inc., of Portland, Oregon.

The subject is required to look into a device with a binocular eyepiece, adjust the interocular distance, fixate upon a crosswire and then press a button to indicate readiness (a more recent version of the device uses a monocular system). Fixation on the crosswire enables the machine to locate the fovea of the subject and to scan, with a low intensity infra-red beam moving spirally, an annular area

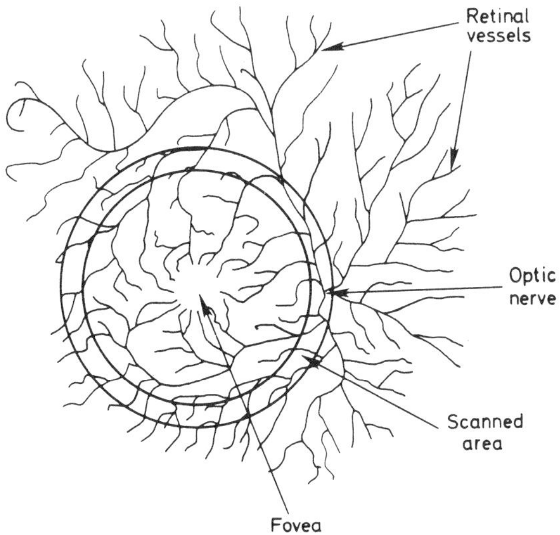

Fig. 7.11 Geometry of retinal scan

around this spot, detecting the nodes and branches of the retinal pattern falling within the scanned area. Figure 7.11 shows the geometry of the retinal scan.

Like the other identification techniques mentioned in this chapter, the subject has an introductory encounter with the recognition equipment, which stores a record of the retinal pattern. On subsequent visits for recognition the measured pattern is compared with the stored pattern. The quality and detail of the measured characteristics are excellent, but the need for careful cooperation of the subject could limit its usefulness.

7.9. THE VERIFICATION PROCESS

All of the identity verification methods that use individual human characteristics have much in common. The design tradeoffs can be studied for these systems in a general way. First consider the two stages of use, introduction of a new user and subsequent identity verification.

Introduction

The subject attends at a verification terminal, usually under the supervision of an operator, identifies himself by name or some other identifier (or is given an identifier by the system) and performs the required act, such as writing a signature, presentation of a finger for scanning of the print or speaking a specified set of words. This is repeated a number of times in order to get a reliable *reference set* of measurements. The system analyses the performance of the subject and stores, either in its files or on a token such as a magnetic-stripe card, its record of the subject's profile. If a magnetic card is involved, the subject is normally given this card to keep for the duration of his authorized use of the system.

Verification

First the subject approaching the system for identity verification claims his identity either by keying in his identifier or by presenting his machine-readable identity card. The identifier must be unique — each individual has a different identifier.

Secondly the verification system finds the profile associated with the claimed identity, either from its files or from the card presented by the subject.

Finally the subject is invited to perform the act on which identity verification is based. The verification system then constructs a profile from the measured response of the subject and compares this with the profile corresponding to the claimed identity. The measured profile and the reference profile each contain a number of parameters. These are the results of calculations made over the set of measurements taken from the subject. For example, a dynamic signature measurement could be a set of coordinates at 20 ms intervals and a parameter might be the average speed of movement or the average vertical travel between changes of vertical direction. The parameters are designed to be difficult for an impostor to control and as invariant as possible for one subject, but varying between subjects.

Exact match of measured and reference parameters is unlikely — it would be a ground for suspicion. The decision to accept or reject the verification requires a measure of 'distance' and a tolerance of error beyond which the decision is 'no'.

Each parameter can have its own tolerance and the tolerances can be decided using the multiple measurements made at the introduction of the subject.

Both the tolerances and the reference values can be allowed to adapt. This is valuable immediately after the introduction because the subject is uncertain during the learning phase and will settle down to a better performance later. For the more objective methods such as fingerprinting and retinal patterns the learning process is less significant. Adaptation must take place only when the identity is verified. Allowing for the possibility of an impostor succeeding, the adaptation works in his favour. For this reason, adaptation must be very gradual and the degree of widening of the tolerance must be limited to the extent that impostors do not get the initial entry into the system.

Tradeoffs

The setting of the error tolerance is critical to the performance of the system. If the tolerance is large, then the discrimination between subjects is reduced and the impostor pass rate (Type II error) is high. On the other hand, if the error tolerance is low, then genuine individuals may have difficulty in gaining the verification which they are entitled to expect from the system; in this case the false alarm rate (Type I error) is high. Generally speaking the tolerance level is adjustable and there is therefore a trade-off between the two error rates. Figure 7.12 shows a typical characteristic for a verification system, plotting how the pass rates vary with the error tolerance level. The way in which 'error' and 'tolerance' are determined is part of the system design and there are many possibilities. Before describing the characteristic we shall discuss the testing process and the ways in which the tolerance is made adjustable. From the raw measurements (such as pen position for a signature or energy in a section of the speech spectrum)

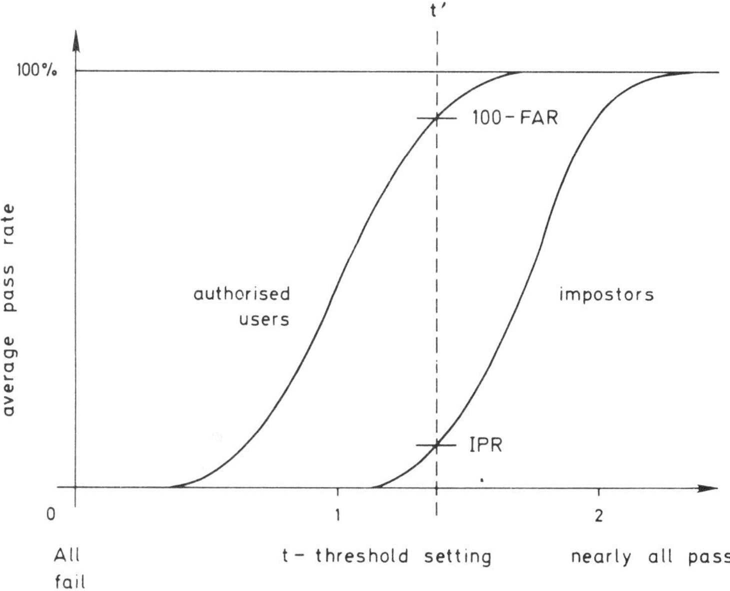

Fig. 7.12 Typical pass rates (averaged over a population)

a number of parameters are computed. Often these are computed by averaging over the measurements, to reduce their variability. In the research leading to Verisign, dozens of parameters were invented and tested on a large body of signatures and, from these, ten were selected to meet the requirements mentioned above and also to be as uncorrelated as possible. For each parameter the *reference* set of measurements is used to find values of mean and variance. Using the standard deviation as a unit of measurement, the 'error' of each parameter for a given signature can be normalized.

A single measure of error could be obtained, for example from the sum of the squares of the normalized parameter errors. This allows a pass when one parameter is wrong by more than normal, provided that the others are closer than normal. In fact, the Verisign system did not work in this way but required for a pass that each parameter be within certain bounds. In order to make the tolerance adjustable, a single threshold value (t in the figure) is used to set the tolerance. For example if x is a parameter measurement, m the mean value obtained from reference values and thus $|x-m|$ the error, then $|x-m|/s$ is the normalized error where s is the standard deviation of the reference set. The condition for acceptance could be chosen to be

$$|x-m|/s < t.$$

Then, if $t = 0$, no tests are passed, but for t large enough, nearly all tests will give a pass decision. It must be emphasized that other ways of defining 'error' and 'tolerance' are equally good. All that is needed is a threshold adjustment to assist in setting up the system.

Figure 7.12 plots average pass rate against t for two populations, a set of authorized users who are doing their best to get accepted according to criteria set by their own reference set of measures and a set of impostors trying to get accepted according to a reference set which is not their own. The characteristics are averaged over many trials and many persons. The characteristics will include some people with very variable performance and others with better consistency. A full analysis should try to separate and study these sub-populations.

An ideal characteristic would have a value of t which separates the two populations, passing the genuine and rejecting the impostor, but in practice this rarely happens. A compromise like the value t' is chosen and it leaves two kinds of error.

Type I — authorized users who fail at t'. The pass rate for authorized users is 100 − FAR where FAR is the 'false alarm rate', given as a percentage.

Type II — impostors who succeed at setting t'. The pass rate for impostors is the 'impostor pass rate' (IPR).

Sometimes the characteristic of Figure 7.12 is plotted with the ordinate measuring error rate instead of pass rate. The effect is simply to invert the curve for authorized users, as shown in Figure 7.13, and otherwise the curves are identical. The nature of the trade-off is made clearer by the overlapping error curves. A setting t'' is shown which improves IPR, a more cautious selection criterion, at the expense of rejecting more genuine users.

Another way to show the trade-off is to plot FAR against IPR, as shown in Figure 7.14, and mark the threshold settings t as a parameter on the curve. In some discussions, FAR + IPR is used as a figure of merit for the system as a whole.

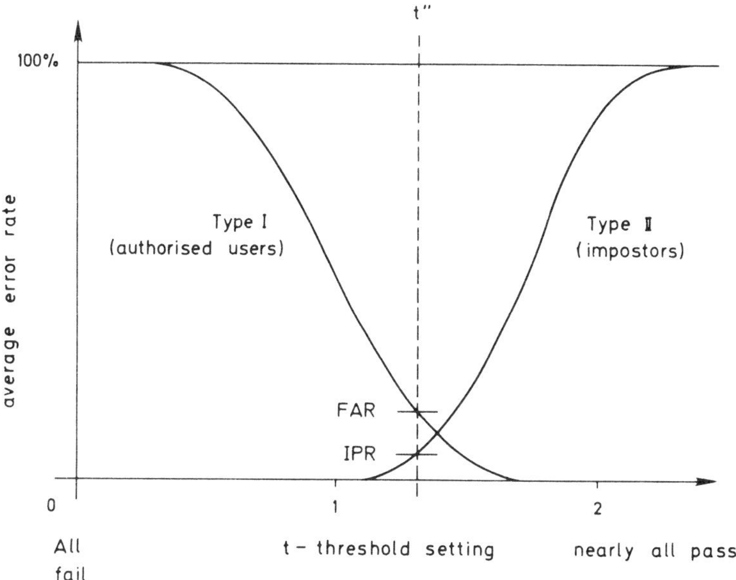

Fig. 7.13 Average characteristic plotted as error rate

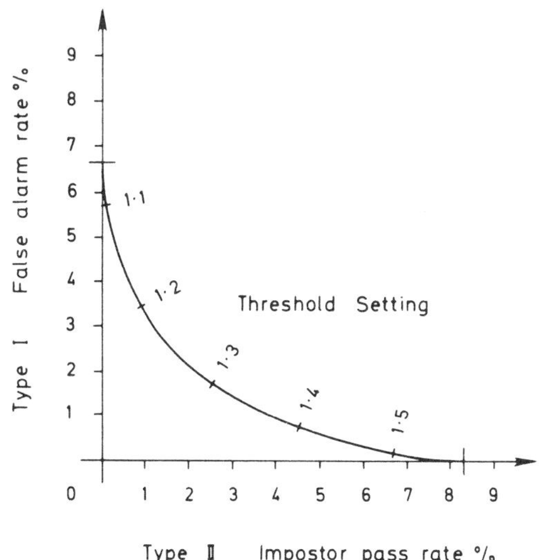

Fig. 7.14 Typical trade-off between error rates of two types

Roughly speaking, this is minimized when FAR and IPR have comparable values.

In many identity verification schemes, the user is given two or three chances and is accepted if any of them meets the pass criteria. The characteristics of the kind shown in our figures should then be plotted for the composite test. Failure to pass then requires several tests to fail and is less lilkely than in a single trial. Correspondingly, the threshold can be set lower so that the single test has a higher FAR and lower IPR. The exact performance on repeat tests after a failure is a matter of measurement. If the tests were independent, we could calculate the results from the characteristics of single tests, but this is not a safe assumption. Under these conditions of independent tests, 10 per cent FAR and 1 per cent IPR would become 1 per cent FAR and 2 per cent IPR after two trials, then 0.1 per cent FAR and 3 per cent IPR after three trials. Practical tests are needed to get true figures and the improvement is much less than these figures suggest.

The characteristics plotted in Figures 7.12 to 7.14 are averages for all users, but the experience of each individual can be different. The system depends on the differences between the parameters of individuals, and these differences extend also to variability of parameters. A small proportion of subjects show large variability. Beyond a certain value, the IPR is excessive. A clever enemy could find ways to identify (or guess with some success) these variable subjects. For example the handwriting of the very old might be variable. Using one of these subjects' identity, an impostor has higher change of success. To avoid this danger, a small proportion of subjects may have to be told that the method will not work for them. Many operators could not afford to say this to valued clients, so their choice of verification method is limited.

7.10. ASSESSMENT OF IDENTITY VERIFICATION TECHNIQUES

Many technical papers have appeared describing verification systems in development and claims have been made concerning the performance (Type I and Type II error rates) obtained with each system. Unfortunately it is extremely difficult to make an objective comparative evaluation from such claims, because the conditions under which the systems were tested were not standardized. For this reason we have not attempted to go into detail concerning any of the systems outlined in the preceding sections, though we shall later give an indication of the levels of performance claimed by the designers of particular systems. It is important to seek results where at least some attempt was made to maintain consistency of test conditions. One of the more significant studies of this kind was carried out by the Mitre Corporation, when, under contract for the US Air Force, several identity verification systems were evaluated in a comparative study. The results of the Mitre survey have been published in a paper by Fejfar and Myers.[11]

The Mitre evaluation studies

The terms of reference of the Mitre evaluations were that the Electronics Systems Division of the US Air Force, under its Base Installation Security System (BISS) programme, required identity verification systems to achieve a Type I error rate no greater than 1 per cent and a Type II error rate no greater than 2 per cent; it was further required that the system should be capable of handling four subjects per minute.

In the Mitre work three identity verification systems were evaluated, voice verification (a system made by the Texas Instruments Corporation), signature verification (a system made by the Veripen Corporation) and fingerprint verification (a system made by the Calspan Corporation). The evaluation was carried out in two laboratory-based phases, followed by field trials held at a US Air Force base. In phase one of the laboratory trials only the voice and signature systems were assessed (the fingerprint system was not available in time), the results being fed back to the manufacturers in order that they might improve the performance of their systems. In phase two all three systems were compared, followed by field trials in which all three were again compared. At each stage over 200 people took part in the experiments. At each visit to the experiment each person attempted to gain verification by each of the three systems. In the laboratory tests the subjects were Mitre staff of all types, engineers, programmers, secretaries and technicians. In the field trials the staff were US Air Force personnel authorized to gain access to a weapon store. Most of the comparative results are given for male subjects because only seven women took part in the field trials; however, some comments are made regarding the relative performance of the two sexes. In each stage of the evaluation the error rates were measured and the time taken in each part of the process was recorded. The measurement of Type I errors needs no comment, being simply a matter of failures to gain verification, but for Type II errors, involving false acceptance of subjects, the experimental philosophy was mainly directed at accidental false matches with the profiles of other subjects and not to deliberate attempts to masquerade as other subjects. The published paper mentions that limited work was done on deliberate masquerades, but the results were not published due to their military security classification.

It has been found[12] in the work at International Business Machines (IBM) that deliberate attempts at signature forgery give a Type II rate greater by an order of magnitude than that measured with accidental false matches. The Type II error rates quoted by Mitre and reproduced here are based on a comparison of each measured subject profile against all the other stored subject profiles and, therefore, refer only to accidental false matches. The experience at IBM is contrary to that at NPL in the development of Verisign, which preceded IBM's work and is apparently a similar system. Data from deliberate forgery attempts were compared with that derived from a comparison of one signature with another's reference data and very little difference was found. This is very convenient because it provides a large quantity of material for the 'impostor' characteristics. Evidently this is not a feature of all identity verification systems.

Voice

The voice verification system involved three operational phases, enrolment, post-enrolment and normal. At enrolment the subject stated his identity and was then required to speak a series of at least twenty, non-repetitive, four-word phrases, in response to aural prompts from the system, so that each of the sixteen words in the system vocabulary was registered five times. The phrases were processed to produce a reference file of formant frequency structure against time for each vocabulary word.

The post-enrolment phase, consisting of the four verification attempts made following enrolment, required the subject to pronounce four phrases chosen by the system. In the normal phase, after post-enrolment (and using a decision strategy different from that of post-enrolment), the system required the subject to utter as many phrases as were necessary to gain verification, with a maximum of eight attempts allowed. In order that the opportunity to try again after rejection should not make things easier for an impostor, the system was programmed

	Phase I	Phase II	Field test
Type I errors %	0.92	0.20	1.06
Type II errors %	0.99	4.40	3.26
Verification time (secs)	6.54	5.85	6.21

Fig. 7.15 Summary of Mitre results for voice verification

to require progressively better correlation between measured and stored profiles as successive attempts were made. After eight failures to gain verification, a Type I error was registered and verification was denied. To allow for long-term changes in voice characteristics, the system amended the stored profile on a one-sixteenth-weighted basis after each verification. It was observed that serious difficulties were experienced by subjects who were suffering from a bad cold or laryngitis.

We quote in Figure 7.15 only the results obtained in normal operation, after the post-enrolment phase, because these are the results that are really significant for an operational system. The particularly poor figures for Type II errors in phase two are due to over-correction by the manufacturer for faults observed at phase one. We see that in the field trials about 1.1 per cent Type I errors were observed, with about 3.3 per cent Type II errors. Considering that the latter were measured by correlation of measured profiles against all stored profiles, and not as a result of deliberate masquerading, the figure is surprisingly high. The evaluators noted that about 11 per cent of the experimental population had a Type I error rate of over 5 per cent and that about 9 per cent of the subjects, regarded as targets, produced a Type II error rate of over 10 per cent. These people were unable to produce an acceptable consistency of speech and were therefore poor subjects for inclusion in a real-life voice verification system. It was also noted that women were usually subject to greater error rates of both types than men, though the difference was not large enough to be statistically significant. On the other hand little correlation was detected between performance and such parameters as area of country where educated as a child, education level and age. The mean time taken for verification by voice was measured to be 6.2 s. The additional actions, entering the enclosure, keying in an identity number, leaving the enclosure etc., accounted for an additional 12.3 s, making a total of 18.5 s.

Signature

In the evaluation of the signature verification system the method used to sense the writing during the Mitre laboratory tests was by measuring the pressure exerted by the ballpoint pen; the pen itself was equipped with a strain gauge; the writing surface did not play any part in the measurement. For the field trial the method of sensing was altered because it had been found that the adapted pen was too fragile; the pressure-measurement transducer was transferred to the writing surface.

At enrolment each subject was required to give a total of six signatures (later increased to nine), five of which were taken to construct the standard reference set. In all, nine measures of the signature made up the reference set for each subject. As for the voice verification evaluation, a post-enrolment phase followed enrolment; during each post-enrolment visit three signatures were required of

	Phase I	Phase II	Field test
Type I errors %	6.25	3.20	1.88
Type II errors %	2.97	1.71	5.63
Verification time (secs)	14.42	13.03	13.49

Fig. 7.16 Summary of Mitre results for signature verification

	All signers	Light signers
Type I errors %	1.88	1.45
Type II errors %	5.63	4.58

Fig. 7.17 Comparison of performance of light and all signers

each individual, all three being used to update the reference set of a verified subject. Acceptance of only one of the three signatures was sufficient to class the subject identity as verified. In the normal phase of verification, following post-enrolment, the subject was asked to sign until accepted, with a limit of three failed attempts when verification was denied. In the normal phase the reference file was updated only with an accepted signature.

The results for male subjects in the laboratory and field trials of signature verification, normal period only, are shown in Figure 7.16. The verification algorithm was changed between phases one and two, as a result of which a very worthwhile improvement in performance was achieved; recall also that the pressure sensor was changed for the field trial. In the latter, the Type I error rate was nearly 1.9 per cent and the Type II error rate just over 5.6 per cent. The experimenters also observe that in the laboratory tests women and left-handers had a higher Type I error rate than the rest of the population. Time of day was not significant, but the performance of some subjects was worse on a Friday than on other days of the week. About 20 per cent of the population required more than six signatures at the enrolment stage (hence the increase from six allowed to nine). The average time taken for the signature verification process was 13.5 s, but the additional actions accounted for another 1.23 s, giving the total time per person as 25.8 s.

The change to a different transducer for the field trials eliminated trouble with the special pen, but was accompanied by other difficulties associated with the pressure-sensitive writing surface. It was found that the surface sensor exhibited hysteresis properties and was also prone to reach its limiting level for some signers. The results for light signers were compared with those for all signers (see Figure 7.17) and it was shown that better results were obtainable for the light signers.

Fingerprints

In the fingerprint verification tests the equipment was designed to scan a presented finger and record the position and orientation of the ridge endings and bifurcations (the minutiae); these parameters were digitized and stored to form the reference profile of the subject. At enrolment all fingers were scanned by the system, the system then deciding which finger to choose as the primary finger, scanning this ten times before forming the reference file. An alternative finger

	Phase II	Field test
Type I errors %	4.58	6.54
Type II errors %	2.18	2.32
Verification time (secs)	8.53	8.94

Fig. 7.18 Summary of Mitre results for fingerprint verification

(second best), chosen by the system, was also scanned ten times and a reference file created for this. At subsequent verification the subject was instructed to present his primary finger for scanning; if the correlation of the scanned minutiae with the reference file was inadequate then the subject was required to re-present his finger. If three failures to verify the primary finger occurred, then the subject was instructed to present his alternative finger, three attempts being allowed to achieve verification with this finger. Six failures produced a rejection. As distinct from the voice and signature systems, no update of the reference file was made. There was no post-enrolment phase between enrolment and normal verification in this case.

The Mitre results for fingerprint verification are shown in Figure 7.18, normal operation only. There was no phase one. Many of the subjects had great difficulty in gaining verification, even though re-enrolment was allowed. For many people only one finger produced consistent results with the system; the rejection rates for the primary and alternate fingers were respectively 14 per cent and 43 per cent, a very dramatic difference. The evaluators suggest that better results might be obtained if update of the reference file were carried out after each successful verification; this goes against the theory that fingerprints do not change. The field trials were significantly worse than the laboratory results; the evaluators suggest that this might be due to the greater number of re-enrolments allowed in the laboratory and to the very cold conditions under which the field trials were conducted. The laboratory trials showed greater error rates for women than for men. Type I error rates, i.e. rejections, were found to be generally higher in the morning than in the afternoon; this might be due to hands being relatively clean and oil-free. The designers and the evaluators agreed that dry and clean skin made the verification process that much more difficult. On the other hand the system also had difficulty in verifying manual workers (and also people with significant hobby DIY activities). The mean time taken for verification of identity by fingerprint was 8.9 s; to this we must add the time for the additional actions, 12.3 s, giving a total of 21.2 s.

Comparison of systems

Since the Mitre evaluators appear to have had no control over the trade-off between Type I and Type II errors in the verification systems, it is very difficult to make an accurate comparative evaluation of the three systems. However, it is possible to make an approximation which allows a rough comparison by adding together the Type I and Type II error rates for each system. By this means we discover a total error rate for voice of 4.4 per cent, for handwriting of 7.5 per cent and for fingerprint 8.8 per cent. The requirement of the US Air Force was for

Type I rate of no more than 1 per cent and Type II rate of no more than 2 per cent. Thus we see that no trade-off of the two types of error would bring any one of the systems within the design requirement. On the grounds of throughput all three systems also fail. However, since the throughput requirement was four per minute and the 'additional actions' took 12.3 s for each system, this left only 2.7 s for the verification process. It is possible that this is too short a time for a practical verification system. Therefore, attention should be given to improving the system environment so as to reduce the time spent in the actions which did not directly involve verification; a visual prompt might be faster than an aural prompt, for example. The Mitre evaluators concluded that the voice verification system was sufficiently promising to warrant further work to improve its performance.

7.11. PERFORMANCE OF OTHER IDENTITY VERIFICATION DEVICES

Though we know of no other published account of a comparative evaluation of identity verification devices, we can refer to the claims of systems designers in order to augment the information available on systems performance; these claims have not been assessed by independent evaluators.

Speaker verification

A collaboration[13,14] between Philips GmbH, of Hamburg, and the Heinrich Hertz Institute GmbH, of West Berlin, has produced the AUROS automatic speaker recognition system, employing a flexible combination of parameter extraction and classification in realistic environmental conditions. The performance of the AUROS system depends on the particular classification procedures in use. Averaged over all the procedures the Type I error rate is 1.6 per cent and the Type II error rate 0.8 per cent. Selection of the best procedures on the basis of minimizing computation time and storage requirements gave a system with Type I error rate of 0.87 per cent and Type II error rate of 0.94 per cent; with this system the computational effort was smaller by a factor of approximately fifteen than that of the other procedures investigated.

A research project at Purdue University[15] sought to determine whether an unknown utterance came from a set of speakers for whom known utterances were on file. Utterance in this context could mean a phrase or just one word. As might be expected the results were not as impressive for this experiment as they would be if chosen utterances were used for identity verification. In an experiment where 6 speakers were involved and 1522 tests were made, the speaker was correctly identified on 1095 occasions, giving a success rate of 72 per cent; this is equivalent to a Type I error rate of 28 per cent. The system could be useful in clustering subjects where large groups of people are required to use a voice verification system. It cannot, without great improvement, be used for identification.

Further information on the Texas Instruments speaker recognition system is given by Doddington.[16] This distinguishes performance against the number of attempts allowed for subjects. Setting the Type II error rate at 1 per cent, Doddington shows that for the first utterance the Type I error rate was 13.0 per cent, for the second 2.3 per cent, for the third 0.8 per cent and for the fourth 0.3 per cent, so that the overall Type I rate was 0.3 per cent if four attempts were allowed. The same results were also expressed in terms of the way in which the impostors progressively gained verification with successive attempts. At the first attempt 76.7 per cent of true speakers and 0.3 per cent of false speakers gained verification, at the second 19.6 per cent and 0.3 per cent, at the third 2.9 per cent and 0.2 per cent, and at the fourth 0.5 per cent and 0.2 per cent. The impostors involved in this experiment were not professional mimics and Doddington suggests that the Type II error rate would be about twice as great if professionals were involved. The Texas system has been in practical use with 180 enrolled users; about 400 verifications are made per day. The overall claimed performance is Type I error rate of 0.2 per cent; the average number of attempts required per user is 1.2.

A recent account of a speaker verification research project comes from de George[17] of Threshold Technology Inc. The study is chiefly concerned with comparison of performance with different speaker training techniques. The overall result gave better than 98 per cent correct verification performance (i.e. sum of Type I and II errors). The errors were also classified according to the word utterances used; for the word 'sing' the error rate was 21 per cent, for 'Pennsylvania' it was 2 per cent; oddly enough, for 'down' there seem to have been no errors at all. From these results it was possible to select the words giving best speaker separation.

Signature verification

Results from the IBM handwriting verification project may be found in Herbst and Liu[18] and Liu, Herbst and Anthony.[12] The initial IBM work was based on acceleration measurements, but was found to lack discrimination. A measurement of writing pressure was therefore added to the experimental technique and the results obtained were vastly improved. System performance then showed a Type I error rate of 1.7 per cent and a Type II error rate of 0.4 per cent. Type II errors in this experiment were the results of deliberate forgery attempts. In the early results of 1977, IBM found that deliberate forgeries were an order of magnitude more successful than the Type II errors logged on a random basis as in the Veripen and other projects mentioned.

More recent results in signature verification at IBM are reported by Worthington *et al.* in 1984.[19] This work was based on the use of accelerometers and strain gauges in the 'pen'. Type II errors were assessed for deliberate attempts at forgery. 4700 signatures were checked. 4625 user sessions yielded only 9 session Type I errors (the term 'session' is used in this work to indicate an attempt to gain verification which may include more than one signature attempt). 2133 forgery attempts were made in 1068 sessions, with only six successful forgeries. Four forgeries attacked four different target subjects. In the measurement of Type II errors no comparison was made with the full profile data base for all users; therefore it is not possible to compare the results published for this work with the analysis produced by Mitre, where the Type II assessment included a comparison with the full user data base. This more recent IBM work includes a degree of adaptivity, which is intended to allow the system to track changes in signature behaviour on the part of genuine registered users. If it were not for this degree of adaptivity, it is asserted that the forgery rate would be even lower.

The results obtained for the NPL Verisign signature verification system[10] were better than those measured by Mitre for Veripen, but not as good as those reported by IBM — certainly not as good as the recent IBM results. Performance evaluation of Verisign was carried out by the Inter-Bank Research Organization of London. It was noted that for new users the average FAR was as high as 25 per cent on first attempt, with 11 per cent of authorized users passing on second attempt and 3 per cent on third attempt, giving an overall FAR for three attempts of 11 per cent. At the end of 4 weeks usage performance was significantly improved; the average FAR for first attempt was reduced to 20 per cent and the overall FAR over three attempts had fallen to 4 per cent. The global IPR was less than 0.1 per cent, but for one individual the IPR was 1.7 per cent; the next worst was 0.2 per cent. The one very poor IPR was due to an individual who had difficulty in producing a consistent signature. The average time to write a signature was measured as 4.75 s; once the signature was complete the system required 100 ms processing time. Thus total identity verification time was less than 5 s.

Another project of interest, already mentioned in this chapter, is that of Nagel and Rosenfeld,[8] who set out to use a computer to detect 'freehand' forgeries, defined as attempts to forge a signature to a known name without any knowledge of how the true signature was written. This is the kind of forgery attempt that may be made by someone finding a cheque book without any indication of the nature of the owner's signature. The detection technique was applied to signatures already written on bank cheques, thus the dynamics of signature production were not measured. Dimension ratios and slant angles were measured for the signature as a whole and also for specific letters within it. The system was found to detect all forgeries, but also rejected an embarrassing number of genuine signatures, giving Type I error rates of 8 per cent and 12 per cent in different tests.

The performance currently achieved with the NPL static signature verification system[9] is 2 per cent FAR (assuming that five specimen signatures are available for comparison) and 3 per cent IPR. The device is not yet ready for commercial exploitation.

Fingerprint verification

Studies have been made of methods of classifying fingerprints for storage in a computer retrieval system,[20] but not a great deal has been done towards automatic identification of fingerprints. One study is by Kameswara Rao,[21] who observed the vital part played by accurate registration of the fingerprint. A further study by Rao and Balck[22] showed that for prints of 'reasonable quality' correct identification was always achieved; in an experiment with sixty prints, fifty-five were correctly classified. The time taken was appreciable, 10 to 15 s for scanning and approximating the print and 5 s to analyse the sampling matrix and classify the print. It is suggested that in a dedicated system the times achieved would be much less.

Retinal patterns

The retinal blood-vessel recognition system of Eyedentify is claimed to give an extraordinarily good discrimination. The manufacturer's literature asserts that for an enrolled population less than 200 million the probability of either Type I or Type II identification errors is effectively zero. The time taken for the recognition process is said to be 'seconds'.

The claimed recognition performance is so much in advance of the measured or claimed results by other methods that this method may become very important for highly secure applications. Whether it is any more acceptable to the subjects is a moot point. Having to look through a binocular instrument and make adjustments to this may not prove convenient to the user, which probably accounts for the development of the monocular version of the device. The system is not appropriate for recognition of blind subjects.

Profile verification

A system for recognition of profile views of the human face from photographs has been developed by Harmon *et al.*[23,24] For over 100 'good' subjects the system gave 100 per cent correct identification; however, it must be said that 'poor' subjects were deliberately excluded from the experiment. For an 'untailored' population of subjects, identification of individuals was almost perfect, confirmation of asserted identity was highly satisfactory and rejection of strangers was excellent. The work is being extended to measurement of ear shape and encouraging results have been obtained.

7.12. SELECTION OF AN IDENTITY VERIFICATION SYSTEM

Selection of an identity verification system is far from easy. Manufacturers' performance claims need validation and there is a need for a wider study than that carried out by Mitre for the US Air Force, described in the 1977 publication. The user must decide what error levels are tolerable, bearing in mind the nature of the relationship between user and system; users who are employees or under command in a military environment can be made to conform to defined procedures, but bank customers are not likely to be as amenable (they can transfer their accounts to other banks).

A useful guide to principles of assessing and choosing identity verificaton systems has been published by the National Bureau of Standards, entitled *Guidelines on Evaluation of Techniques for Automated Personal Identification*.[25] This publication lists twelve points which should be taken into consideration.

1. Resistance to deceit.
2. Ease of counterfeiting an artefact.
3. Susceptibility to circumvention.
4. Time to achieve recognition.
5. Convenience to the user.
6. Cost of the recognition device and of its use.
7. Interfacing of the device for its intended purpose.
8. Time and effort involved in updating.
9. Processing required in the computer system to support the identification process.
10. Reliability and maintainability.

11. Cost of protecting the device.
12. Cost of distribution and logistical support.

We have touched on some of these issues in our discussion in this chapter. The features can be split into three groups — the strength of the system as a security device, the acceptability to the user and the cost of the system. Whilst all of these aspects are important, most of this chapter has been devoted to an examination of security strength; to some extent acceptability depends on convenience and speed of response, users being unlikely to tolerate a slow and awkward system. If the users find the system unacceptable, then they will occupy themselves in finding means to circumvent the control which the system provides. Where cost is concerned, a risk analysis should show whether a proposed system is worth its cost.

REFERENCES

1. *Personal identification number management and security*, X9.8-1982, American National Standards Institute, 1982.
2. Wilkes, M.V. *Time-sharing Computer Systems*, Elsevier, New York, 3rd edition 1975.
3. Morrison, R. and Thompson, K. 'Password security: a case history', *Comm. ACM*, **22**, 11, 594, November 1979.
4. Gasser, M. *A random word generator for pronounceable passwords*, Mitre Corp., Bedford, Mass., Report MTR-3006, November 1976.
5. Azzarone, S. 'Safety PIN: can it keep card systems secure?', *Bank Systems and Equipment*, November 1978.
6. *Identification cards — physical characteristics* ISO 7810, International Organization for Standardization.
7. Boltz, C.L. 'Secrecy and security in one magnetic strip', *Private communication*, Emidata/Malco, 1977.
8. Nagel, R.N. and Rosenfeld, A. 'Computer detection of freehand forgeries', *IEEE Trans. on Computers*, C-26, 9, 895, September 1977.
9. Brocklehurst, E.R. 'Computer methods of signature verification', *Journal of the Forensic Science Society*, **25**, 6, 445, November 1985.
10. Pobgee, P.J. and Watson, R.S. 'Signature verification', *IEE Colloquium 'Pattern Recognition — Fact or Fiction?'* Digest 7, January 1976.
11. Fejfar, A. and Myers, J.W. 'The testing of three automatic identity verification techniques', *Proc. International Conference on Crime Countermeasures*, Oxford, July 1977.
12. Liu, C.N., Herbst, N.M. and Anthony, N.J. 'Automatic signature verification: system description and field test results', *IEEE Trans. on Systems, Man & Cybernetics*, SMC-9, 1, 35, January 1979.
13. Bunge, E., Hofker, U., Hohne, H.D., Jesorsky, P., Kriener, B. and Wesseling, D. 'The Auros project — automatic recognition of speakers by computers', *Frequenz*, **31**, 11, 345, November 1977.
14. Bunge, E., Hofker, U., Hohne, H.D., Jesorsky, P., Kriener, B. and Wesseling, D. 'Report about speaker-recognition investigations with the Auros system', *Frequenz*, **31**, 12, 382, December 1977.
15. Kashyap, R.L. 'Speaker recognition from an unknown utterance and speaker-speech interaction', *IEEE Trans. on Acoustics, Speech and Signal Processing*, ASSP-24, 6, 481, December 1976.
16. Doddington, G.R. 'Speaker verification for entry control', *Proc. Wescon*, 31-3, 1975.
17. De George, M. 'Experiments in automatic speech verification', *Electronic Engineering*, **53**, 653, 73, June 1981.
18. Herbst, N.M. and Liu, C.N. 'Automatic signature verification based on accelerometry', *IBM Journal of Research and Development*, **21**, 3, 245, May 1977.

19. Worthington, T.K., Chainer, T.J., Williford, J.D. and Gundersen, S.C. 'IBM dynamic signature verification', *Proceedings of IFIPSec '85*, Elsevier Science Publishers, 1985.
20. Rao, T.Ch.M. 'Feature extraction for fingerprint classification', *Pattern Recognition*, **8**, 3, 181, 1976.
21. Rao, K. 'On fingerprint pattern recognition', *Pattern Recognition*, **10**, 1, 15, 1978.
22. Rao, K. and Balck, K. 'Type classification of fingerprints: a syntactic approach', *IEEE Trans. on Pattern Analysis and Machine Intelligence*, PAMI-2, 3, 223, May 1980.
23. Harmon, L.D., Kuo, S.C., Ramig, P.F. and Raudkivi, U. 'Identification of human face profiles by computer', *Pattern Recognition*, **10**, 5, 301, 1978.
24. Harmon, L.D., Khan, M.K., Lasch, R. and Ramig, P.F. 'Machine identification of human faces', *Pattern Recognition*, **13**, 2, 97, 1981.
25. National Bureau of Standards *Guidelines on evaluation of techniques for automated personal identification*, Federal Information Processing Standards Publication 48, 1977.

Chapter 8 PUBLIC KEY CIPHERS

8.1. THE PRINCIPLE OF PUBLIC KEY ENCIPHERMENT

Most ciphers in use today employ a secret key which is known to both the sender and the receiver of the information. These are *symmetric* ciphers because the knowledge of the secret key by both A and B allows secret communication either from A to B or from B to A. Nothing distinguishes A and B in this cryptosystem and, therefore, it provides a protected channel in both directions.

A surprising new development was revealed in 1976 with the publication of a paper 'New directions in cryptography' by Diffie and Hellman.[1] They proposed a new kind of cipher system in which the sender and receiver used different but related keys, only one of which need be kept secret. The receiver of data holds a *secret* key with which he can decipher but a different key is used by the sender to encipher and this can be made public without in any way compromising the system. This is an *asymmetric* system, providing secure communication only in one direction. To set up secure communication also in the other direction a second pair of keys is needed.

The great value of the public key cryptosystem is that it makes it unnecessary to carry a secret key between A and B in order to establish the protected channel. Transporting this secret key is risky and, because of the many precautions needed, it can be troublesome and expensive. Storing a secret key also involves risks and in some applications of public keys these risks are drastically reduced. When a key must be stored in many places, removing the need for secrecy is an immense improvement.

Figure 8.1 is a block diagram showing the relationships. Encipherment employs an algorithm E and an encipherment key *ke*. Decipherment employs the algorithm *D* and the decipherment key *kd*. Both algorithms are public and are unchanging features of the design of the system. The two keys are specially created for each new protected channel and can be changed, like all keys.

In order to make the decipherment process an inverse of encipherment, the two keys must evidently be related yet at the same time they are chosen from a large universe of possible keys. For this reason they are shown in the figure as both derived from a seed or starting key *ks*, which is chosen at random. Two public algorithms, F and G, are used for calculating the keys. In order to distinguish the two parties in this asymmetric cipher we call them a sender and a receiver.

Only the intended receiver of the information should be able to decipher it, therefore the decipherment key *kd* is kept secret by the receiver. The other key, *ke*, is made public, enabling anyone to encipher data for the one receiver to whom that key belongs. In order to keep the secret, the receiver must himself carry out

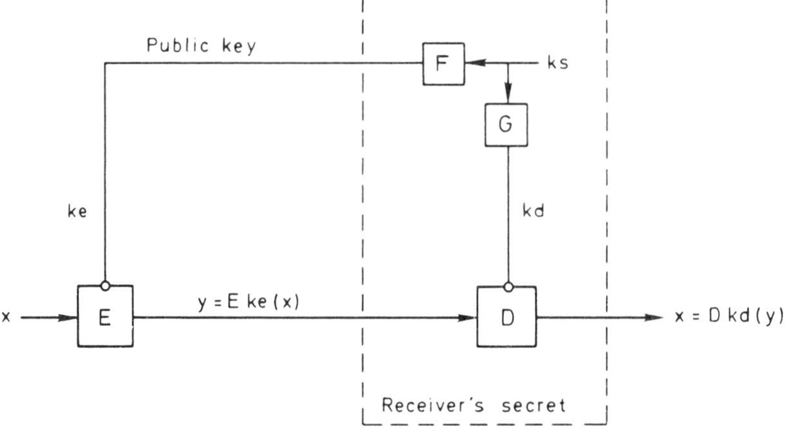

Fig. 8.1 Principle of the public key cipher

the calculations F and G to create both keys,of which he keeps *kd* strictly for his own use. In the ideal situation, the receiver's operations F, G and D would be carried out in one micro-electronic chip and the value *kd* need never leave that chip, which makes it more secure. Having discarded the random starting number, *ks*, the receiver has only to ensure the physical integrity of the chip and he is provided with a decipherment capability which no one else has and which enables him alone to decipher the messages that have been enciphered with his public key.

The best kind of public key cipher would be one that did not expand the data when enciphering it. We can then think of the ciphertext y as a number lying in the same range of values as x, the plaintext. Later we describe an important cipher which has this desirable property and it also happens that, for this cipher, the two functions E and D are identical. Their effect differs, of course, because they use different keys.

Since the encipherment key is public and anyone can use it, no kind of authentication of the received message is provided. An enemy could generate bogus messages and the system (in its simple form) cannot warn the receiver that they have come from an enemy. This contrasts with the symmetric cipher in which the sender A and the receiver B share a secret key which 'unites' them. Then, assuming that the key has been kept from all others, either A or B can assume that a message which deciphers correctly with this key has come from the other. Since we speak of 'deciphering correctly' we are assuming that there is sufficient redundancy in the message for the receiver to distinguish genuine messages from random output with certainty.

The lack of authentication in public key encipherment can be overcome in more than one way. One of the methods, called the digital signature, is of great importance as an exceptionally strong method of authentication. This is the subject of chapter 9. When A and B have each produced keys for a public key cipher and each knows the other's public key, authentication of messages is possible. Let A place into his enciphered message to B a random number R of his own choosing. Only B can decipher this message and he returns R in his reply. A can decipher the reply and he checks that the correct R has been returned. A can then regard the returned message as authentic if he knows B's public key correctly and can assume that B knows his. Obviously two-way authentication is possible if B also sends a random number and A returns it in his enciphered reply.

There is another potential problem, arising from the completely public nature of encipherment. If an enemy can guess what the message might have been, he can encipher the guess and check the result against the enciphered text to discover if his guess is correct. He can go on guessing and testing and he may hit on the right plaintext. With large message blocks (which are necessary for other reasons) the chance of a good guess is low, but not if the owner of the channel has been exceptionally careless, for example sending fixed-format messages with only a small variable part. It would be better to use procedures which guard the data in all cases, in the same way that cipher chaining with a random initializing variable (IV) protects against repeated starts to messages. One way to increase the difficulty of the 'guessing' strategy is to put into each block a large enough field of random digits. The enemy would have to guess this random number correctly and this can be made hard enough by choosing a long enough random number. Comparison with the DES suggests that a 64-bit random field would be long enough for all purposes. This random number field can if necessary contain the random numbers that we suggested earlier for authentication, in fact 32 bits of it can be used by A and 32 bits by B. In his very first message, A can put in 64 random bits to fill the whole field, but only 32 are used for authentication in the reply, since B uses the rest of the space.

In the public key cipher, the need for transporting a secret key has been avoided but in its place is the requirement that the key used by the sender shall be *authentic*. Otherwise, an enemy might be able to fool the sender into using a key of his own choosing for which he knows the decipherment key. All the messages generated by that sender would then be decipherable by the enemy. If he is to maintain the apparent normal communication between sender and receiver, the enemy would have to decipher them, re-encipher them with the correct public key and send them on to the intended receiver. It would be difficult for an enemy to establish a bogus key and keep up the masquerade, which requires a continuously successful active tap on the communication line. Nevertheless, success, even for a short time, could be troublesome. Therefore the sender must be very certain that he has the correct public key. A requirement of authenticity replaces the requirement of secrecy.

There have been discussions of protocols for secure systems in which authentication and secrecy were treated as though they were similar requirements in practice. Conceptually they are closely related — perhaps 'duality' is the best description of their relationship — but practically they are very different. If we want to keep a message secret, the risk of its discovery is increased by almost everything extra that is done with it, such as processing or copying it. But if we want to ensure that a message is authentic (a false value has not been substituted) there are many ways to do this, such as making it very public, storing it in many data bases and transmitting it to anyone who enquires. The more that is done the safer it becomes. Furthermore, if a secret number has been discovered, there need be no trace of the incident left by the thief, but if a public, authentic number has been changed the evidence is clear and public. The authenticity can be ensured by a system of checking.

Access control with an asymmetric cipher

The essential feature of the asymmetric cipher is that two different keys are used, one for encipherment and one for decipherment and that knowledge of one does not imply knowledge of the other. There is no reason why either should be made public, so the *public* key cipher can, from this point of view, be regarded as only one way to employ the asymmetric cipher.

In any cipher, knowledge of the key gives access to data. Suppose, for example, that data are stored enciphered by the Data Encryption Standard (DES). Whoever knows the key can both read the stored data and write new data. It would be possible to control access to the data by controlling distribution of the key. There is one problem. Even those without the key, if they are allowed access to the stored data, can damage and destroy it; possession of the key controls only the ability to make *controlled* or *meaningful* changes to the data.

This same principle, using an asymmetric cipher, permits the rights of access to be refined even more, since read and write access are separated. Some users can be given the encipherment key, enabling them to write new data in the store, others can be given the decipherment key so that they can read it and yet others can be given both, which confers read/write access. This idea, though interesting, has not been put to real use and perhaps the reason is that, in most applications, control of write access should be more positive and should completely prevent the overwriting of stored data.

Constructing a public key system

The early publications which introduced the idea of public key cryptography contained ingenious arguments which indicated that such systems should be possible. They did not specify the functions D, E, F and G which make a working system. It was later, with the publication of a concrete example by Rivest, Shamir and Adleman[2] that public key systems were put on a firm footing. By this time, the 'knapsack' public key cipher had also been evolved.[3] The scheme published in 1978 by Rivest, Shamir and Adleman, now known as the RSA method, has a number of advantages and will be the main example in this chapter.

Referring to Figure 8.1 we can see how difficult it is to devise a satisfactory method. The various functions (or algorithms) must satisfy a number of conditions. Since *ke* is known, the function E using this parameter is a public function, generating y as a public function of x. Yet in order to work as an encipherment method, it must be impossible for an enemy to calculate x given y. This is the property usually described as making a 'one-way function'. The function F must also be a one-way function, otherwise it will be possible to deduce from *ke* the value of *ks* and thus calculate the value *kd* which should have remained secret.

With regard to the function D, this must have all the properties normally required of a cipher function. For example, given the values of some ciphertext blocks y and their corresponding plaintext blocks x, it must not be feasible to calculate the key *kd*. The chosen plaintext criterion is especially realistic for this kind of cipher because an enemy can choose any plaintext x and calculate the corresponding cipher for himself using the public key *ke* and the known function $y = \mathrm{E}ke(x)$.

The requirements we have just stated seem inconsistent in that E is to be a one-way function which cannot be inverted, yet $\mathrm{D}kd(y)$ is precisely that inverse.

Without a knowledge of *kd*, the function E is not invertible but the secret key *kd* is a kind of 'trapdoor', the knowledge of which makes the inverse of E very simple to implement. For this reason, the encipherment function was called by Diffie and Hellman a 'trapdoor one-way function'. E*ke*(x) is the one-way function with a trapdoor D*kd*(y) which allows it to be inverted and can be 'sprung' by a knowledge of *kd*.

A public key cryptosystem comprises encipherment and decipherment functions, together with methods for generating pairs of keys, using random information (seed values) to make the guessing of keys impossible. It has proved extremely difficult to devise satisfactory public key cryptosystems. Testing the security of a system has followed the same general pattern as in symmetric ciphers, requiring all kinds of attacks to be proposed and, if we are fortunate, shown to be ineffective. The RSA cipher has stood up well to this testing, though full certainty is as yet unattainable.

One-way functions revisited

A one-way function $y = \mathrm{f}(x)$ has the property that if x is given it is easy to calculate y but if y is given it is, in general, difficult to calculate $x = \mathrm{f}^{-1}(y)$. Such a function can be used to protect a file of reference passwords against passive attack, as we described in chapter 7 (page 172). If we know a set of corresponding (x,y) pairs, these make the function easy to 'compute' for just these y values. The hard problem is to compute x for an arbitrary y, without depending on the lucky chance that y is a value for which x is already known.

Any regularity or smoothness in the function can be exploited to invert it. For example a set of tabulated (x,y) pairs can be used for interpolation of a 'smooth' function and any error in x can be corrected by finding $\mathrm{f}(x)$ and repeating the interpolation. So a good one-way function should appear to generate random values, like a good cipher function.

The most useful kind of one-way function is a one-to-one mapping or 'bijection'. A bijection $y = \mathrm{f}(x)$ maps a range of N different x values into the same range N of y values and each x maps into a different value of y. Consequently, each of the N values of y maps into a different value of x.

There can be no absolute sense in which the function $y = \mathrm{f}(x)$ is non-invertible. It is possible, in principle, to calculate y for each of the N values of x, store all the pairs (x,y) and sort them into increasing sequence of y. The result is a tabulation of x as a function of y — an explicit inversion of the given function f.

The one-way property of a function is a matter of computational complexity. The computation of $y = \mathrm{f}(x)$ should be relatively easy, carried out in comparatively few steps. On the other hand, given y the computation of x should be very much harder, requiring an extremely large number of operations. In the extreme case, tabulating (x,y) values and sorting these may be the best method of inversion and the sorting process requires of the order of $N\log N$ steps. It is clear that N must be a large number to make a one-way function.

Number theory and finite arithmetic

Public key cryptosystems lean heavily on the branch of mathematics which is called number theory. For many readers this will be unfamiliar territory though a few of its concepts are generally known. It is fortunate for our purpose that only the most basic ideas of number theory are needed to explain the RSA and knapsack ciphers. It is worth the effort to understand the underlying mathematics because it gives an idea of the sources of the security of these methods, the practical problems of implementing them and their limitations.

The ideas which are needed belong to 'finite arithmetic' which manipulates numbers according to rules different from ordinary numbers, with the useful result that arithmetics can be constructed using only *finite* sets of numbers instead of the infinite set of integers 0, 1, 2, . . . , which are the subject of ordinary arithmetic.

It is, of course, only the familiar kind of arithmetic that is useful in everyday counting, measuring and calculating. Finite arithmetic is nevertheless called 'arithmetic' because it is so very closely related to normal arithmetic and it is convenient to use the same names and symbols for operations such as addition and multiplication and the same notation for the numbers themselves. The arithmetic uses integer values like 0, 1, 2, . . . , and it has no need for separate negative or fractional numbers. A finite arithmetic can be constructed with just two integers, 0 and 1, or with three integers or four . . . , a set of any size we like. For cipher purposes the arithmetics will need to be very large, though finite. For explaining the theory, small sets can be used.

Ciphers operate on message units (characters, blocks) of fixed size, which take potentially any one of a finite set of values. The result of enciphering should be to produce a unit of the same size, taken from the same finite set of values. Therefore, finite arithmetics are just what is needed for encipherment.

The appendix (page 245) introduces finite arithmetics with a general modulus m where m is the number of different integer values in the arithmetic. It defines addition and multiplication and their inverses, subtraction and division. It shows that subtraction does not require negative numbers. The properties of multiplication are simplified when the modulus is a prime number, that is, a number p which has no factors except the trivial factors 1 and p. For a prime modulus, the arithmetic has the very useful property that division is always possible, even though there are no fractional numbers. In the arithmetic we normally use, subtraction can only be made universal by introducing negative numbers and division can only be made universal by introducing fractions, so finite arithmetics are much neater and simpler.

We shall introduce, in sections 8.2 and 8.3, arithmetic operations from which useful cryptographic functions can be derived. Useful cipher applications are at once demonstrated but these fall short of our main purpose which was to describe a public key cryptosystem. The RSA cipher requires a further step, which is introduced in section 8.4. In this way we are able to develop the arithmetic properties in easy steps, and show practical usefulness at each stage. In section 8.5 the knapsack public key system is introduced, still depending on finite arithmetic. In section 8.6 a third form of public key cipher, using error correcting codes, is described. Finally, in section 8.7 we consider the organization required for the wide use of public key cryptosystems.

8.2. THE EXPONENTIAL FUNCTION AND KEY DISTRIBUTION

The basic operations of ordinary arithmetic: counting, adding and multiplying, are developed one from the other. Counting 0, 1, 2 . . . , is most basic. Addition of b to a is defined by counting from the value a for b steps. If a is added into a zeroized register b times, the result is a multiplication of a by b. This sequence of increasingly complex arithmetic functions can usefully be taken one stage further.

The next stage is to start with 1 and repeatedly multiply by a, doing this n times. This generates the 'nth power of a'. The construction is

$$a^n = a.a \ldots a \quad \text{(with } n \text{ terms).}$$

This construction is used in finite arithmetic with the same notation.

To develop the idea of the exponential function further, let us confine our attention, for the present, to moduli which are prime numbers, denoted p. In the appendix it is shown that a prime modulus ensures that multiplication always has an inverse (omitting zero). The multiplication function defines a 'group' which ensures elegant properties.

This operation of forming a^n combines two numbers a and n which cannot be interchanged, so there are two ways of describing the function, according to which of the numbers is regarded as the independent variable. Regarded as a function of n it is the *exponential function*. Regarded as a function of a it is the *power function*, to be more specific, the nth power. For the present, we shall investigate the exponential function, a^n as a function of n. The independent variable is n and there are two parameters which are held constant, the modulus p and the base a.

In order to understand this function, an example is shown in Figure 8.2 for arithmetic with modulus 11. For each value of a, the exponential function a^n is the corresponding column of this table. The value $a = 0$ has been omitted because we are using the multiplication group and 0 does not belong to that group. The table might have been extended to the right, but values of a greater than 10 do not belong to modulo 11 arithmetic. On the other hand, the table can genuinely be extended downwards (increasing values of n) because n simply counts the number of terms in the repeated product $a \times a \times a \ldots$ There is no reason to believe that n can be treated as a number in modulo 11 arithmetic. In fact, the table shows something else because the row $n = 10$ reveals that a^{10} is always equal to 1 (except of course when $a = 0$). Beyond this point, the table simply repeats since each row is obtained by multiplying the row above by a. This new fact can be expressed thus,

$$a^{n+10} = a^n \quad \text{(modulo 11).}$$

It illustrates that, for a prime modulus, p, except for $a = 0$

$$a^{p-1} = 1 \quad \text{(modulo } p)$$

and for all a,

$$a^p = a \quad \text{(modulo } p).$$

These are alternative ways of expressing *Fermat's theorem*, which is a basic part of number theory. Nearly all the applications of number theory in cryptography depend on it.

The exponential function a^n is periodic as a function of n with a period $p-1$. Therefore n cannot be regarded as a number in modulo p arithmetic, it belongs instead to modulo $p-1$ arithmetic. To be more precise, we can reduce the complexity of the expression a^n in modulo p arithmetic by the two following rules.

n \ a	1	(2)	3	4	5	(6)	(7)	(8)	9	10
0	1	1	1	1	1	1	1	1	1	1
1	1	2	3	4	5	6	7	8	9	10
2	1	4	9	5	3	3	5	9	4	1
3	1	8	5	9	4	7	2	6	3	10
4	1	5	4	3	9	9	3	4	5	1
5	1	10	1	1	1	10	10	10	1	10
6	1	9	3	4	5	5	4	3	9	1
7	1	7	9	5	3	8	6	2	4	10
8	1	3	5	9	4	4	9	5	3	1
9	1	6	4	3	9	2	8	7	5	10
10	1	1	1	1	1	1	1	1	1	1
11	1	2	3	4	5	6	7	8	9	10

Fig. 8.2 Table of powers and exponentials, modulo 11

1. Reduce the value of a to its residue, modulo p.
2. Reduce the value of n to its residue, modulo $p-1$.

Because of these rules, the whole of the arithmetic of the exponential function, modulo 11, is contained in the square section of Figure 8.2 above the broken line.

In the figure there is a feature almost as striking as the row of ones at $n = 10$. This is a row which is half way through the period at $n = 5$ where there are two different values, 1 and 10. For modulus p, the row contains $a^{(p-1)/2}$, which is the square root of $a^{p-1}=1$. The square root of 1 is either $+1$ or -1 and the value of -1 in modulo p arithmetic is $p-1$, in our example 10.

Four of the columns in the table have been picked out, those corresponding to $a=2$, 6, 7 and 8, because their columns of figures contain the numbers 1, 2, ... 10 permuted. In other words, for these values of a, the exponential function a^n is a one-to-one mapping or bijection. When we use the exponential function it will usually be with such values of a. In practice, nearly half the values have this property and they are called 'generators' because they generate all the numbers in the multiplication group by means of the multiplication operation.

When we apply the function a^n to the set of values $n=0, 1, 2, \ldots p-2$ and when a is a generator, this function is a one-to-one mapping onto the set of values $1, 2, 3, \ldots p-1$. It follows that a unique inverse can be defined under these conditions. The function and its inverse are: the exponential, $y=a^x$ and the logarithm, $x=\log_a y$.

The logarithm function is named by analogy with the same function in ordinary arithmetic but here the result of the logarithm is always an integer. In the equations, a is a generator, x is less than $p-1$ and y is non-zero.

	8473	8474	8475	8476	8477	8478	8479
211	388600	125442	482038	713725	472600	893438	014498
212	663764	013579	341495	635933	298302	696122	930616
213	167980	070401	219323	259738	749030	822633	827194
214	318731	588206	794011	576705	635243	401132	897147
215	653663	542372	357618	234676	046456	854896	038732
216	580745	138460	864060	147589	814193	931487	414204
217	736025	329981	032991	985614	031395	281035	095420
218	445837	306526	603468	206282	139937	655224	079933
219	641110	545473	478238	475948	266111	083507	763416
220	217374	416776	135951	203826	861282	984365	114305
221	841199	819851	204309	658535	211631	588335	208568
222	600286	535473	548202	837537	026485	988909	478128
223	309740	675331	090932	084295	517153	113030	116230
224	471628	852168	661790	496558	980492	284626	530915
225	171976	394389	765586	897144	771971	100249	704802
226	177417	109200	451646	321812	109415	925455	117750
227	279792	376525	764909	724871	526714	140872	419216
228	717906	727080	713969	111044	030483	333114	592882
229	920932	380657	990125	224941	408777	188500	131937
230	189487	742243	452022	632318	261534	130166	712829

Fig. 8.3 Part of a table of powers and exponentials, modulo 999 983

The exponential as a one-way function

The exponential function a^x with generator a and modulus p is very valuable as a one-way function. In order to demonstrate this we have to show that calculating the exponential is easy but calculating the logarithm is hard.

There are some exceptions to this one-way property that are apparent in Figure 8.2. The logarithm of 1 is 0 and the logarithm of $p-1$ is $(p-1)/2$. These are not significant because it is always possible to calculate a part of the table of a one-way function and wait for values to appear which can be inverted using the table. We have to rely on the improbability of hitting one of the chosen values when the modulus (hence the size of the table) is very large. We might also choose to avoid those ranges of the function for which the inverse is easily seen. A better idea of the function is given by using a larger modulus. Figure 8.3 is a small section of the table for $p=999\ 983$ the largest prime less than 1 million. The table has a random appearance, which is expected, since any regularity could be exploited to help with inverting the function. Figure 8.4 shows one of the corners of this table, where the non-randomness is clear. Values near to zero or p must be avoided. For the very large values of p used in cryptography the proportion of the table which is not random is extremely small, less than 10^{-60}.

For practical use of the exponential, the numbers could have a length around 500 bits or more. This would be an impossible calculation if it required 2^{500} multiplications, or anything like that number, to produce the exponentials. In fact, the number of multiplications required lies in the region 500 to 1000. The method of calculation was known to Indian mathematicians more than 2000 years ago and is illustrated in Figure 8.5 for the exponent 41.

The exponent x is expressed as a binary number 101001 and this sequence is used as a 'program' for a series of multiplications. Each stage comprises

	999976	999977	999978	999979	999980	999981	999982
1	999976	999977	999978	999979	999980	999981	999982
2	49	36	25	16	9	4	1
3	999640	999767	999858	999919	999956	999975	999982
4	2401	1296	625	256	81	16	1
5	983176	992207	996858	998959	999740	999951	999982
6	117649	46656	15625	4096	729	64	1
7	176440	720047	921858	983599	997796	999855	999982
8	764886	679633	390625	65536	6561	256	1
9	645696	922117	046841	737839	980300	999471	999982
10	480043	467196	765778	048593	59049	1024	1
11	639631	196773	171042	805611	822836	997935	999982
12	522498	819328	144773	777488	531441	4096	1
13	342446	083947	276118	889980	405643	991791	999982
14	602827	496301	619376	440012	783037	16384	1
15	780126	022143	903052	239918	650838	967215	999982
16	539016	867125	484655	040311	047452	65536	1
17	226820	797148	576674	838739	857627	868911	999982
18	412226	217027	116579	644976	427068	262144	1
19	114367	697804	417088	420045	718762	475695	999982

Fig. 8.4 A corner of the same table as Figure 8.3

41 = 101001

binary	a^{41}	7^{41} modulo 97
1	$1.a$	$1.7 = 7$
0	a^2	49
1	$a^4.a = a^5$	$73.7 = 26$
0	a^{10}	94
0	a^{20}	9
1	$a^{40}.a = a^{41}$	$81.7 = 82$

Fig. 8.5 Calculation of a^{41} and 7^{41}, modulo 97

(a) forming the square of the previous number,
(b) multiplying the result by a if the binary digit is 1.

With the starting value 1 this rapidly generates a^{41}. The number of multiplications needed for the exponent x will lie between $\log_2 x$ and $2\log_2 x$. The figure shows also the specific example 7^{41} modulo 97, using seven multiplications.

The complexity of the logarithm

To complete the demonstration that $y=a^x$ is a one-way function we must show that the logarithm function is exceptionally hard to calculate. For a long time the best method of calculation was a time-memory trade-off which reduced the

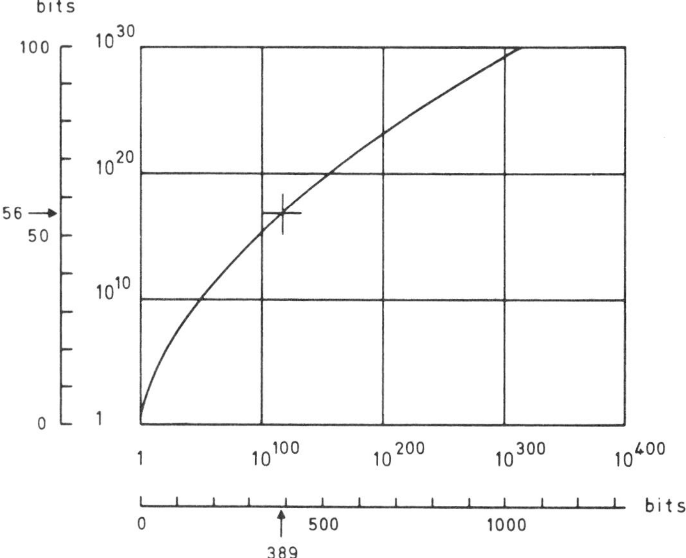

Fig. 8.6 The function $y = \exp\sqrt{[\ln x.\ln(\ln x)]}$

number of operations from p (testing each value of x until $y=a^x$ was solved) to approximately $\sqrt{p}$ operations using $\sqrt{p}$ storage cells. Then the interest in this function for cryptography led to two new developments. First, Pohlig and Hellman[4] described a quicker method for finding the logarithm whenever $p-1$ had no large factors. The complexity of his method depends on the sum of the factors of $p-1$ and it can be very large if there is a large factor. Since $p-1$ always has a factor 2 the hardest case that can be achieved is for $(p-1)/2$ to be a prime. It is not difficult to find primes of this kind. At the other end of the spectrum, if we wanted to make it easy to calculate the logarithm we could choose a prime of the form 2^n+1, so that Pohlig's method gave a complexity proportional to n.

If we use a prime for which $p-1$ has a large factor, the best logarithm method available is one due to Adleman,[5] one of the authors of the RSA Algorithm. The number of elementary operations required to calculate the logarithm for modulus p using Adleman's method is approximately

$$\exp\sqrt{[\ln p.\ln(\ln p)]}.$$

The complexity of the logarithm according to this criterion is shown by a graph in Figure 8.6. The logarithm calculation can be made sufficiently difficult for the most determined attack. We shall return to the practical question of computing exponential functions when we look at the RSA cipher, because similar operations are required.

Key distribution

In their classic paper, 'New directions in cryptography',[1] Diffie and Hellman described a method by which a secret key can be distributed using messages which were not secret. With the right conditions this can be an effective method of solving the key distribution problem, which was one of the purposes of public key cryptosystems. The possibility of distributing secret values by public messages is as surprising as encipherment by public keys.

The principles of the method are shown in Figure 8.7. To begin with, the two parties, whom we call X and Y, choose a suitable, large prime number, p, such that Pohlig and Hellman's method for calculating the logarithm is not effective and the complexity of the logarithm is sufficiently high. They also choose a generator a for modulo p arithmetic. These choices are communicated between the parties in messages that need not be secret, so they are shown in the figure as the public values a and p.

Next each participant chooses a secret value, making the choice at random among the entire set of values in modulo p arithmetic (but avoiding values too close to 0 or $p-1$). These secret values are called x and y. The participants X and Y then form the exponentials a^x and a^y respectively.

In the next stage, the two values a^x and a^y are exchanged, so they then become public values, because there is no secrecy in the communication network.

In the final stage, each participant computes a further exponential. In the case of X the received value a^y is raised to the power x. In the case of Y, the received value a^x is raised to the value y. As a consequence, both participants have computed the value a^{xy} and they now possess the same secret number. All the computations are carried out in modulo p arithmetic so the result is a number lying in the range 0, 1, ... $p-1$.

Although the two participants share the common secret number, neither of them controls its value. Consequently this is not a method of communication. Nevertheless, the secret value can be used to form a common key for a symmetric cipher like the DES.

The security of the scheme depends on the difficulty of computing the logarithm. An enemy knows the modulus p of the arithmetic and the value of the base a. From the exchanged values a^x and a^y he can compute x or y only by the

	Known to X	Public	Known to Y
Initially	x	a , p	y
Exchange	a^x → a^y ←	a^x , a^y	← a^y → a^x
Calculate	$(a^y)^x$ $= a^{xy}$		$(a^x)^y$ $= a^{xy}$

Fig. 8.7 Key distribution using the discrete exponential function

logarithm calculation, which we know to be extremely complex. There are no manipulations with these public quantities which will enable the enemy to deduce the secret value a^{xy} which now belongs to X and Y.

Unfortunately, the exponential method of key distribution suffers from the important practical limitation that it depends on authentic communication. The clearest indication of this problem is that X and Y have exchanged no information prior to the interactions shown in Figure 8.7, therefore neither has any means of establishing that he is exchanging information with the other. It might be thought that this difficulty could be overcome, after secure communication had been established, by the exchange of passwords. This unfortunately is not so, because an enemy can already have established an effective 'tap' on the communication path. To show this we need only imagine a third party Z who intervenes in the communication and can alter messages as they passed from X to Y and back again. Z will not interfere with the exchange of p and a but intercepts the quantities a^x and a^y. He presents both X and Y with his own quantity a^z. The result is that he shares a secret quantity a^{xz} with X and another secret quantity a^{yz} with Y. From this point onwards, he is able to relay messages from X to Y and Y to X, changing their encipherment en route in such a way that neither X nor Y knows that their communication is being shared with Z.

If there is a community of users who want to set up common keys by this method they could make all their public values (such as a^x) known in advance by some other authentic method of communication, so that establishing the common key between any two partners does not require any further communication and cannot be interfered with. This method of proceeding would require that the whole community use the same value of p and a. It would require careful investigation of its security, because it changes the environment in which the enemy works. If he can rely on the use for a long period of standard values for the modulus p and base a, he might embark on a lengthy computation to produce a large set of stored tables which enable the logarithm calculation to be done with relative ease for each value a^x that he captures. If after careful investigation this kind of attack is found not to be important, the exponential method can be used among a group of people, any pair of whom might want to communicate. The values like a^x become public keys, distributed in a way that ensures their authenticity. Any pair such as X and Y can establish a common secret key in using their own secret keys x and y and the public keys of their partners. They are ready to communicate without any further exchange of messages, after the public values have been broadcast effectively.

Another approach to the authentication problem is to use an independent method of communication to authenticate the numbers used in the exponential method. It is a characteristic of the attack by active linetap that the two parties attacked have different results a^{xz} and a^{yz}. If they form a one-way function of these values and compare the results over a reliable channel, equality assures them that their values are authentic. The check digits can be read over the telephone between people who recognize each other. This rests on the assumption that it is not possible to organize both an active line tap and a simulated telephone conversation. If a group of stations need to set up master keys, a one-way function of all their public values a^x, a^y, . . ., can be checked between them all by telephone calls.

Authentication and transparency

If X and Y have exchanged no information before they start the key distribution procedure of Figure 8.7, they cannot authenticate each other. An enemy Z using an active line tap can intercept and change all messages, impersonating X to Y and Y to X, as we described, but there is a limit to what this enemy can do. At first sight it may seem that Z is in complete control. If X and Y tried to exchange passwords P_x and P_y under encipherment after the common keys had been established, Z could translate the passwords from one false key to the other, so that both X and Y believe the authenticity of the other, yet Z remains able to read or alter any subsequent messages.

Yet by a change of protocol, X and Y can prevent this — an idea due to Shamir and Rivest. The exchange of passwords, or of any other messages, is done by breaking the enciphered values of P_x and P_y into two halves. Let Q_x and Q_y be the enciphered passwords and divide them each into halves so that Q_x is $Q_x(1)$ followed by $Q_x(2)$ and Q_y is $Q_y(1)$ followed by $Q_y(2)$. The exchange is by alternating messages; X sends its first half $Q_x(1)$, Y replies with $Q_y(1)$, then X sends $Q_x(2)$ and Y sends $Q_y(2)$. It is not possible for Z to translate $Q_x(1)$, knowing only half of the cipher block, yet Y will not reply until he receives something, so Z is forced to concoct a value P_x in order to receive P_y. Also, Y's reply cannot be translated by the enemy and passed on at the correct time to X. After the exchange it is possible for both X and Y to verify what they have received, using prior information.

This 'half-word exchange' protocol does not itself provide authentication, because Z is able to masquerade to both X and Y. It prevents Z from providing a transparent path from X to Y, that is, one that immediately communicates at least some part of X's message to Y. If every message between X and Y contains a one-time password, this would provide authenticity — at a cost. If all the interaction is conversational with each reply related to the previous message yet having some unpredictable content, this cannot be simulated by Z. But the protocol needs careful handling if it is to give authenticity.

8.3. THE POWER FUNCTION

Figure 8.8 illustrates the power function $y = x^n$ in modulo 23 arithmetic. The nth power function is represented by row n of the table.

Whereas the exponential function could be made difficult to invert, it is a characteristic of the power function that the inverse is itself a power function. This can be illustrated in Figure 8.8 by the two functions with $n = 13$ and $n = 17$, each of which is the inverse of the other. For example

$$5^{13} = 21 \quad \text{and} \quad 21^{17} = 5 \quad \text{(modulo 23)}.$$

To check that this relationship holds for all values of x notice that if $y = x^{13}$ then $y^{17} = x^{13 \cdot 17} = x^{221} = x$. The exponent 221 reduces to 1 because $x^{22} = 1$ $(x \neq 0)$ according to Fermat's theorem, hence $x^{220} = 1$.

In the general case, for power functions modulo p, the exponents m and n produce mutually inverse power functions if $y = x^m$ implies $x = y^n$, i.e. $x^{m \cdot n} = x$

		1	2	3	4	5	6	7	8	9	10	11	12	13	14	15	16	17	18	19	20	21	22
						*		*			*	*			*	*		*		*	*	*	
0		1	1	1	1	1	1	1	1	1	1	1	1	1	1	1	1	1	1	1	1	1	1
1	*	1	2	3	4	5	6	7	8	9	10	11	12	13	14	15	16	17	18	19	20	21	22
2		1	4	9	16	2	13	3	18	12	8	6	6	8	12	18	3	13	2	16	9	4	1
3	*	1	8	4	18	10	9	21	6	16	11	20	3	12	7	17	2	14	13	5	19	15	22
4		1	16	12	3	4	8	9	2	6	18	13	13	18	6	2	9	8	4	3	12	16	1
5	*	1	9	13	12	20	2	17	16	8	19	5	18	4	15	7	6	21	3	11	10	14	22
6		1	18	16	2	8	12	4	13	3	6	9	9	6	3	13	4	12	8	2	16	18	1
7	*	1	13	2	8	17	3	5	12	4	14	7	16	9	19	11	18	20	6	15	21	10	22
8		1	3	6	9	16	18	12	4	13	2	8	8	2	13	4	12	18	16	9	6	3	1
9	*	1	6	18	13	11	16	15	9	2	20	19	4	3	21	14	8	7	12	10	5	17	22
10		1	12	8	6	9	4	13	3	18	16	2	2	16	18	3	13	4	9	6	8	12	1
11		1	1	1	1	22	1	22	1	1	22	22	1	1	22	22	1	22	1	22	22	22	22
12		1	2	3	4	18	6	16	8	9	13	12	12	13	9	8	16	6	18	4	3	2	1
13	*	1	4	9	16	21	13	20	18	12	15	17	6	8	11	5	3	10	2	7	14	19	22
14		1	8	4	18	13	9	2	6	16	12	3	3	12	16	6	2	9	13	18	4	8	1
15	*	1	16	12	3	19	8	14	2	6	5	10	13	18	17	21	9	15	4	20	11	7	22
16		1	9	13	12	3	2	6	16	8	4	18	18	4	8	16	6	2	3	12	13	9	1
17	*	1	18	16	2	15	12	19	13	3	17	14	9	6	20	10	4	11	8	21	7	5	22
18		1	13	2	8	6	3	18	12	4	9	16	16	9	4	12	18	3	6	8	2	13	1
19	*	1	3	6	9	7	18	11	4	13	21	15	8	2	10	19	12	5	16	14	17	20	22
20		1	6	18	13	12	16	8	9	2	3	4	4	3	2	9	8	16	12	13	18	6	1
21	*	1	12	8	6	14	4	10	3	18	7	21	2	16	5	20	13	19	9	17	15	11	22
22		1	1	1	1	1	1	1	1	1	1	1	1	1	1	1	1	1	1	1	1	1	1
23	*	1	2	3	4	5	6	7	8	9	10	11	12	13	14	15	16	17	18	19	20	21	22
24		1	4	9	16	2	13	3	18	12	8	6	6	8	12	18	3	13	2	16	9	4	1

Fig. 8.8 Table of powers and exponentials, modulo 23

(modulo p). Since calculations in the exponent are modulo $p-1$ the condition for m and n to generate inverse power functions is

$$m \cdot n = 1 \quad (\text{modulo } p-1).$$

An unrestricted choice of m and n is not allowed because not all power functions are one-to-one. We need functions that have an inverse. The functions with this property are marked in Figure 8.8 by the symbol $*$ to the left of the row of figures. They are 1, 3, 5, 7, 9, 13, 15, 17, 19, 21. The rejected values are all those with factors 2 or 11, which are factors of $p-1=22$. The general condition is that m and n must have no factors in common with $p-1$. If n is relatively prime to $p-1$, the congruence $mn=1$ (modulo $p-1$) has a solution for m which is also prime to $p-1$.

The example of modulus 23 is small enough that all the inverse pairs of power functions can be listed. The significant ones are

$$3,15 \quad 5,9 \quad 7,19 \quad 13,17.$$

The trivial cases are 1,1 which means $y=x$ and 21,21 which means $y=1/x$ and $x=1/y$, since 21 is equivalent to the power -1.

The solution of the equation $mn=1$ to find m when n is given is very simple. It is described on page 249 and in Figure 8.21.

The power function $y=x^n$ is a useful cipher if the power n is kept secret, for n is the key. To discover n means calculating the logarithm $n=\log_x y$ and we know that this can be made difficult if the modulus p is large enough and $p-1$ has a large factor. The decipher function is the same function (the power function) with a different key $m=1/n$, modulo $p-1$. Both m and n must be kept secret because each can be derived easily from the other. The power function provides a cipher system with two different, but secret, keys and it can be used in all respects like a symmetric block cipher. The modulus should be a prime of about 500 bits in length, so the computations are not easy. This cipher is a stepping stone to the RSA cipher and has its own novel possibilities, to be described next.

Encipherment without key transport

An elaboration of the power-function cipher enables messages to be protected by encipherment without having to transport a key between sender and receiver. This was first described in a report entitled *Mental Poker*[6] where the problem was to shuffle and deal cards when the only communication between the players was over telecommunication channels.

There is a good physical analogue of this method in which secret papers are to be transported in a locked box without transporting the keys. The method is illustrated in Figure 8.9. The box is like a cipher which protects the message contained in it and can be locked by padlocks, for each of which there is a key. It is first locked by the sender M and transported to the receiver U, who adds his

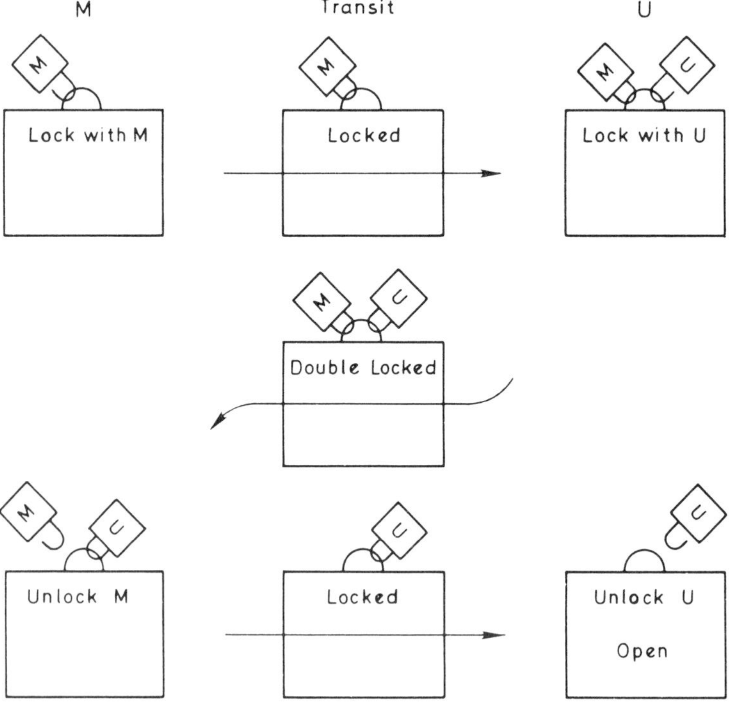

Fig. 8.9 Secure transport with keys remaining behind

own padlock which he locks with his key. The double-locked box returns to M who removes his padlock. The box is then sent again to U, who can remove his padlock and retrieve the message. So the contents of the box have been protected but the keys have not been transported. Locking the box with padlocks can be likened to encipherment by the power function. The two padlocks correspond to double encipherment. Figure 8.10 shows the cipher procedure. The two ciphers applied to the message commute, so that they can be taken off in the same order as they were applied instead of the normal method of removing super-encipherment, which is to remove the last-applied cipher first.

For the sender M, locking and unlocking consists of applying the power function with the exponents m and n which satisfy the relationship $m \cdot n = 1$ (modulo $p-1$). A similar relationship allows the receiver U to lock and unlock his cipher with the exponents u and v. The first transport carries the message x in the enciphered form x^m. The message is then returned, double enciphered, in the form x^{mu}. The sender M can remove his encipherment with the aid of the exponent n reducing the enciphered message to x^u. When this is sent again to the receiver U he can decipher it with the power function $x^{uv} = x$.

In order to apply this to the shuffling and dealing of cards, the names of the fifty-two cards are first enciphered then shuffled by S who hands them to the dealer D who shuffles and deals these enciphered card names. Each player (except S, who can decipher his own) enciphers the received messages with his own key and returns them to S who removes his encipherment and returns the results to their owners. The owners can then decipher the names and discover what cards they have.

This method of secret communication is closely related to the exponential method for key distribution. It suffers in just the same way from the authentication problem, because there is nothing to tell the sender *who* has returned the message x^{mu} and, therefore, to assure him that his message reached its intended receiver.

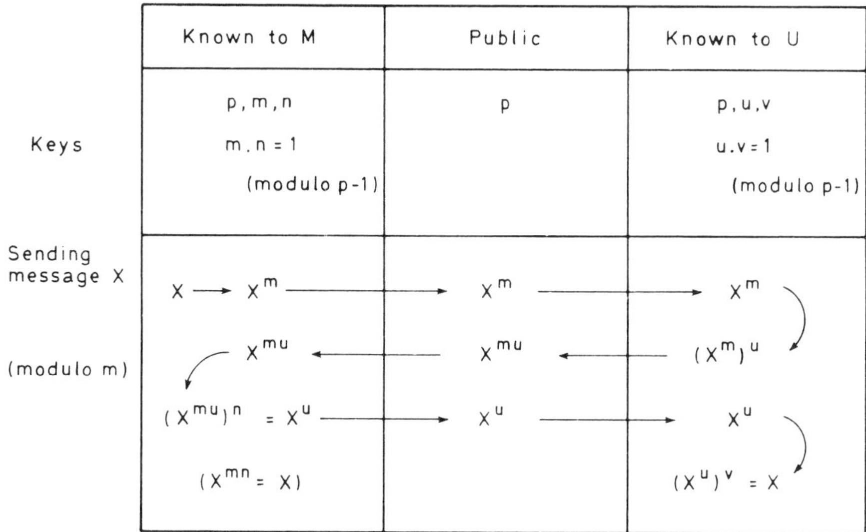

Fig. 8.10 A cipher analogue of Figure 8.9

The important public key cipher of Rivest, Shamir and Adleman also uses power functions, with only one difference, that the modulus is no longer prime.

8.4. THE RIVEST, SHAMIR AND ADLEMAN PUBLIC KEY CIPHER

Among the first of the public cipher systems that was published[2] was the one devised by Rivest, Shamir and Adleman. Encipherment and decipherment in the RSA method employ the power function, thus

$$y = x^e \quad \text{and} \quad x = y^d, \quad (\text{modulo } m)$$

where x is the plaintext, e is the public key used for encipherment, y is the ciphertext and d is the secret key used for decipherment. It was not possible to use power functions with a prime modulus as a public key cipher because the encipherment and decipherment keys could be deduced from each other by an easy calculation. The new idea in the RSA cipher is to employ a non-prime modulus.

To keep the system as simple as possible, the modulus is the product of two large prime numbers, different from each other. We call the primes p and q and the modulus $m = pq$. Actually, the method can be extended to the product of more than two primes but there seems no advantage in doing so. As in all public key systems, the receiver of information generates the keys so it is his task to find large primes p and q, then announce their product m as part of the public key, because encipherment as well as decipherment requires a knowledge of the modulus. The essential point of the RSA scheme is that the receiver can keep p and q secret while revealing m.

The factors p and q could not be kept secret if an enemy could factorize m. For small numbers, factorization is very easy but it has been known for a long time that the factorization of large numbers is a difficult problem. It is a famous problem in number theory, so it has been given a lot of attention. By making the values of p and q large enough, the system can be made secure against any attack by factorizing m.

The RSA cipher can be explained by looking at a simple example using the modulus $m = 15$ with the factors 3 and 5. Figure 8.11 shows the table of powers. To make the method work we need to find a periodic structure down the table and this is easily seen in our example, except that the first row, consisting entirely of ones, never repeats. The repetition period is 4. For the general case, the symbol for the repetition period is λ. If the example of $m = 15$ is a good guide, there is an analogue of Fermat's theorem which states that for all values of x

$$x^{\lambda+1} = x \quad (\text{modulo } m).$$

It is also evident that arithmetic in the exponent of these expressions is arithmetic modulo λ except that it is not correct to deduce that $x^{\lambda} = x^0 = 1$. But it is easy to find corresponding exponents e and d for encipherment and decipherment. Decipherment restores the plaintext value if

$$x = y^d = (x^e)^d = x^{ed} \quad (\text{modulo } m).$$

	1	2	3	4	5	6	7	8	9	10	11	12	13	14	
0	1	1	1	1	1	1	1	1	1	1	1	1	1	1	
1	1	2	3	4	5	6	7	8	9	10	11	12	13	14	1
2	1	4	9	1	10	6	4	4	6	10	1	9	4	1	
3	1	8	12	4	5	6	13	2	9	10	11	3	7	14	
4	1	1	6	1	10	6	1	1	6	10	1	6	1	1	
5	1	2	3	4	5	6	7	8	9	10	11	12	13	14	$\lambda+1$
6	1	4	9	1	10	6	4	4	6	10	1	9	4	1	
7	1	8	12	4	5	6	13	2	9	10	11	3	7	14	
8	1	1	6	1	10	6	1	1	6	10	1	6	1	1	
9	1	2	3	4	5	6	7	8	9	10	11	12	13	14	$\phi+1$

Fig. 8.11 Table of powers and exponentials, modulo 15

This requirement is met if

$$ed = 1 \quad (\text{modulo } \lambda).$$

In our example with $m=15$, a useful cipher function is $y=x^3$ and the decipherment which corresponds ix $x=y^3$ since $x=x^9$ because $9=1$, modulo 4.

The method is complete when we know the way to calculate λ for any modulus m. The function for general m is a little complex[7] but for $m=pq$ where p and q are different primes, the period λ is given by

$$\lambda = \text{lcm } (p-1, q-1).$$

The least common multiple (lcm) is easily calculated by dividing the product $(p-1)(q-1)$ by the greatest common divisor and the greatest common divisor can be obtained by the Euclidean algorithm. We therefore have an easy method, given p and q, of finding a pair of corresponding keys e and d for encipherment and decipherment respectively. The method remains secure only if p and q can be kept secret and this depends on the extreme difficulty of the factorization problem.

If we began by choosing a random value for e, there would be some doubt whether $y=x^e$ was a one-to-one mapping. When the modulus was a prime p it was necessary to choose an exponent having no factors in common with $p-1$. In the present case, the one-to-one mapping can be ensured by choosing a value of e which has no factors in common with $p-1$ or $q-1$; a prime number that does not divide $p-1$ or $q-1$ would be suitable, for example. The same condition ensures that the equation $ed=1$ (modulo λ) has a solution. The finding of keys could begin with either e or d and generate the other and there may sometimes be an advantage in having one of these quantities small. But it would be unwise to have a very small value of d in case the enemy, knowing this, could obtain the value by a search.

The RSA cipher method is summarized in Figure 8.12 which shows first the calculations required to generate a pair of keys and then the algorithms of encipherment and decipherment. The first step in calculating keys is to find two

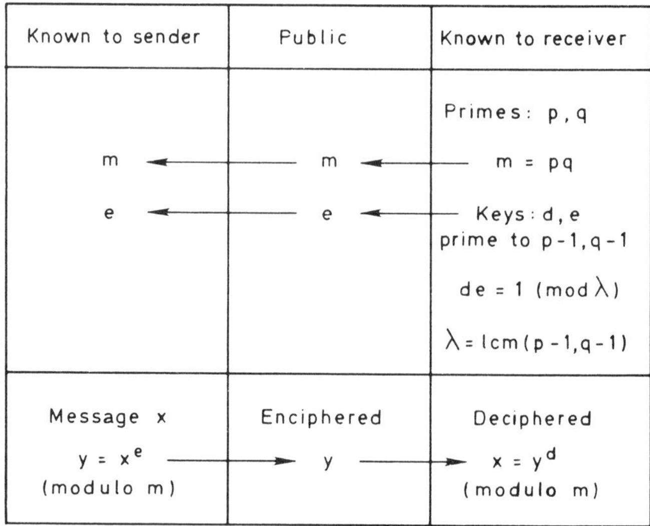

Fig. 8.12 Summary of the RSA cipher

prime numbers of sufficient size. Primes can be found by searching odd numbers of about the size required and testing them to discover whether they are prime. The simplest test for primality[8] is a probabilistic test which rejects composite numbers with a probability of about 0.5 at each test. Many independent tests can be made and, the more tests a number survives, the more probable it is to be prime. Very few composite numbers will survive 10 tests, for example, and those that do can be subjected to 100 more tests, by which time the probability that they are prime amounts to a practical certainty. There are other strategies for finding suitable primes and although it is not a trivial matter, primes of the size required in the RSA algorithm can be found in a reasonable time.

It has sometimes been suggested that users of this method should buy pairs of primes from an organization set up to find and sell large primes. This cannot be recommended because the vendor of primes would be able to break any of the ciphers used by his clients. The purchaser would need a great degree of trust in the vendor.

An attack by iteration and a defence

When the RSA method was first published, many people tabulated the power function for small examples, like our Figure 8.11 and saw, in these examples, methods of attacking the cipher. For example, if the plaintext x has one of the primes p or q as a factor, then since the modulus m has these factors, the ciphertext y must have the same factor. It is easy to calculate x,y pairs, knowing only m and e, then if the greatest common factor of x and y is very large, it may point to the value of p or q. The enemy who noticed this property would have determined the factorization of m. This method fails in practice because the proportion of plaintexts which has this property is so extremely small. Searching for such a coincidence is no better than trying to factorize m by testing its divisibility with a sequence of large primes. Both approaches would take a hopelessly long time.

There were published comments[9] about a more significant potential weakness in the RSA method. The method of attack that was proposed is instructive because it can potentially find the plaintext for a single example of ciphertext without finding the key, and without factorizing m. It illustrates that breaking a cipher and finding its key are not always the same thing.

The method of attack depends on knowing the ciphertext which is to be deciphered, and the public key (the modulus, m, and exponent, e) and from these data finding the plaintext by a search which is much shorter than p or q steps. Since the encipherment method is public, an enemy can repeat it and do so indefinitely. Starting with the ciphertext y he first forms y^e. He repeatedly applies this encipherment process to the result until the result is equal to y again. Since the operation y^e has an inverse and only a finite number (m) of different values can be generated, eventually it must come back to its starting point. Then the value which was last enciphered to produce y is x, the plaintext.

This attack is best illustrated with a numerical example, but the modulo 15 case is too small to be a useful illustration. We can illustrate the attack with modulus 23 using the table in Figure 8.8, even though this was not an RSA example. Assume an encipherment key $e=9$ and a ciphertext $y=3$. Then the repeated encipherment takes the following values

$$3^9=18, \quad 18^9=12, \quad 12^9=4, \quad 4^9=13, \quad 13^9=3.$$

Only five steps are needed to return to the ciphertext $y=3$. This completes the attack, because the last step of the cycle was $13^9=3$ and this demonstrates that 13 was the plaintext. If the iteration completes its cycle in a reasonable number of steps, the iterative attack will have succeeded.

In our example, useful plaintexts and ciphertexts can take any value between 2 and 21. These 20 values can be grouped into 4 cycles each of length 5 as follows

$$3,18,12,4,13 \quad 5,12,19,10,20 \quad 6,16,8,9,2 \quad 7,15,14,21,17.$$

In a reply to these proposals Rivest[10] analysed the cycle length for this iteration and showed that there were simple conditions on the primes p and q which would ensure that the method was impractical. Consider just the prime p because the same conditions apply to q. Given any encipherment $y=x^e$ and the inverse $x=y^d$, the value of d can be found very easily if the logarithm calculation $d=\log_y x$ is feasible. We know that in modulo-p arithmetic this can be prevented by making p large enough and avoiding values for which $p-1$ has only small factors, so that the Pohlig and Hellman method cannot be used. With the modulus $m=pq$ the same condition holds — to avoid weakening the cipher by the logarithm calculation $p-1$ should have at least one large factor, and so should $q-1$. Suppose that the large prime factor of $p-1$ is p'. The condition which must be satisfied to ensure that the iteration attack fails is that $p'-1$ also has at least one large prime factor. For q there is a similar condition. We can try applying this to our example, although the modulus is a prime, $p=23$. In this case $p-1$ has the 'large' prime factor 11 and $11-1$ has the 'large' prime factor 5. This was the period which we observed in the repeated application of encipherment in modulo 23 arithmetic. Larger periods are also observed, for example if we had chosen $e=13$ we would have found two cycles of size 10.

Since this early correspondence, very careful analyses of the cycles of this iteration have been made[11] and they confirm that the RSA method is proof against iterative attack provided that the values of p and q are chosen with the precautions stated above. A more complex set of attacks of the same general nature has been described by Herlestam[12] but they have not been demonstrated to work in examples of practical size. Rivest[13] has replied denying the value of these attacks but Herlestam was apparently not convinced of the completeness of Rivest's argument. This controversy appears to have reached almost a philosphical disagreement about what is necessary to demonstrate a successful attack. Herlestam's method encompasses an enormously large range of possible methods, too large to be analysed completely, yet if the quoted examples are typical, some of these attacks may succeed. The problem for the enemy is to discover which method to use in each case.

Practical aspects of the RSA cipher

The size of the numbers which must be used to make the RSA method secure is determined by the complexity of the factorization problem. The method of factorization quoted by Rivest *et al.* was a variation described by Schroeppel on a method published by Morrison and Brillhart.[14] The complexity of this method for a prime of a given size is of the same order as the complexity of the logarithm problem which was plotted in Figure 8.6. The same criterion for the size of m ensures that the cipher cannot be attacked either by factorization or by the logarithm calculation, $d = \log_y x$, except at enormous computational cost.

The RSA cipher has given practical importance to the complexity of the factorization problem, which can at present be determined only by finding the best method and computing its complexity. As a 'difficult' calculation on which a public key cipher is based, factorization is well-chosen because it has received close mathematical study for a long time. The better methods yield complexities of the form

$$L(x) = \exp\sqrt{[a \ln x . \ln(\ln x)]}$$

where a is constant and x is the number to be factorized. In recent years a series of refinements has reduced the value of a from 6 to a little greater than 1. The unit operation which is counted in these estimates is the multiplication of numbers of size $\sqrt{x}$. The complexity of factorization has the same form as the complexity of the discrete logarithm (see page 219). An easy solution for the logarithm would also upset the security of the RSA cipher.

The factorization problem has been studied intensively since the time of Gauss. The related discrete logarithm problem has had less attention. Many of the theoretical studies of complexity do not give a practical guide to the magnitude of the calculation; information that we need to understand the strength of the RSA cipher or the size of modulus needed for a given strength — in some cases they are vague about the unit of measurement and they rarely consider the size of store or the complexity of controlling a calculation, which can be a significant factor in a program for implementing an algorithm.

The complexity of factorization determines the strength of the RSA method which is the best available public key cipher. It is also the only public key cipher that can also function as a method of digital signature — a technique described in chapter 9. Because of its central place in public key ciphers and signatures, we need to know as much as possible about the difficulty of the factorization problem.

The theory of factorization is very sophisticated and a feel for the difficulty of the problem can better be obtained from a consideration of the various practical approaches now being worked upon. Six methods are worthy of note:

(a) the elliptic curve algorithm of Lenstra,[15]
(b) the class-group algorithm of Schnorr and Lenstra,[16]
(c) the linear sieve algorithm of Schroeppel,[17]
(d) the quadratic sieve algorithm of Pomerance,[18]
(e) the residue list sieve algorithm of Coppersmith, Odlyzko and Schroeppel,[17]
(f) the continued fraction algorithm of Morrison and Brillhart.[14]

The worst-case running time of method (a) and the typical running times of all the other methods are given by the expression $L(x)$ above. For algorithm (a) we state the worst case time because this algorithm performs best when the two primes being sought are very different in size, which is not the case when the primes are about the same size as is recommended in RSA implementations.

In early 1988 the largest composite number made up of 'RSA primes' which had been factored was of 90 digits length. This was factorized by Silverman, of the Mitre Corporation, into 41 and 49-digit primes; the smaller of these factors was said to be the largest penultimate factor yet found for any number. The method used was the multiple polynomial quadratic sieve and it was implemented on a parallel network of 24 Sun-3 workstations; the running time of the whole system, working in parallel, was about 625 hours or somewhat under four weeks.

In September 1988 there was an announcement from A. Lenstra and M. Manasse of the factorization of the 96-digit product of two primes (each of 48 digits). The technique was again the multiple polynomial quadratic sieve, but the implementation was on a large number of cooperating machines, connected by a combination of local area networks and electronic mail. The elapsed time for the result was 23 days, but well over 10 years of CPU time was expended. At the time of writing the most recent information on success in factorization comes again from Lenstra and Manasse, announced in October 1988, with the achievement of the factorization of the 100-digit product of primes of 41 and 60 digits; this was also achieved using a massively parallel approach with many cooperating machines.

Future progress in factorization is considered by Pomerance, Smith and Tuler[19] by taking present achievement of some of the methods cited above and extrapolating on the basis of the expression for $L(N)$. Since the Pomerance *et al* paper was published in 1988, but written and accepted for publication rather earlier, the base of extrapolation does not take into account the most up-to-date achievements. It is nevertheless an interesting indication of what we may expect in the next few years. Pomerance *et al.* ask the question 'what size composite number can be factorized in one year given equipment that can be purchased for 10 million dollars?' They take as one of their base examples the achievement of Davis, Holdridge and Simmons,[20] Sandia National Laboratories, in 1984, who factorized a 71-digit number in 9.5 hours using the quadratic sieve algorithm on a Cray XMP computer (it is assumed for the purposes of comparison that 10 million dollars would buy a supercomputer). Using this method it is projected that any 101-digit number could be factorized in one year. The other example is based on Silverman's achievement of factorizing an 81-digit number in one week on a small network of 9 Sun workstations. For 10 million dollars it is proposed that 2000 Sun workstations could be bought; the projection of performance with such a constellation is factorization of any 126-digit number in one year.

Pomerance, Smith and Tuler devote the main part of their paper to describing a machine costing 50,000 dollars, which should yield the factors of 100-digit numbers in two weeks. The machine uses the quadratic sieve method and is constructed with a pipeline architecture. With a larger machine based on the same assumptions, but costing 10 million dollars, they project that a 144-digit number can be factorized in one year.

On the basis of very different assumptions Wang[21] suggested in 1983 that, using the method of Schnorr and Lenstra on a parallel machine costing 10 million dollars with many LSI devices, a 125-digit number could be factorized in 50 years. In deciding what is a safe size of modulus to restrict factorization, the gradual improvement of technology must be taken into account.

The disparity between the projections we have discussed here underlines the danger in placing too much faith in this method of estimating progress. However, it seems certain that, with no improvement in algorithms, but with greatly increased parallelism, much larger composite numbers can be factorized than the present level of achievement. The fact that all the methods proposed thus far have approximately the same running time is itself significant, but too much should not read into this. The most that one can say is that it seems rather unlikely that a totally new algorithm will yield a dramatic reduction in running time.

There are special values of the modulus $p \cdot q$ which can be factorized easily, for example if p and q are nearly equal or if $p-1$ or $q-1$ have only small factors or if $p+1$ or $q+1$ have only small factors. It is quite possible that further special forms of p and q will be found which give easy factorization. These special cases create an extra problem in finding suitable primes.

A discussion of selection of primes for use in RSA implementations may be found in Gordon.[22] In Gordon's notation, a prime p to be used as one of the factors of the RSA modulus should be such that $p-1$ has a large factor r and $p+1$ has a large factor s. Also to prevent the iteration attack, $r-1$ should have a large factor t. Gordon has shown how suitable primes can be found by finding t first, then r, finding s and, with r and s, making a search for a suitable value of p. This adds to the computation, compared with finding an unrestricted prime p, by a factor 19/16, which is not excessive.

The nature of Lenstra's elliptic curve method of factorization throws light on the $p+1$ and $p-1$ conditions. In this method an elliptic curve is chosen and the factorization will succeed if the group under addition of points on the curve, modulo p, has a size Np which is smooth, i.e. has only small factors. To do the factorization a number b must be chosen such that all the prime factors of Np are less than b. If b is chosen too large, the computation is excessive, but if b is chosen too small it is very unlikely that Np will be smooth to the extent b. An optimal choice of b can be found which minimizes the total computation, estimated as computation per chosen curve divided by probability of success (the numbers of curves to be tried, on average, before success is the inverse of this probability). It turns out that Np is similar in size to p.

The $p-1$ and $p+1$ 'smooth' condition is of exactly the same kind and the measure computation/probability of success is just as large as for Lenstra's method. So if the prime is big enough to avoid all the best factorization methods, which have similar levels of complexity, the $p-1$ and $p+1$ attacks must be infeasible.

It is clear that the $p+1$ condition can be dropped, but the $p-1$ condition defines r, for which $r-1$ must have a large factor t. Applying the same probability of smoothness to this condition it is found that the iterative attack may indeed be feasible for the size of p which prevents factorization. If the $r-1$ condition is necessary, so is the $p-1$ condition, and we must also remember Pohlig's discrete logarithm method, which is avoided if $p-1$ has a large factor. Gordon's method of finding suitable primes (with or without the $p+1$ condition) gives factors r and t that are plenty large enough.

Choice of modulus size in the RSA algorithm must be made by system implementers bearing in mind the value of any resource which the algorithm is being used to protect. Many systems at present use 512-bit (154 decimal) moduli and it may be considered that the cost of factorizing such moduli greatly exceeds the benefit that a cryptanalyst would gain by such an achievement.

The RSA cipher is a block cipher, which treats message units of a fixed size. The value of each block can be in the range 0 to $m-1$, so there is an advantage in choosing a large value of m, but how large? Messages are coded as strings of bits, and a block of n bits has 2^n values, so the m values of the RSA block do not exactly fit a convenient message block. Suppose that we choose $2^{511} \leq m < 2^{512}$, in other words the 512-digit binary representation of m has a 1 in its most significant digit — it fills the available space. The ranges of values of m is 2^{511} and certainly large enough to give a sufficient choice of the primes p and q. A convenient message size would be 511 bits to ensure that the plaintext x is less than m, but the ciphertext y would have to be a 512-bit field. This expansion of the message field by 1 bit in 511 is not serious, but it prevents the result of one RSA operation being used as input to another with a different value of m. In the signatures described in the next chapter there will be applications for such double RSA operations. Messages in 512-bit blocks could be reblocked into 511-bit blocks, but this is awkward, so other solutions of the 'reblocking problem' are needed.

In chapter 9, methods of avoiding the reblocking problem are proposed. Here we shall consider how the RSA cipher method itself might be modified to ease the reblocking problem. If the modulus m can be chosen very close to a power of 2, the need for reblocking is minimized and modulo m arithmetic is simplified. Louis Guillou, in a paper contributed to an ISO working group, proposed, among other possibilities, that m should lie in the interval $2^{512} \leq m < 2^{512} + 2^{400}$. This means that m takes the form of a 513-bit number with an initial 1, then 112 zeros. To specify the value of m requires only the remaining 400 bits. Values of p and q giving the product $m = pq$ in this range are easily found. The first prime p can be chosen with little restriction, then m/p is calculated for each of the extreme values of m. A starting point for q is found at random within the range given by m/p. Primality testing finds a suitable prime in the neighbourhood of this value and finally the resultant m is checked against the required size.

With p and q chosen in this way, no help is apparently given to an enemy trying to factorize m. Two advantages flow from this choice of m. First, reblocking is hardly ever necessary. A 512-bit plaintext enciphers into a number which has a probability less than 2^{-112} of exceeding 512 bits. Secondly, the calculation of the residue modulo m is simplified. The residue is the remainder of a division by m. Division by a number so close to 2^{512} is very simple.

With numbers of 512 bits, arithmetic operations have to be carried out by multiple-length routines using the much shorter word length of the available processor. The time taken for multiplication is proportional to the square of the word-length ratio. The number of multiplications needed for a 512-bit RSA operation $y=x^e$ lies between 512 and 1024. The result is that encipherment or decipherment operations range in time from seconds on a fast main frame to many minutes on a small microprocessor. Because of the cost of devoting a fast processor to this task, it can be said that this method is not economic for any reasonable data rate. The DES algorithm can be implemented to operate in a simple microcomputer in less than 100 ms. For anything but slow operation, the DES is implemented in a special semiconductor chip. For the RSA algorithm, special hardware is needed for almost any practical embodiment. The essentials of a hardware RSA device are an arithmetic unit which operates on long words and a controlling microprocessor. Various purpose-built chips to implement the RSA algorithm are now becoming available, both in the USA and in Europe. The Cylink Corporation, of Sunnyvale, California, has a chip (CY1024) which gives operation time of 320 milliseconds with 1024-bit modulus (20 MHz clock); this is equivalent to a throughput of 6.4 Kb/s. The Cylink chip is cascadable, so that the maximum modulus that can be handled is 16,384 bits, though the throughput rate for such modulus is not stated. British Telecom has a chip (Meteor) capable of 10 Kb/s throughput at 256-bit modulus (one RSA operation takes 25 milliseconds (10 MHz clock)); this chip is not cascadable. The Belgian organization, Cryptech, has a range of multi-chip implementations (PQR3, PQR5 and PQR6) carrying out RSA with operation times of less than 10 milliseconds at 256-bit modulus (PQR3), less than 40 milliseconds at 512-bit modulus (PQR5) and less than 70 milliseconds at 672-bit modulus (PQR6). For the PQR6 implementation with 672-bit modulus, the throughput rate is said to be in excess of 9.6 Kb/s.

An account of a chip design has been published by Sedlak[23]; this cryptographic processor contains two 'ciphering units', one capable of handling 340 bits and the other 440 bits. The two ciphering units collaborate as a single 780-bit unit and achieve speeds of up to 200 Kb/s in general and 3 Mb/s with exponent of the fourth Fermat number, 65537. At the time of writing no production model of the chip described by Sedlak had been announced.

In one further respect the RSA calculations can be simplified; this is also a proposal by Louis Guillou. The condition $ed=1$ (modulo λ) does not greatly restrict the choice of e. The smallest practical value for e is 3, since e must have no factors in common with $p-1$ and $q-1$, which are even numbers. If p and q are chosen suitably, this value of e can be used and it gives the simplest possible RSA encipherment; only two multiplications. The value of d cannot be constrained in this way, since it has to remain secret. In some applications this choice of e

allows simple microprocessors to encipher very fast, but the complexity of decipherment cannot be avoided. Implementers should note that, though the use of exponent 3 for encipherment is acceptable in many circumstances, it is not always safe. In particular, if the same message is enciphered with exponent 3 and three different co-prime moduli, then a straightforward application of the Chinese Remainder theorem will yield the plaintext (the secret key is not compromised). The advantage of an exponent with a small number of non-zero bits can safely be obtained if an exponent such as the fourth Fermat number, $2^{16}+1$ (or 65 537), is used. Note that neither $p-1$ nor $q-1$ should be divisible by 65 537, but this requirement will reject very few primes.

Public key cryptosystems can be used to simplify the problem of key distribution and make it more secure. This can be achieved if RSA encipherment is employed to transmit session keys, which are then used in a faster symmetric cipher, such as the DES algorithm, to encipher and decipher data. Since the session keys can be contained in a single RSA block, which also has space for authenticating information (such as identities, date, time and initializing variables if these are needed), a time of about 1 s for RSA operations is fast enough.

Speeds of operation change rapidly as the technology improves. The significance of the RSA method extends beyond encipherment and includes public key signatures, employing the same computations and the same keys. Its elegance and versatility make it a strong contender for widespread practical use.

8.5. THE TRAPDOOR KNAPSACK

One of the first ideas for constructing a public key cryptosystem was to use puzzles of a known complexity.[24] The puzzles would be constructed by the sender of the message and the receiver would solve the puzzles to reveal the plaintext. In some way, the receiver must have special knowledge which makes it easier for him to solve the puzzles than for an enemy. The trapdoor knapsack cipher[3] employs as its puzzle the 'knapsack problem' which is known to be very difficult to solve in the general case. Figure 8.13 illustrates the knapsack problem with a simple example. The puzzler is given a set of 8 weights of assorted integer sizes, in this case 9, 33, 112, 118, 203, 250, 269 and 361 units. With these weights he has to try to balance a 'knapsack' containing an unknown selection of the same weights. The puzzle is specified by giving a *knapsack vector*, and a *knapsack total* (357) which

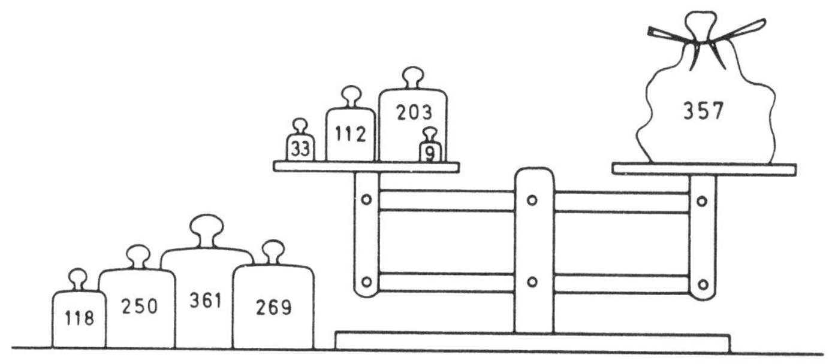

Fig. 8.13 The knapsack problem

has to be matched. The puzzle is to find the subset of weights which adds up to the total.

The way in which the knapsack problem was first formulated was slightly different. Given the vector and the knapsack total, the puzzle was: 'Can this exact total be obtained by adding together any subset of the numbers in the vector?' This is known to be among the category of hard problems called 'NP complete'. General algorithms to solve NP complete problems are computationally complex but that does not mean that every example of a problem in this category must be hard. This can be seen in the knapsack problem because it is very easy to solve if the knapsack vector is the geometric series, 1, 2, 4, 8, 16 . . ., for then we can decide whether the largest weight must be used without knowing anything about the rest because the largest weight is greater than the sum of those of lesser size. If the largest weight is part of the solution, we subtract it from the total and continue the process with the next lower weight. The subset is thus found, step by step. This can be done not only for the vector 1, 2, 4, 8, 16 . . ., but for any vector where each weight is greater than the sum of the previous ones — a so-called 'super-increasing sequence'.

The trapdoor knapsack cipher uses special cases of the knapsack problem. Therefore the NP completeness of the general case does not necessarily indicate that the cipher problems will be hard to solve. It is now known that many of the ciphers based on the knapsack problem are weak, but they still have some theoretical interest.

Using our example we will first show how it can be used to carry messages and then how the receiver can possess a special trick which enables him to solve the knapsack problem and reveal the plaintext without having to search all the possibilities. The searching method can be made unprofitable by using enough knapsack elements, for example 100, since a search of 2^{100} possibilities is too difficult.

In our example we have 8 weights in the knapsack vector which enables it to carry messages of 8 bits. The 8 bits of the plaintext are used to select which of the weights is to be included in the table. For the case of a 100-element knapsack this can be expressed by treating the plaintext block, x, as a binary array of 100 elements $x_1, x_2 \ldots x_{100}$. The knapsack vector is the set of weights $w_1, w_2 \ldots w_{100}$. To encipher the plaintext block, the inner product of these vectors is formed, namely

$$S = \sum_{1}^{100} x_i w_i$$

The knapsack sum, S, is the ciphertext and this is transmitted to the receiver. The receiver of the message has to solve the knapsack problem to discover which selection of weights was used in forming S. This reveals the binary array, x_i, which is the plaintext. The supposed difficulty of the knapsack problem ensures that an enemy should not be able to solve it in a useful time.

The intended receiver of the plaintext must have some special knowledge by which he can arrive at the solution. This relates to the 'secret structure' illustrated in Figure 8.14. Two knapsack problems are shown, an easy one and an apparently hard one which is just a transformation of the easy one. The easy problem at

Key generation	a = 203			m = 491					
Easy knapsack	1	3	5	11	23	46	136	263	A Total 488
Transformed	203	118	33	269	250	9	112	361	B

Encipherment

Public key	9	33	112	118	203	250	269	361	C
Plaintext	1	1	1	0	1	0	0	0	
Ciphertext	9	+33	+112	+0	+203	+0	+0	+0	S = 357

Decipherment	1/a = 387 m = 491	
Transformed ciphertext	357 × 387 = 188	(modulo 491)
Easy vector solution	188 = 136 + 46 + 5 + 1	see A
Corresponding hard vector	357 = 112 + 9 + 33 + 203	see B
Plaintext	1 1 1 0 1 0 0 0	see C

Fig. 8.14 Trapdoor knapsack example

A in the figure has a super-increasing sequence

1, 3, 5, 11, 23, 46, 136, 263, total 488.

With this vector, any knapsack problem can be solved by a trivial calculation. This problem is transformed to conceal this structure. The numbers are multiplied by $a=203$ in a finite arithmetic with the modulus $m=491$. The result of this multiplication is shown at B in Figure 8.14; it happens to be the set of weights we used in Figure 8.13. To make sure the structure is concealed, it is best to present this new vector in an increasing sequence, see C.

In this way, a simple problem has been transformed into one that is apparently complex but with a knowledge of a and m the inverse transformation can be performed. All that is necessary is to find the inverse of the multiplier, $a=203$, in modulo 491 arithmetic. Using the Euclidean method described in the appendix on page 249, the inverse is found to be 387. In the figure the ciphertext S has the value 357. The first step is to transform it from the hard knapsack problem to the easy one by multiplying by 387 in modulo 491 arithmetic, giving 188. This is the knapsack sum in the easy problem.

Its solution clearly does not include the weight 263 (too large) but must include 136. The remainder $188-136=52$ must include weight 46 and the remainder $52-46=6$ includes 5 and 1. The weights 1, 5, 46, 136 correspond to 203, 33, 9 and 112 respectively in the hard vector, and it is easily checked that their sum is 357, the ciphertext value, so the problem is solved.

The 'hard' vector was 9, 33, 112, 118, 203, 250, 269, 361 so the binary plaintext was 11101000 to select for these values. We have to imagine that the number of terms is big enough that solving the knapsack without knowing the transformation is too difficult for any enemy. But the receiver of the ciphertext, who knows the values of a, $1/a$ and m and the weights of the easy knapsack vector, can easily find the solution.

Like any public key system, the trapdoor knapsack cipher comprises a method of forming the secret and public keys, a method of encipherment and a method of decipherment. The receiver generates the keys by first constructing an easy, super-increasing, knapsack vector. Adding the vector elements gives the largest number that this knapsack can produce. The next step is to choose, in some random fashion, a modulus m larger than this maximum sum and a multiplier a such that the modulus and the multiplier have no factors in common. Then the receiver can compute the inverse of the multiplier using the Euclidean algorithm. The receiver then generates the hard knapsack vector by multiplying each element of the easy vector by the multiplier a, using arithmetic with the chosen modulus m. The hard knapsack vector given in increasing sequence is the *public key* and the receiver keeps secret the easy knapsack vector, the modulus m and the multiplier a, which together form the secret key.

Encipherment consists of using the plaintext blocks as a binary array with which to select some of the elements of the knapsack vector for adding together. The resultant sum is the ciphertext which is sent to the receiver. The decipherment operation consists of multiplying this knapsack sum by the inverse multiplier to get the corresponding sum in the easy knapsack problem. Solving the easy knapsack problem gives the plaintext, which is a binary vector describing which weights (taken in the sequence of the hard vector) were used in the solution.

Practical aspects of the trapdoor knapsack

None of the calculations required for the trapdoor knapsack is as complicated as those for the RSA method. Put in the simplest terms, the RSA cipher, with n bit numbers, has a complexity proportional to n^3 whereas the knapsack method has a complexity proportional to n^2. But the secret key is much bigger for the trapdoor knapsack method. The knapsack cipher expands the message. In the example, a plaintext of 8 bits produces a ciphertext with a maximum value of 1355 and this requires an 11-bit ciphertext. A 'compact' knapsack would be one that had minimum expansion but it has been found that, in this type of cipher, compactness is not compatible with security.

Suppose, for example, that a 100-element knapsack vector is used. We first have to estimate the sum of the easy knapsack and the size of the modulus. The most compact knapsack problem would have for its vector the geometric series 1, 2, 4, 8 . . . , but we require an easy vector that the enemy cannot guess and this makes the numbers necessarily somewhat larger. Merkle and Hellman suggest choosing the smallest element in the range 1 to 2^{100}, the next not greater than 2^{101}, the next not greater than 2^{102} etc. and the hundredth element not greater than 2^{199}. Their sum is less than 2^{200} so a modulus m less than 2^{202} is suggested, and the elements of the hard vector are numbers of 202 bits. The ciphertext requires a field of 209 bits. This means a data expansion from plaintext X to ciphertext S by a factor 2.09. The public key consists of all the elements of the hard vector, 20 200 bits in all.

The encipherment calculation is extremely simple because each of the 100 bits of plaintext selects one of the knapsack elements and the selected ones are added together to form the ciphertext. This is as complex as a multiplication of 100 bit numbers, which compares favourably with the average of 750 multiplications of 500 bit numbers typical of the RSA method. Even if the number of elements had to be increased, encipherment would still be exceptionally easy. Decipherment begins with a multiplication followed by a division by m in order to find the residue. The solution of the easy knapsack is a matter of 100 test subtractions followed by decisions whether to include a particular element. This is similar in complexity to a division. Approximately, decipherment is three times as complex as encipherment by this straightforward method. A simple 8-bit microprocessor can perform the encipherment and decipherment operations for 100-bit blocks in a few tens of milliseconds. This would give a useful data rate, but if higher speeds are needed, special LSI logic can be designed and, with an improved decipherment algorithm,[25] speeds in the region of 10 Mbit/s can be achieved.

The security of the trapdoor knapsack method depends on the difficulty of solving the knapsack problem for the hard vector. The straightforward solution of the 100-element problem by testing all 2^{100} subsets of the elements can certainly be improved. A time-memory trade-off has been described[26] which reduces the number of operations to about 2^{50} at the cost of requiring storage for about 2^{25} numbers. Although it would be extremely expensive to solve the problem this way, it is feasible, whereas 2^{100} operations are not. The trapdoor knapsack method can be strengthened by making more than one transformation, using different moduli. It is well known that repeating calculations with different moduli increases the complexity of a problem and this seems a useful precaution, but each transformation of the 100-bit knapsack requires a bigger modulus than the previous one and adds approximately 7 bits to the size of ciphertext, so 20 or 30 transformations might be used, with data expansion factors of 2.79 or 3.49, respectively.

Shamir and Zippel[27] described an attack on the trapdoor knapsack assuming that only one transformation had been made and that the enemy knew the modulus m. If he can identify the knapsack vector elements corresponding to the two smallest elements of the easy knapsack and if they are small enough, the multiplier a can be found and the cipher has been broken. Using large elements increases the data expansion. The remedies are to keep m secret, use more than one transformation and shuffle or sort the published knapsack elements.

In reference 27 a new kind of knapsack is described, due to Graham and Shamir. Figure 8.15 shows how they construct the elements of the easy knapsack. The elements are shown in a binary array with the most significant bits on the left. The two blocks of random digits are separated by a diagonal matrix and a strip of zeros. When a selection of these elements is added together, the carries from the right-hand random numbers do not reach the diagonal region because there are $\log_2 n$ zeros, where n is the number of knapsack elements. Consequently, in the knapsack sum, the bits in the diagonal region 'read out' the identities of the elements that have been added together, i.e. the plaintext.

As in the Merkle–Hellman knapsack, the problem for an enemy can be made more difficult by several transformations using different moduli. When the

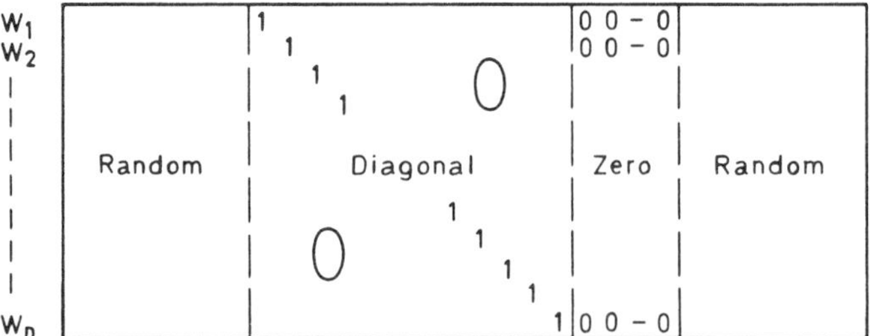

Fig. 8.15 The elements of an easy knapsack

receiver, knowing the a and m values, strips off the transformations, he can read the plaintext from the n bits in the diagonal part of the sum. Decipherment is almost as easy as encipherment.

Almost from the beginning there have been questions about the security of the knapsack cipher, proposed attacks and improvements to resist these methods. Then in 1982 Shamir[28] described a method which can defeat the basic Merkle–Hellman knapsack cipher. At about the same time others described promising alternative methods and Adleman[29] demonstrated a method that solved the Graham and Shamir variant. Also, ways to attack the knapsack when it has two multiplicative transformations were being discussed. To be strictly accurate, though polynomial in complexity, Shamir's algorithm is not able to solve trap-door knapsacks of large size in a useful time, so a Merkle–Hellman cipher could still be secure. There are selected key values which are apparently safe at the level of 100 or so elements. A comprehensive study[30] concluded that the security of knapsack ciphers was still an open problem but that a prudent cryptographer would not use them; a more recent survey[31] reached the same conclusion. New designs of knapsack cipher continue to be produced. If one of them proves to be secure it could serve a useful purpose. Encipherment and decipherment will be much easier than for RSA, but the public key will be larger and there will be data expansion.

8.6. A CIPHER BASED ON ERROR-CORRECTING CODES

A third kind of public key cipher was described by McEliece. It exploits an error-correcting code by requiring the sender to encode his data with a large number of errors which only the receiver knows how to remove.

There are many coding methods designed to allow error correction and one large class of these, known as Goppa codes, is exploited here. The general form of the Goppa code is shown in Figure 8.16. The message $\boldsymbol{u}$, which has k bits, is enciphered by modulo 2 matrix multiplication with an $n \times k$ matrix denoted $\boldsymbol{G}$. The result is a code word of length n, which is larger than k in order to provide error correction. This is very similar to a knapsack calculation in which the rows of the matrix are knapsack elements and the rows to be added are selected by the message bits. In this case, however, the addition is modulo 2, whereas knapsack elements are added as ordinary integers. The usual way in which $\boldsymbol{G}$ is expressed has on its left a square matrix of size $k \times k$ with units on its diagonal and otherwise zero. The effect of this is that the first n bits of the code word repeat

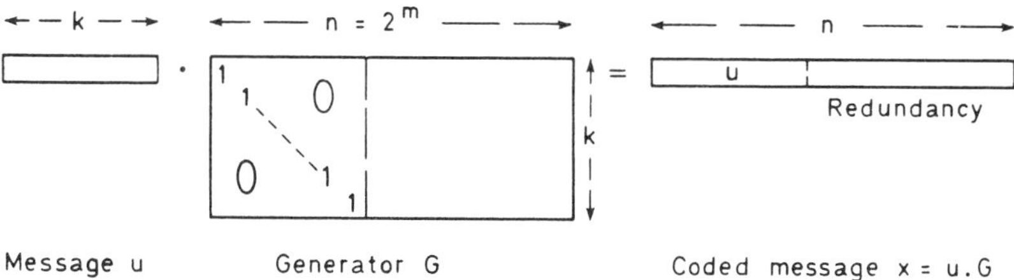

Fig. 8.16 General form of a Goppa code

the message and the remaining $n-k$ bits are added for error correction. Any matrix **G** can be put into this form by taking linear combinations of the rows, using the elimination process which is used to solve simultaneous equations.

Such a Goppa code can correct t errors and the code word length n is a power of 2, namely 2^m. The position of an error requires m bits to specify it, therefore all the error positions can be specified in about tm bits, actually rather less because errors are interchangeable. The redundancy $n-k$ is equal to or a little less than tm. The example given by McEliece has $m=10$ and $n=1024$ and contemplates using $t=50$ errors so that the message length k is 524 bits or slightly greater.

There is an efficient algorithm for removing these errors and restoring the original k message bits provided the design of the code is known. There is also an efficient way to produce many varieties of code using the fact that any polynomial of degree t in the Galois field GF(2^m) generates such a code. In order to make it difficult for an enemy to discover which code is being used, the matrix **G** can be modified in a secret way, known only to the receiver. Combining the rows of **G** linearly is equivalent to pre-multiplying **G** by a non-singular matrix **S**, but this process could easily be undone to restore the diagonal form of the first part of the matrix, as we have seen. This can be prevented by permuting the columns of the **G** matrix, by post-multiplying **G** by a permutation matrix **P** of dimensions $m \times m$. Thus the generator matrix which the sender of the message actually receives, namely the public key, is

$$\boldsymbol{G}' = \boldsymbol{SGP}.$$

To encipher a message **u** the sender computes the vector

$$\boldsymbol{x} = \boldsymbol{u}\boldsymbol{G}' + \boldsymbol{z}$$

where **z** is a locally generated random vector of length n containing exactly t ones and the rest zeros. Thus **z** introduces the required number t of errors.

In order to apply the efficient algorithm for removing these errors, the receiver can compute $\boldsymbol{xP}^{-1}$ and this is a code word in the Goppa code of generator **SG**, which allows error correction. Having found the corresponding message value $\boldsymbol{u}'$ the receiver can compute the real message **u** as

$$\boldsymbol{u} = \boldsymbol{u}'\,\boldsymbol{S}^{-1}.$$

It is doubtful that the security of this cipher has been adequately tested since the only discussion of proposed attacks seems to have come from the author. The best attack proposed by McEliece is to select k digits of the n-digit code word in the hope that there is no error in them and solve the equations for **u**. For the example given, he estimates a complexity of about 10^{19} multiplications. In general, there is a great deal of knowledge about error-correcting codes, which leads us to wonder whether better attacks will be found.

A big advantage of the McEliece cipher is the ease of implementation both for encipherment and decipherment. He estimates that communication rates near 10^6 bit/s should be posible. Clearly it cannot be used for signatures because of the large data expansion. Such expansion is a disadvantage, but the analogy with the knapsack system indicates that data expansion may be essential for security.

One curious new feature of the method is that the sender could choose the errors apparently at random but actually according to a strict rule which provides a concealed data channel.

8.7. THE REGISTRY OF PUBLIC KEYS

The sender of an enciphered message must know the public key of the receiver. The security of his communication depends on using the authentic key because any other key might have been planted by an enemy who can then read his message. In practice, the conversational nature of most communication would make it difficult for an enemy to exploit a planted key but the danger remains and must be guarded against. People who frequently communicate can usually establish authentic keys by exchanging them by hand. All our business depends on the authentication of people by meeting them. We use a large number of clues to identify people and the organizations which they represent. Public keys can be carried on magnetic-striped cards or stored electronically in microcircuit cards, therefore much of the public key distribution and authentication can take place by direct exchange. The values received in this way must be stored carefully to avoid fraudulent changes by an enemy.

As well as frequent communication with well-identified people there is a need for secure communication in an 'open network' where calls are made to people or organizations identified only by their name. The requirement of authenticity in public key distribution can be met by a public register of keys run by a trustworthy organization. The establishment of the key registry needs a legal basis, similar to that of a register of companies. In addition to the public key registry there can be key registries operated within organizations for internal traffic.

A public key may be identified with an individual or his function in an orgnanization or both. Any large organization will have many public keys, each corresponding to a different domain of security within the organization. Someone communicating with the company can be sure whether he is in contact with the sales office, the research department, the managing director or one of the company's computer systems.

When a new public key is presented to the registry, its authenticity can be tested by the registry only by traditional means, based on meeting people, credentials, telephone calls etc. The same is true whenever a new public key is introduced to replace an old one. The old public key could be used to authenticate the new one, for example by the registry sending the owner a message of the form: 'Do you agree that the new value K should replace your registred public key? If you do agree, include in your reply the random number R and encipher your message with the public key of this registry.' But it is not a good idea to use an old key to validate a new one, because if the enemy has undermined it (by finding its corresponding secret key) he can falsify the new one.

Requests to the registry for the values of public keys will be sent through a data communication network, using messages which are themselves authenticated. The reply from the registry can use a public key signature and its authenticity then rests on the public key of the registry itself. The public key of the registry must be authenticated outside the registry system using wide publicity to prevent falsification. For example, it can be printed in bar-coded form in an official gazette and in public newspapers. Each user of the system is able to monitor the value of his key stored in the public registry, checking it from time to time. This monitoring would detect any carelessness in the registry or attempts to falsify keys. It is the ability to monitor the published keys which makes public key systems more effective than those based on secret keys. The effectiveness of monitoring depends on assuming that the registry gives the same answer to all enquiries and cannot falsify a key to one caller while reporting it correctly to its owner. The problem of the 'two-faced registry' can be met by dividing its equipment into two separate units, one of which stores and retrieves keys while the other identifies callers and answers their enquiries. Communication between the two units is made by messages identified only by random numbers created for the purpose. With this arrangement, the strict security requirements concerning the registry's software and its method of operation can be concentrated mainly on the unit which answers the calls and the method of communication between the two units.

There is no possibility of designing a system which functions correctly whatever treachery is happening. All systems depend on some degree of trustworthiness. Public key ciphers reduce the amount of trust needed.

8.8. COMPLEXITY THEORY AND CRYPTOGRAPHY

Practical ciphers are at most *conditionally* secure — they can be broken if enough computational effect is available. So the analysis of the strength of ciphers, if it can ever be made rigorous, will depend on the mathematical discipline called 'complexity theory'. This aims to measure the difficulty of computational tasks and to find methods with the least computational effort.

Complexity theory has had a lot of success and one of its achievements has been to demonstrate that a large class of problems which are apparently rather different are, in fact, computationally equivalent. This class is called 'NP complete' and it includes the general knapsack problem. Since none of these problems has yet been solved by a general method which is in any sense 'easy' they can be regarded, in our present state of knowledge, as hard problems. If one of these problems could be solved more easily then all the others would become correspondingly simpler to solve.

To measure complexity we need a unit of measurement, a 'scale of difficulty' and a way to characterize the problem. Consider the problem of multiplying two numbers, a and b. They can be expressed by binary representations and the number of bits would be approximately $\log_2 a$ and $\log_b b$ respectively. Suppose we ask how complex it is to multiply, in the conventional way, two numbers of n digits each. This could be done with n additions, but since the addition of n

bit numbers requires a number of logical operations proportional to n, it could more fairly be stated as proportional to n^2 logical steps. From another point of view, if the elementary steps were 8-bit additions such as a simple microcomputer provides, the multiplication of large numbers would still be proportional to n^2 but with a different factor of proportionality.

Complexity is always measured in terms of an assumed 'unit operation' but constants of proportionality are ignored. It is usually sufficient, for theoretical purposes to know if the complexity is proportinal to n or n^2 or e^n or such. Since hard problems are the ones studied, the criterion most often used is whether complexity is proportional to a polynomial in n or something which increases faster, such as e^n.

When a computational problem is studied it has a measure of size, such as n in our example. The size of a number is such a measure and this usually enters as the logarithm. Recall the expressions given earlier (pages 218 and 230) for the discrete logarithm and factorization problems. Sometimes the complexity is the size of an array, such as the number of components in the knapsack.

For any problem, new computational methods might be found which reduce the known complexity. The obvious methods are not usually the best. Multiplication with a complexity proportional to n^2 is a case in point since it can be shown[7] that the complexity is actually $n^{1+\epsilon}$ where ϵ can be made as small as we please. A calculation of complexity for one method is interesting but a more fundamental result is a proved complexity, a statement that the general problem cannot be solved in less than $f(n)$ operations, where n is the size of the problems. These proofs of complexity are hard to find.

Complexity theorists are inclined to ignore the unit of measurement since they are concerned with very broad questions (exponential or polynomial) rather than precise measurement. For the cryptanalyst the unit can be a vital factor.

The limitations of complexity theory for cryptography

The first concern of the theory has been: 'How hard are certain general problems?' For the purpose of the cryptanalyst the results have been disappointing, for two reasons.

The first reason is that the subject studied is the general problem and how it can *always* be solved. Therefore, it is the hardest example that sets the complexity measure and it may be difficult or impossible to decide how hard is each individual case. For a problem of size n, the problem is not fully defined until it is expressed in actual numbers, such as the actual knapsack vectors and the sum. We know that a vector with superincreasing elements makes a trivial knapsack problem yet this problem is NP complete for the general case. Furthermore, we now know, as a result of Shamir's work, that the knapsacks generated by the basic Merkle and Hellman scheme are easily solved even when their method of construction is not revealed.

Cryptanalysis is concerned with the easiest examples of a class of problems. We want the problem to be hard in all cases, but we might settle for a problem with just a few easy examples provided that, (a) the enemy cannot spot easily which are easy problems, and (b) the probability, with a random key, that the cryptanalytic problem can be solved easily is much too small to make it worthwhile to try. The criterion is actually very hard to define but it seems that the complexity of some easy fraction of problems is needed. As a partial approach to this problem, Shamir[32] studied the 'median complexity' of the knapsack problem — such that half the problem instances are harder than the median value.

When the complexity of an algorithm is known, and this is the best-known method, it may not be the answer to the practical problem. The mathematical interest lies in the largest problems, and if n is the size of the problem it is the asymptotic behaviour of complexity as n tends to infinity which is usually studied. As a problem is studied, better and better algorithms are devised but the best algorithm tends to score over the lesser ones only for very large problem sizes.

The advanced methods which give good asymptotic behaviour are necessarily ingenious and rather complicated in their design. Consequently the steps involved in controlling the calculation may become as significant as the arithmetic itself. Yet complexity theory tends to concentrate its attention on arithmetic, except in such stages as sorting items into sequence where decisions are the essence of the calculation. It is possible that this factor pushes even higher the real size of problem when computer running time is the measure of complexity.

So complexity theory does not provide the theoretical basis for cryptography — yet. But it could develop in directions which make it the perfect tool for a theory of ciphers. Indeed there are, even now, ciphers[33] which are provably as complex to break as solving the factorization problem and this puts the complexity of factorization at the heart of the available ciphers based on number theory.

Complexity theory, by its study of a wide range of difficult problems, provides an excellent 'scouting party' for finding new ciphers, since there must be a complex problem behind every cipher. The history of the trapdoor knapsack provides a warning that a hard problem in general is not enough.

8.9. APPENDIX: FINITE ARITHMETIC

There was a young fellow named Ben
Who could only count modulo ten.
He said, 'When I go
Past my last little toe
I shall have to start over again.'

Anon. Quoted by Martin Gardner in *Scientific American*, February 1982.

Arithmetic with integers, rationals or real numbers employs an infinite set of numbers. Ciphers can be constructed more easily with a finite arithmetic.

Finite means that the arithmetic operations (add, subtract, multiply, divide) work on a finite set and give results that belong to the same finite set. The results do not always agree with those of normal arithmetic. The operations of addition, multiplication etc. have such a close relation with their arithmetic counterparts that the things they operate upon are given the usual names, except that there is only a finite set of numbers, $0, 1, 2 \ldots m-1$. We call this 'modulo m arithmetic' and m can take any positive integer value except 1. The result of an arithmetic operation belongs to the same set of m different numbers.

Counting in modulo m arithmetic

When counting with the numbers 0, 1, 2 . . . $m-1$, the next numbers in the series are 0, 1, . . ., and so the sequence continues and repeats. The minutes in the hour count as 0, 1, 2, . . . 59, 0, 1 . . ., which are numbers in finite arithmetic, in this case with modulus 60.

Addition

The operation, $a+b$, means 'start at a and count b times'. The value we reach is the same as in normal arithmetic if $a+b$ is less than m, otherwise the effect of counting on from $m-1$ to 0 subtracts m from the result, so modulo m addition gives

$$\begin{array}{ll} a+b & \text{if } a+b<m \\ a+b-m & \text{if } a+b\geq m. \end{array}$$

Fortunately all the properties of addition are like those of familiar arithmetic. In order to show that we are working with a finite arithmetic, equations are qualified by adding the words 'modulo m', often in brackets.

If we have to evaluate a longer sum, such as $a+b+c+d+e+f$, there are two ways to do it. We can add the numbers in normal arithmetic, then bring it in range by subtracting a multiple of m, or we can keep the number in range at each operation by subtracting m if necessary.

'Bringing the number in range' is of special importance. If the result of a string of operations is x (a normal integer) and it may be outside the range 0 to $m-1$, we need to find the result

$$x' = x-im$$

where integer i is chosen so that $0\leq x'<m$. This can be obtained by dividing x by m, then

$$x/m = i \quad \text{with remainder } x'.$$

The quantity x' is the remainder left when x is divided by m to give an integer result. It is called the *residue* of x, modulo m.

For example,

$$2+4+5+7+8=4 \quad (\text{modulo } 11)$$

because the sum in normal arithmetic is 26 and when divided by 11 this leaves a remainder 4.

Subtraction

The quantity $a-b$ is the solution of the equation $a=b+x$. If we can solve $b+x=0$ (modulo m) this gives a value $-b$ and then $a-b$ can be calculated as $a+(-b)$, modulo m. It is easy to see that $-b$ can be calculated in modulo m arithmetic as $m-b$ in ordinary arithmetic, except for the trivial case $-0=0$.

Now we see one of the advantages of finite arithmetic. Subtraction in ordinary arithmetic requires the introduction of a new kind of number, the negative integer, but in finite arithmetic the set 0, 1 . . . $m-1$ is sufficient.

		a						
		0	1	2	3	4	5	
	0	0	1	2	3	4	5	
	1	1	2	3	4	5	0	
b	2	2	3	4	5	0	1	a + b
	3	3	4	5	0	1	2	
	4	4	5	0	1	2	3	
	5	5	0	1	2	3	4	

Fig. 8.17 Addition, modulo 6

A useful characteristic of finite arithmetic is that the complete table of an arithmetic function can be written down. Figure 8.17 shows addition in modulo 6 arithmetic. Subtraction is always possible because each row and column contains all the values 0, 1 . . . 5.

Multiplication

Addition is derived from repeated counting and similarly multiplication is derived from repeated addition. We can define ab as $a+a+\ \ldots\ +a$ with b terms in the sum (modulo m). Figure 8.18 is a table of multiplication, modulo 6.

The properties of multiplication are like those of ordinary arithmetic, for example

$$ab=ba \qquad (ab)c=a(bc)$$
$$a(b+c)=ab+ac \quad a(-b)=-(ab).$$

Division (except by zero) would be possible in finite arithmetic if each non-zero row and column of the multiplication table contained all the values 0, 1 . . . $m-1$, but Figure 8.18 shows that this will not always be true. The zeros in the body of the table arise because $3.2=0$ in this arithmetic, 2 and 3 are said to be 'divisors of zero'. This could be avoided if the modulus m had no factors, in other words if m were a prime number.

		a						
		0	1	2	3	4	5	
	0	0	0	0	0	0	0	
	1	0	1	2	3	4	5	
b	2	0	2	4	0	2	4	ab
	3	0	3	0	3	0	3	
	4	0	4	2	0	4	2	
	5	0	5	4	3	2	1	

Fig. 8.18 Multiplication, modulo 6

	b \ a	0	1	2	3	4	5	6	
	0	0	0	0	0	0	0	0	
	1	0	1	2	3	4	5	6	
	2	0	2	4	6	1	3	5	
b	3	0	3	6	2	5	1	4	ab
	4	0	4	1	5	2	6	3	
	5	0	5	3	1	6	4	2	
	6	0	6	5	4	3	2	1	

Fig. 8.19 Multiplication, modulo 7

Division

Primes are very important in ciphers based on finite arithmetic. A prime p has no factors except the trivial ones 1 and p. The sequence of primes

$$1, 3, 5, 7, 11, 13, 17, 19, 23, 29, 31, 37, \ldots$$

has no end, though they get more sparse as the sequence continues. There is no formula for generating the nth prime. Testing for primality is possible, even for very large numbers. Figure 8.19 is a multiplication table, modulo 7, and illustrates that when the modulus is a prime, each column except $a=0$ has a single 1 in it. Therefore $ax=1$ always has a solution, (for non-zero a) called $x=1/a$, the reciprocal of a. In other words, for a prime modulus, non-zero numbers have an inverse under multiplication, just as all numbers have an inverse under addition. This is such a valuable property that we shall very often use a prime modulus.

Since any non-zero number y has a reciprocal $1/y$ (modulo p) it follows that division by non-zero y is always possible. The result of dividing x by y is $x/y=x(1/y)$ (modulo p). It can be seen in Figure 8.19 that division is always possible because each row and column, except those corresponding to zero multipliers, contains all the values 0, 1 ... $m-1$, consequently $z=x/y$ can be found as the value of z for which $zy=x$.

The Euclidean algorithm

A classical problem of number theory is to find the largest number which divides exactly into two given numbers, their greatest common divisor (gcd). For example, the gcd of 60 and 84 is 12. An elegant and efficient method is the Euclidean algorithm. As an example we shall calculate the gcd of 323 and 238. If both a and the smaller number b have a divisor D then $a-b$, $a-2b$ ..., also have that divisor D (unless one of these is zero, which means that b divides a and therefore b is the gcd). So we find the smallest number we can by taking $a-nb$ where n is the integer result of dividing a by b and $a-nb$ is the remainder, which is less than b. In our example $n=1$ and the remainder is $323-238=85$.

(a)	323		
(b)	238		
(c)	85		(a – b)
(d)	68		(b – 2c)
(e)	17	g.c.d	(c–d)
(f)	0		(d–4e)

Fig. 8.20 Euclidean algorithm for the greatest common divisor

Now we have two numbers 238 and 85 which both have the divisor D. The process is repeated, and 238/85 gives 2 with remainder 68. We therefore know that 85 and 68 both have the common divisor D. The whole procedure, starting with $a = 323$ and $b = 238$ is tabulated in Figure 8.20, labelling the lines $a, b, c \ldots,$ and showing how each line is obtained from the preceding pair.

Since the numbers are decreasing, eventually one of them is the gcd itself. This divides the previous number exactly, giving a zero remainder. The last number before zero, in this case 17, is the gcd. If the two starting numbers have no other common factor the number 1 will be the result; they are relatively prime.

Calculation of the reciprocal

With modulo m arithmetic, the reciprocal of a number a is a solution of the equation $ax = 1$, modulo m. This can always be solved if a and m have no common factors (except 1). The Euclidean algorithm can be modified to find x, as shown in Figure 8.21 where the reciprocal of 299 is found in modulo 323 arithmetic. The starting equations a and b are self evident, then they are manipulated by finding $a - nb$, just as before. All the right-hand sides of the equation have 299 as a factor. This process continues until the gcd process reaches 1 on the left-hand side

Modulo 323

(a)	323	=	$0 \cdot 299$	
(b)	299	=	$1 \cdot 299$	
(c)	24	=	$-1 \cdot 299$	(a – b)
(d)	11	=	$13 \cdot 299$	(b–12c)
(e)	2	=	$-27 \cdot 299$	(c–2d)
(f)	1	=	$148 \cdot 299$	(d–5e)

Fig. 8.21 Euclidean algorithm for the reciprocal

of the equation. This final line states the required reciprocal relationship. If it gives a negative value, add the modulus (323) to get the correct value.

The examples show how rapidly the Euclidean algorithm converges in practice. Typically the number of steps is roughly the number of binary digits in the starting value. The worst case has 44 per cent more steps than this.

REFERENCES

1. Diffie, W. and Hellman, M.E. 'New directions in cryptography', *Trans. IEEE on Information Theory*, IT-22, No. 6, 644–654, November 1976.
2. Rivest, R.L., Shamir, A. and Adleman, L. 'A method of obtaining digital signatures and public key cryptosystems', *Comm. ACM*, **21**, No. 2, 120–126, February 1978.
3. Merkle, R.C. and Hellman, M.E. 'Hiding information and signatures in trap-door knapsacks', *Trans. IEEE on Information Theory*, IT-24, No. 5, 525–530, September 1978.
4. Pohlig, S.C. and Hellman, M.E. 'An improved algorithm for computing logarithms over GF(p) and its cryptographic significance', *Trans. IEEE on Information Theory*, IT-24, No. 1, 106–110, January 1978.
5. Adleman, L. 'A subexponential algorithm for the discrete logarithm problem with applications to cryptography', *Proc. 20th IEEE Symposium on Foundations of Computer Science*, 55–60, October 1979.
6. Shamir, A., Rivest, R.L. and Adleman, L. *Mental Poker* MIT Laboratory for Computer Science, Report TM-125, 7 pages, February 1979.
7. Knuth, D.E. *The art of computer programming*, 2, *Seminumerical algorithms*, p19, Adison-Wesley, Reading, Massachusetts, 1969; 2nd edition, 1981.
8. Solovay, R. and Strassen, V. 'A fast Monte-Carlo test for primality', *SIAM J. of Computing*, **6**, No. 1, 84–85, March 1977.
9. Simmons, G.J. and Norris, M.J. 'Preliminary comments on the MIT public-key cryptosystem', *Cryptologia*, 1, No. 4, 406–414, October 1977.
10. Rivest, R.L. 'Remarks on a proposed cryptanalytic attack on the MIT public-key cryptosystem', *Cryptologia*, **2**, No. 1, 62–65, January 1978.
11. Blakley, B. and Blakley, G.R. 'Security of number theoretic public key cryptosystems against random attack', Part 1: *Cryptologia*, **2**, No. 4, 305–321, October 1978. Part 2: *Cryptologia*, **3**, No. 1, 29–42, January 1979. Part 3: *Cryptologia*, **3**, No. 2, 105–118, April 1979.
12. Herlestam, T. 'Critical remarks on some public key cryptosystems', *BIT (Nordisk Tidskrift for Informationsbehandling)*, **18**, 493–496, 1978.
13. Rivest, R.L. 'Critical remarks on "Critical remarks on some public-key cryptosystems" by T. Herlestam', *BIT (Nordisk Tidskrift for Informationsbehandling*, **19**, 274–275, 1979.
14. Morrison, M.A. and Brillhart, J. 'A method of factoring and the factorization of F7,' *Mathematics of Computation*, **29**, No. 129, 183–205, January 1975.
15. Lenstra, H.W. 'Factoring integers with elliptic curves', *Annals of Mathematics*, to appear.
16. Schnorr, C.P. and Lenstra, H.W. 'A Monte Carlo factoring algorithm with linear storage', *Mathematics of Computation*, **43**, 298–311, 1984.
17. Coppersmith, D., Odlyzko, A.M. and Schroeppel, R. 'Discrete logarithms in GF(p)', *Algorithmica*, **1**, 1–5, 1986.
18. Pomerance, C. 'The quadratic sieve factoring algorithm', *Advances in Cryptology, Lecture Notes in Computer Science 209*, Springer, 169–182, 1985.
19. Pomerance, C., Smith, J.W. and Tuler, R. 'A pipeline architecture for factoring large integers with the quadratic sieve algorithm', *SIAM J. Computing*, **17**, 387–403, April 1988.
20. Davis, J.A., Holdridge, D.B. and Simmons, G.J. 'Status report on factoring (at the Sandia National Laboratories)', *Advances in Cryptology, Lecture Notes in Computer Science 209*, Springer, 183–215, 1985.
21. Wang, S. S-Y. A study of Schnorr and Lenstra's factoring algorithm. Thesis, in the Department of Electrical Engineering and Computer Science, MIT, Cambridge Mass., September 1983.

22. Gordon, J. 'Strong RSA keys', *Electronic Letters*, **20**, 514–516, June 1984.
23. Sedlak, H. 'The RSA cryptography processor', *Advances in Cryptology, Lecture Notes in Computer Science 304*, Springer, 95–105, 1988.
24. Merkle, R.C. 'Secure communication on insecure channels', *Comm. ACM*, **21**, 294–299, April 1978.
25. Henry, P.S. 'Fast decryption algorithm for the knapsack cryptographic system', Bell System Telephone Journal, **60**, 767–773, May–June 1981.
26. Schroeppel, R. and Shamir, A. 'A $T.S^2 = 0(2^n)$ time/space tradeoff for certain NP-complete problems', *Proc. 20th IEEE Symposium on Foundations of Computer Science*, 328–336, October 1979.
27. Shamir, A. and Zippel, R.E. 'On the security of the Merkle–Hellman cryptographic scheme', *IEEE Trans. on Information Theory*, IT-26, 339–340, May 1980.
28. Shamir, A. 'A polynomial time algorithm for breaking the Merkle–Hellman cryptosystem', *Proc. 23rd IEEE Symposium on Foundations of Computer Science*, 145–152, 1982.
29. Adleman, L. 'On breaking the iterated Merkle–Hellman public-key system', *Advances in Cryptology, Proc. Crypto'82*, Plenum, 303–308, 1983.
30. Brickell, E.F. and Simmons, G.J. 'A status report on knapsack based public key cryptosystems', *Sandia Report SAND 83-0042*, Sandia National Labortories, February 1983.
31. Diffie, W. 'The first ten years of public-key cryptography', *Proc. IEEE*, **76**, 560–577, May 1988.
32. Shamir, A. 'On the cryptocomplexity of knapsack systems', *MIT-LCS TM-129*, MIT Laboratory for Computer Science, March 1979.
33. Blum, L., Blum, M. and Shub, M. 'A simple, secure pseudo-random number generator', *Memorandum UCB/ERL M28/65*, Electronics Research Laboratory, UC Berkeley, September 1982.

Chapter 9 DIGITAL SIGNATURES

9.1. THE PROBLEM OF DISPUTES

Message authentication can assure the receiver that a message is authentic in two respects — the receiver knows the true identity of the sender and can be sure that, if the message has been changed since it left that sender, this interference is detectable. If the sender and receiver of the message never have a dispute about its origin or content, this is all that is needed. Both can be sure that no third party can interfere with a message or try to misrepresent its origin without being detected. But the situation changes when sender and receiver are in dispute.

The receiver could forge a message which gave him an illegal advantage by pretending that the message came from the sender. For example, if a bank cheque were sent through a digital communication network, the receiver could increase the amount of the payment in the message and claim that the cheque for this larger amount had arrived from the sender. Nothing in the authentication methods of chapter 5 prevents manipulation by the receiver. Authentication uses a secret key which is known to both parties so that the sender can create an authenticator value and the receiver can check it. Using this key, the receiver can also construct an authenticator for a message that he has forged.

In many business applications where paper documents are used to create contractual obligations (like a cheque) the receiver of a message is the person most likely to gain by falsifying it. Although authentication is entirely effective against third parties, it does nothing to prevent forgery by the receiver. With paper documents it is the signature and the laws against forging signatures that give them contractual value. A signature can be taken to a third party, such as a court of law and used to 'prove' the origin and correctness of the document so that the sender is confronted with his obligation.

The other side of the picture is that the sender may be dishonest and may be trying to escape a contractual obligation. The message could have contained instructions to a stockbroker for transactions which have turned out badly. The sender dishonestly tries to pretend he never sent the message. If the message is sent through a digital network and protected only by authentication, the powerlessness of authentication against forgery by the receiver makes it possible for the sender to claim that the message was forged. The receiver, who has been wronged, cannot produce any convincing evidence that the message was genuine.

A third party in a dispute cannot distinguish between these two cases. If B takes a received document to a court and A, who sent it through a digital network, claims it was forged by B, the two situations we have described are indistinguishable to the judge. B may indeed have forged the document but on the other hand A may be trying to escape the consequences of a message he actually sent.

The weakness of the authenticator in this problem of disputes lies in its symmetry as between A and B. A symmetric cipher, like the Data Encryption Standard (DES), can be used to encipher messages both ways, since both A and B know the secret key. Likewise it can be used by either party to authenticate messages which the other can check. A true analogue of the signature, which we would like to have, is created by A and checked by B but could not have been produced by B.

9.2. DIGITAL SIGNATURE USING A PUBLIC KEY CIPHER

In the previous chapter we explained that a public key cipher is asymmetric and, while effective for encipherment, gives no help in message authentication. It is, therefore, surprising to find that public key ciphers can be used to provide a digital signature, able to resolve disputes between sender and receiver, hence are capable of a very strong form of authentication.

Figure 9.1 shows the signature principle, using the same notation that was used in Figure 8.1 to introduce the public key cipher. The functions D, E, F and G are the same as before and the pubic key *ke* and private key *kd* are generated in the same way. In this case the public and private keys belong to the *sender*, who creates these keys from the seed value *ks* and makes the value *ke* public.

It is not at once obvious that the relationships shown in Figure 9.1 are valid. We know that if the value x is enciphered to form the ciphertext $y = \mathrm{E}ke(x)$, then decipherment restores the plaintext value in the form $x = \mathrm{D}kd(y)$. Applying these functions in the reverse order creates a quantity $s = \mathrm{D}kd(x)$, which is the signature. The receiver of s can obtain the plaintext by using the encipherment function and generate $x = \mathrm{E}ke(s)$, if our figure is to be believed. This is easy to verify for the RSA cipher because applying the exponential functions in either sequence must give the same result. Thus

$$(x^e)^d = x \quad \text{implies} \quad (x^d)^e = x.$$

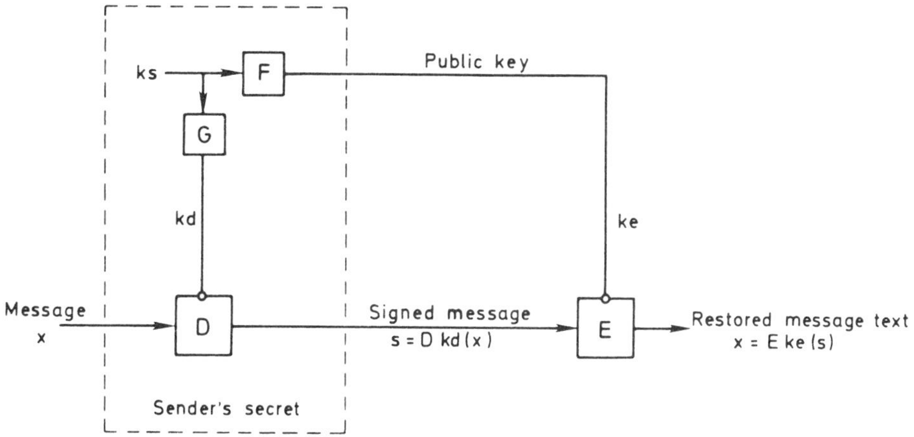

Fig. 9.1 Principle of the public key signature

Mutually inverse functions commute in this way whenever they constitute a one-to-one mapping or 'bijection'. For encipherment functions, which must have a unique inverse, this simply means that there is no expansion of data by encipherment. The RSA cipher can create signatures but the knapsack cipher, in its usual form, cannot, because of its data expansion.

Unlike a normal signature on a document, the value s is a transformed version of the whole plaintext which can be transformed back by the receiver into readable form. But there are other ways of forming digital signatures, which we describe later, in which the signature is not a transformation of the message but an additional and separate value that goes along with the plaintext. Our figure shows the signature process applied to just one block of plaintext, such as 512 bits for a typical RSA cipher. When a longer message is to be signed, transforming it block by block carries a risk. An enemy could change the sequence of the signed blocks and the signatures would still be checkable. So for long messages it is better to have a signature which is a separate value tacked on to the end. But for the present we shall consider s as the signature of just one block.

In order to check the signature, the receiver applies the encipherment function using the public key of the sender. If the result is a plaintext message, the signature process has been successfully completed and the signature is considered to be valid. The argument for its validity is that only by possessing the secret key *kd* could anyone produce the transformed message s which enciphers with the public key to generate a valid plaintext. This rests on some assumptions.

It is assumed that *kd* has indeed been kept secret by the owner of the public key *ke*. The ownership of *ke* is made known through a key registry and both its value and its association with the owner are assumed to be authentic. If an enemy could persuade the receiver to accept a false public key, which he created himself (and therefore knew the secret key which corresponded), this enemy could produce documents which appeared to have a valid signature. Therefore, authenticity of the public key is important, as it was with encipherment.

Another assumption is that any interference with the message in the form s will produce an unacceptable output which the receiver will not accept as a plaintext. This requires enough redundancy in the plaintext message that the probability of a random message being accepted as plaintext is very low. Fortunately, this is true for nearly all the circumstances of practical communication. Redundancy is essential, since in any authentication or signature procedure the checking depends on the fact that only a very small proportion of received bit patterns could be accepted as valid messages. When the signature takes the form of a block which is appended to the plaintext message, the signature itself provides the necessary redundancy, which is perhaps a better scheme.

With these assumptions, it can be argued that only the owner of the secret key could produce a valid signature, but anyone knowing the public key of that sender can verify that a message has his signature. A signature procedure must depend on *all* the bits of the plaintext, so that the text cannot be altered. It is important that no one can make calculated changes to the plaintext and leave the signature unchanged. Signature differs from symmetric authentication because the sender has more information than the receiver (namely the secret key) enabling him to create signatures which the receiver cannot create, yet the receiver can check these signatures using the public key.

The statement of a signature procedure is not complete until the method of arbitrating a dispute is specified. The receiver must be able to prove a genuine signature's validity to a judge, in spite of the assertion by the sender that it was forged. The evidence he produces is the signed form of the message, s, the public key ke and the plaintext x. The judge is able to verify for himself that $x = Eke(s)$ and can also verify that ke belongs to the sender. For the latter, he depends on the authenticity of information coming from the key registry. Concerning s, he has only the word of the receiver that this was the message he received. By issuing this message, which must have been created by the use of the secret key, the sender is giving it legal validity. It can be copied but not altered without destroying the signature property. Merely by possessing this message the receiver can, he hopes, establish that it came from the sender.

The text x is the substance of the signed document and since a judge is concerned with the legal effect of the document he will want to see this in its plaintext form. There must be enough redundancy to ensure that the text presented could not have happened by accident. In order to sign a message with very little redundancy (which is hardly ever the case), the redundancy could be added, for example by forcing 64 bits of the 512-bit block to be zeros. So with the three items, x, s, and ke the judge can made a decision and declare whether or not the message is forged. Because the decision is virtually automatic and does not require the exercise of judgement, it would hardly ever be necessary to take the dispute to arbitration. Presented with the evidence of a valid signature, the sender would be wasting his time and money to begin a formal dispute.

The process of transforming a message into its signed form must not be confused with encipherment. It provides no secrecy because anyone with the aid of the public key can transform s (by encipherment) back to the plaintext form. It seems characteristic of public key systems that they provide either secrecy or authentication in its pure form. The secrecy they provide contains no authentication and the authentication they provide in the manner of Figure 9.1 provides no secrecy. If we need both secrecy and signature, the encipherment and signature schemes must be combined.

Signatures using a public key allow an authentic message to be broadcast to many destinations. Each of the destinations can obtain the public key of the sender and check the authenticity of the message, which would be unwise for authenticators using a secret key since a widespread knowledge of the key would weaken its security. A valuable practical application of this principle is the authentication of software. Preventing unauthorized changes to software is an essential requirement for the security of information systems and copies of the software are often widely distributed. By providing an authentic public key, which is difficult to falsify, the originator of the software can use a digital signature process which enables all the software users to check that it has not been interfered with. The same procedure should be applied to subsequent updates of the software to prevent falsification.

With messages consisting of many blocks, if each block is signed by the transformation we have described, there is a need to ensure that the sequence of blocks is preserved during transmission and this can be done by giving each block a serial number before it is signed. This sequence number can be part of the redundancy if extra redundancy is needed.

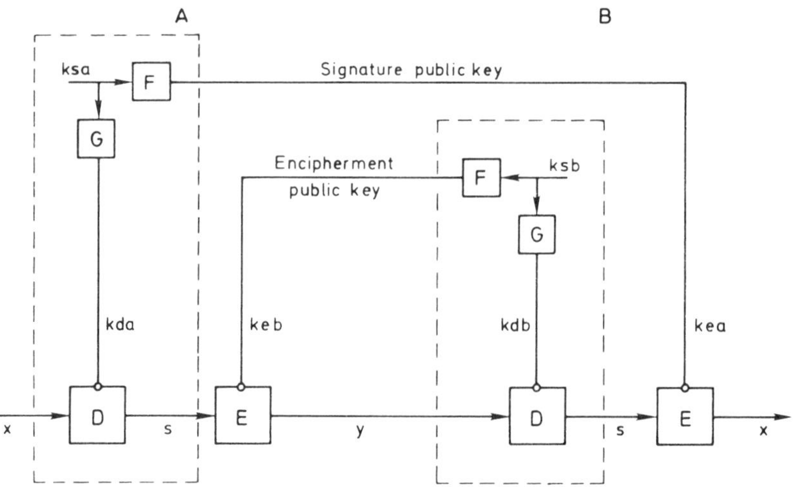

Fig. 9.2 Public key encipherment of a signed message

Signature and encipherment combined

By combining the two transformations of public key signature and public key encipherment, a message can be protected against both unauthorized change and unauthorized reading. That is to say, reading is prevented by encipherment and unauthorized changes are detected when the signature is checked. The two transformations are 'nested' one inside the other. Figure 9.2 shows the principle of the method. A message is being sent from A to B using first the signature transformation then the encipherment. The public and secret keys of both parties are employed so we have labelled the secret keys *kd* and the public keys *ke*, according to their ownership, *kda* and *kea* for the sender A and *kdb*, *keb* for the receiver B. For the signature transformation, A uses his secret key *kda* and then enciphers the signed message by a transformation with the public key of the receiver, *keb*. At the receiving end, these transformations can be stripped off, first decipherment with *kdb*, then the signature transformation with *kea*.

This way of nesting the operations is the most convenient for two reasons. First, it is easier to present the evidence for the signature to a third party, should this be necessary because, after the encipherment has been stripped off, the signed message, its verification and the data used as evidence for a judge are the same as before. Secondly, the signed message, unencumbered by encipherment, can be stored as evidence for use in the future or given to others whom it may concern. If the message were enciphered before signature, the result would be of no value to a third person. Figure 9.2 shows messages travelling from A to B. For a message to pass in the other direction the same set of keys is used but B signs with *kdb* and enciphers with *kea*, then the receiver A deciphers with *kda* and checks the signature with *keb*.

Signature using the RSA cipher

In RSA encipherment, the public key consists of the modulus m and the encipherment index e. The encipherment and decipherment functions can be expressed very simply as

$$y = x^e \quad y^d = (x^e)^d = x \quad \text{(modulo } m\text{)}.$$

Signature transformation and the inverse transformation used for signature checking take the form

$$s = x^d \quad s^e = (x^d)^e = x \quad \text{(modulo } m\text{)}.$$

The RSA cipher easily produces signatures because it is a bijection — it maps from the $m-1$ values 1, 2, 3, ... $m-1$ onto the same set. On the other hand, the knapsack cipher produces a message expansion and this makes it unsuitable for signature. If a 100-bit message transforms into a 208-bit ciphertext, only a minute fraction of the 208-bit messages offered for signature by the inverse transformation will correspond to a 100-bit signature. There have been proposals for constructing 'dense' knapsack transformations so that signatures can be obtained by varying part of the message and making repeated trials. This is unsatisfactory, because it is now known that the security of the knapsack transformation depends on its message expansion property. Shamir[1] proposed a variation of the knapsack method for digital signatures which he called 'a fast signature scheme' because of the much lower computational complexity of the knapsack calculations. Instead of the knapsack sum being a selection of the weights from the knapsack vector, these weights can be multiplied by small numbers within a certain range. But although Shamir's fast signature scheme seems secure, the basis for its security is difficult to establish and Shamir considers it to be an open question. Unless or until this method becomes better understood the only true signature procedure we have, based on a public key cipher, is the RSA method.

Unfortunately, the RSA method presents a computational awkwardness because each set of encipherment keys uses a different modulus for its arithmetic. The generation of each set of keys begins with finding a pair of primes, p and q, using a random seed to make sure these values cannot be predicted by an enemy. In the situation shown in Figure 9.2, the keys used for encipherment and decipherment belong to B and have one modulus which we shall call m_b and those keys used for signature and signature checking belong to A and use the modulus m_a. The result of the first operation using A's secret key *kda* is to produce a signed message block which is virtually a number chosen at random in the range, 1, 2, ... m_a-1. This quantity enters the second operation, which is encipherment using *keb*, the public key of B, an operation in arithmetic with modulus m_b. This creates no problem if m_a is less than m_b but in a general population of users where each determines his own modulus, this can be true only half the time. In the awkward cases, the number resulting from the signature is too large to fit into one plaintext block for encipherment. If a string of blocks is first signed, then enciphered with a smaller modulus, the signed blocks will have to be rearranged into shorter blocks for this purpose. This is rather awkward.

There are several ways to avoid this reblocking problem. It might be thought that all the RSA cipher users in a closed community could accept their keys from a single key distribution authority which employs the same modulus for all sets of keys. But the need to distribute the secret keys to the users makes this no more convenient than the use of a symmetric cipher like the DES. Also, a group of users might get together to find the value of λ, enabling them to attack other user's ciphers or forge their signatures. Given a set of values of $de-1$, which are multiples of λ, forming their least common multiple (lcm) might allow λ to be found.

Kohnfelder[2] suggested an ingenious way to solve the blocking problem. The operations of encipherment and signature can be nested in either of two ways, of which one was shown in Figure 9.2 and the other is shown in Figure 9.3. By choosing either of these arrangements according to the size of their moduli, we can always ensure that the result of the first operation is less than the modulus for the second. Whichever operation has the smallest modulus is taken first. Since both moduli are part of the public keys, the sender and receiver both know which sequence of operations is being used. But it is not immediately clear how the receiver could present evidence to a judge on which the judge could verify the correctness of the signature. The received signature s together with the public key of A and the received ciphertext y will enable the judge to verify the signature transformation, but y contains no obvious redundancy, being apparently just a random number. Furthermore, the judge is not usually interested in a ciphertext; he needs to enforce a contractual obligation expressed in plaintext terms. The solution to this problem is that B also gives the judge the plaintext x which the judge can encipher using B's public key *keb* and thus generate for himself the ciphertext y. If this agrees with the value given him by B, then both the plaintext and its association with the owner of the public key used for signature is proved. If B wants to pass on the signed document to another person whom it may concern, he has to include the plaintext x and a note of his own identity or his own public key so that the transformation from x to y can be made whenever the document has to be verified. In fact the quantity y is not an essential part of the evidence because it is produced as part of the verification process. To summarize, the evidence in this case comprises the plaintext x, the signed, enciphered message s and the public keys of both the sender and the receiver.

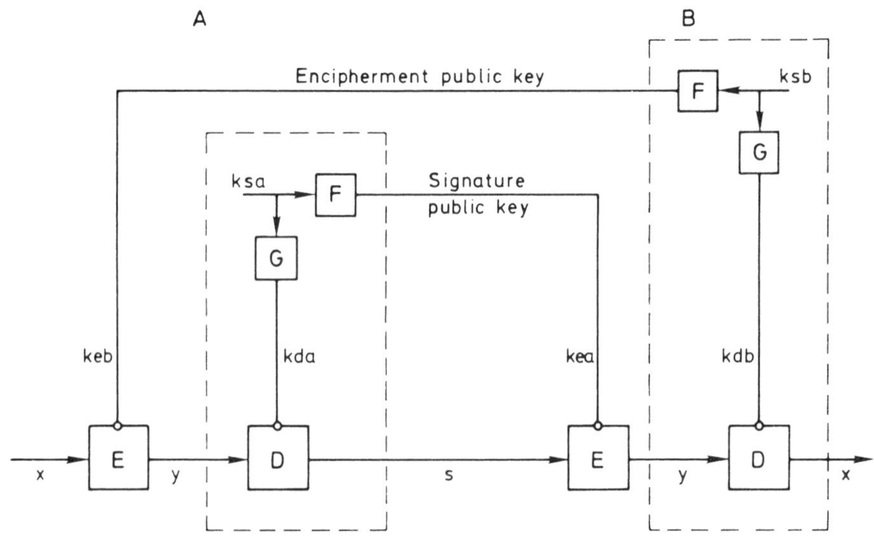

Fig. 9.3 A second way of combining encipherment and signature

In chapter 8 we discussed the possibility of selecting the primes p and q so that the modulus $m = pq$ is very close to a power of 2 such as 2^{512}. It seemed possible that moduli could be constructed with, for example, 112 ones in their most significant bits while still leaving enough choice of primes to prevent any guessing by an enemy. If this can be done, the blocking problem takes a different form. Each plaintext or ciphertext or signed block can be regarded as a number of 512 bits. The output of an encipherment or signature process can be regarded as a random number for which the probability of exceeding the modulus of the next RSA process is very small. For the unlucky cases the text of one block expands to two blocks. It should be possible to develop practical systems of this kind.

The asymmetric use of DES as a signature substitute

The use of DES keys inside a tamper-resistant module (TRM) enables their function to be controlled and allows the DES cipher to be used asymmetrically, rather like public key systems. This was illustrated in the 'Key notarization' method of handling keys, described in chapter 6, page 166.

The asymmetric method of authentication is illustrated in Figure 9.4. All the TRMs are loaded with the same master key, *km*, which is inaccessible to the users. For calculating the authenticators of messages, user A employs in his TRM the key *ka*. This key is generated in the TRM by a random or pseudo-random process and distributed to other TRMs as the enciphered value *kpa* = E*km*(*ka*, A), which is like a public key for sender A. The identifier A is included in the enciphered chain. Any user with a TRM containing *km* can employ *kpa* to check the authenticator of a received message, but the design and the software of the TRM does not allow it to release an authenticator value calculated with a key, like *ka*, which it has received from elsewhere. Thus B can create authenticators with his key *kb* but only check them if the key used is *ka*.

This has some of the virtues of a public key signature because only the owner of the public key *kpa* can 'sign' messages with the secret key *ka*. If a judge is required to settle a dispute, he can do so with the aid of a TRM containing *ka*, the disputed message and its authenticator and an authentic public key, *kpa*, of the sender. The scheme depends on the security of the TRM, its keys and software

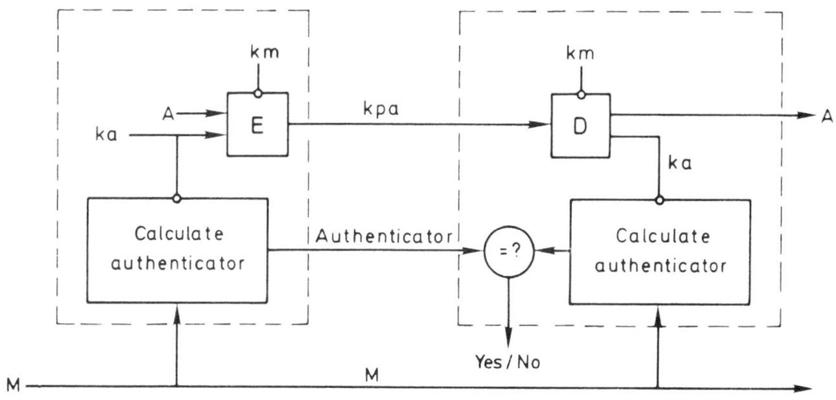

Fig. 9.4 Asymmetric use of DES authentication

and the secure distribution of the common key *km*. It is a workable method for a small, controlled group but not for general public use.

9.3. SEPARATION OF THE SIGNATURE FROM THE MESSAGE

A signature requires sufficient redundancy in the transmitted message that unauthorized changes can, with a high probability, be detected by the receiver. This redundancy can either be part of the message itself, as we have described, or it can be added to the message by appending a signature. When both signature and encipherment are needed, they can be applied either signature first or encipherment first. It is best to apply the signature first (to the plaintext) so that the signed message can be stored for future use or given to a third party for his use. For transmission in secret, the plaintext message with its appended signature can be enciphered by any method; a public key system or a symmetric cipher.

The principle used in generating a separate signature is shown in Figure 9.5. Sender A is transmitting a message *M* and signing it, using his secret key *kda*. In this case the message is not enciphered so the identity of the receiver is unimportant. It could be a broadcast message or a signed document that is open for public inspection. The construction of the signature is shown by the upper line in the figure. A one-way function $H(M)$ is formed using all the bits of the message and then this quantity is transformed by the public key signature method, to form the signature of the whole message. At the receiving end, the message is obtained in plaintext form and, in order to check the signature, it is enciphered with A's public key *kea* in the usual way. This should return the value $H(M)$. The receiver applies the one-way function to the message and this should also produce $H(M)$. If these two values are equal, the signature is pronounced valid and the message is authenticated.

The one-way function must be made public and should be a standard function

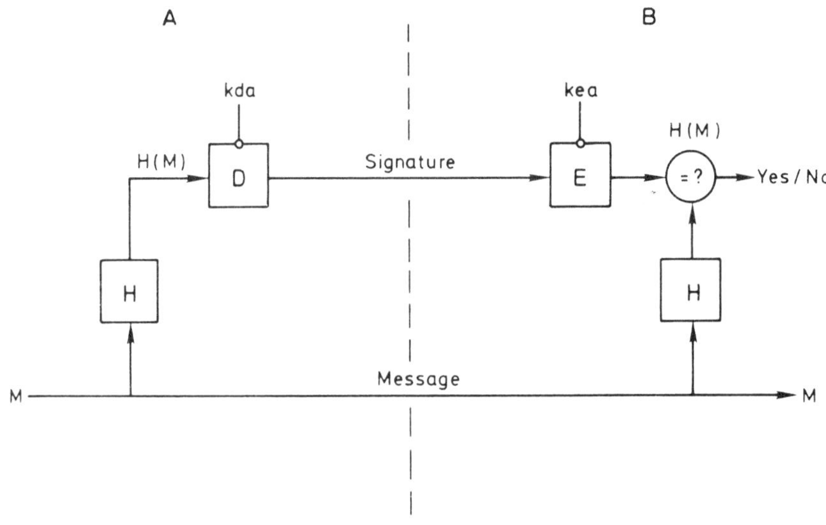

Fig. 9.5 Signature separated from the message

that anyone can use in generating and checking signatures. The value of the one-way function should not be such a large number that it exceeds one RSA block nor should it come from so small a range of values that there is a non-trivial probability of guessing it. We might propose a one-way function of 64 bits, which means that the signature generated from a randomly chosen value has a probability 2^{-64} of appearing to be a valid signature. This probability is small enough to make it worthless to guess at the value of $H(M)$.

An enemy can take an arbitrary signature and calculate, by RSA encipherment with *kea*, the corresponding value of $H(M)$. If he can then find a useful message which fits this value, he has produced a forgery. It should certainly not be possible, given the value of $H(M)$, to enumerate all the corresponding messages and find which matches $H(M)$. A function $H(M)$ would not be effective if there were a way of modifying a valid message such that its $H(M)$ value did not change. Because of the many requirements which the function has to satisfy, choosing a satisfactory function is not easy and, in particular, functions which are suitable as authenticators are not usually suitable for $H(M)$. Suitable functions will be described later.

Encipherment of a signed message presents no new problems because the plaintext with the signature appended can be enciphered as a plaintext chain. In principle the signature provides error checking, since any change to the message or signature (whether enciphered or not) would be recognized when the signature is checked. But sometimes signatures are not checked until someone is ready to take action on the message contents, so there may be an advantage in providing a cyclic sum check as part of the chain, which includes both message and signature.

It is, therefore, possible for the message to have three different forms of checking. A sum check may be used by the communication system to verify the correct receipt of the enciphered chain. This enables the communication system to call for a re-transmission if there have been line errors. But an enemy can easily change the message and reconstruct this communication sum check. Therefore the second level of checking used is a cyclic sum check for the plaintext chain which detects 'enemy activity' on the line. A failure of this check when the communication check succeeds could indicate an attempted active attack but it might also be caused by using the wrong decipherment key. The third level of protection is the signature itself and the purpose of this is to prevent forgery by the intended receiver, that is, changes to the message after it has been deciphered.

It is not strictly necessary to encipher the signature. Figure 9.6 shows an example of a signed message which has been enciphered using a public key system. Such a message is specific to one sender A and one receiver B and uses both their public keys. The plaintext signature is transmitted and the message is enciphered by the receiver's public key *keb*. After decipherment, checking the signature proceeds as before. This method of operation (which is not necessarily recommended) would allow an enemy to substitute messages and signatures but he would gain nothing because the probability of a signature match is no better than the random case. If the receiver finds that the signature and the message produce matching $H(M)$ values using the public key *kea*, then he can assume that the correct signature arrived for that particular message and the message came unaltered from A.

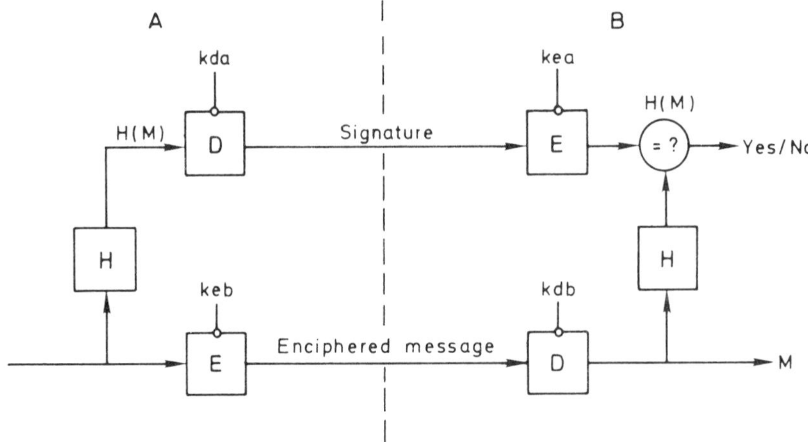

Fig. 9.6 Encipherment of a message with a separate signature

There is little practical advantage in sending the signature in clear as Figure 9.6 suggests. Since RSA encipherment is more complex than the DES, it will often be the case that the message with its appended signature is treated as one message for encipherment by DES. The advantage of the signature based on the one-way function is that the RSA signature process is used only once to sign a message of any length.

Falsifying a signed message by the 'Birthday' method

If the one-way function has 64 bits, it would seem impossible to find a bogus message which has the same function as a genuine one. The probability of any particular bogus message having the required function is 2^{-64} and it is impossible to make 2^{64} trials. Searching a variety of messages to find one that fits is impossible in practice. Like many attacks on cryptographic systems, this one can be changed to a rather different plan which becomes within the bounds of possibility making use of a variety of the Birthday problem. This problem is described in an appendix to this chapter, page 279.

The new plan is as follows. The enemy must be able to present a document of his own choosing for signature by the person who is to be tricked. With the aid of this document he will be able to create another document, damaging to that person, which will have the same signature. By signing an apparently harmless document which has been chosen by the enemy, one could be giving that enemy a method for forging a harmful document which has the same signature. Suppose that the enemy can construct, by a systematic method, as many as 2^{32} different documents all of which are likely to be signed without question. Suppose also that he can generate, by a similar systematic method, 2^{32} documents of the kind he would like to use as forgeries. For each of these 2^{33} documents he calculates the value of the one-way function $H(M)$. Assume that every one of these values is a random choice out of the 2^{64} possible values. Then the Birthday problem shows us that there is a probability $1-e^{-1}$ that one of the values from the good documents equals one of the values from the bad documents. If the enemy is unsuccessful after all these attempts, he can go on trying and he should succeed without many more than 2^{32} documents of each kind.

The reason for the high probability is essentially that there are 2^{64} different ways of comparing a value from the first set with a value from the second set and though each of these has a probability 2^{-64}, there is a reasonable chance that one of these 2^{64} comparisons will give equality. Of course the method would be worthless if it required 2^{64} comparisons, but there is an easier method which is to sort the whole collection of 2^{33} items according to the value of $H(M)$, after which sets of coincident values can be examined to find equalities between values taken from both sets. Sorting would take of the order of 33.2^{33} steps. Though 2^{38} steps in a calculation take a long time or a lot of processing power, it is certainly feasible with today's technology.

The plan will work only if 2^{32} message variants can be found, but this is quite easy. Figure 9.7 shows a letter of a kind that a trickster might use to give himself a signed authority which will enable him to steal money belonging to his employer. This letter has 37 words or phrases where a choice can be made among two

Dear Anthony,

This letter is	to introduce	you to	Mr.	Alfred	P.
I am writing		to you	---		--

Barton, the	new	chief	jewellery buyer for
	newly appointed	senior	

our	Northern	European	area	He	will take	over
the		Europe	Division.		has taken	

the	responsibility for	all	our interests in
---		the whole of	

watches and jewellery	in the	area	Please	afford
jewellery and watches		region.		give

him	every	help he	may need	to	seek out	the most
	all the		needs		find	

modern	lines for the	top	end of the market.	He is
up to date		high		

empowered	to receive on our behalf	samples	of the
authorized		specimens	

latest	watch and jewellery	products,	up	to a
newest	jewellery and watch		subject	

limit	of ten thousand pounds sterling.	He will	carry
maximum			hold

a signed copy of this	letter	as proof of identity.	An order
	document		

with his signature, which is	appended	authorizes	you to
	attached	allows	

charge the cost to this company at the	above	address.	We
	head office		

fully	expect that our	level	of orders will increase in the
-----		volume	

following	year and	trust	that the new appointment will	be
next		hope		prove

advantageous	to both our companies.
an advantage	

Fig. 9.7 A fraudulent letter with 2^{37} variations

alternatives and, as far as we can tell, all these are independent so there are potentially 2^{37} different letters which would serve the enemy's purpose. These letters can be generated systematically by a processor which calculates the one-way function of each message and stores it, along with a note of the 37-bit number which specifies the variant. With another bit to distinguish the functions from good and bad messages, the items which have to be sorted contain a total of $1+37+64$ bits.

A simple defence against this plan is to place into each signed document a number which is chosen at random just before the signature is generated. All the effort of the enemy would then be wasted because he cannot predict the $H(M)$ value which will actually be generated by any document. A random number of 32 bits is sufficient to frustrate the Birthday method of forgery. An alternative precaution, very similar in its effect, is to introduce a randomly chosen initializing value into the one-way function and include this value in the block which is to be signed. A signature method of this kind has been described in detail.[4]

A one-way function for signature or authentication

The function $H(M)$ reduces a message of arbitrary length to a number of a convenient and standard length for the purpose of signature. It could equally well be used as a preparation for calculating an authenticator. It can even be used by itself to authenticate a large message. Suppose that M is to be transmitted from Ann to Bill by a communication path that might be subject to active attack. It can only be authenticated if another, reliable, communication method is available. In the case of an authenticator this is the path by which the secret key is transported. But in another case, the message might be too long to read over by telephone but Ann and Bill, knowing each other well, could communicate authentically but not secretly by telephone a short number, say 16 hexadecimal characters or 64 bits. They use $H(M)$ by telephone to verify M sent by insecure network. The same kind of function will serve in all these applications.

The essential quality of $H(M)$ is that it is a one-way function of M. It employs no key and is public. For a given message, the enemy can calculate $H(M)$ and the security can be broken only if he can find another message M' such that $H(M') = H(M)$. Since the range of $H(M)$ is much smaller than that of M there must be a great number of solutions, but there is no systematic way to find one.

It is well known that $\mathrm{E}k(x)$ is a one-way function of k but not of x. Nor is it a one-way function of k and x jointly. There are many easy solutions of $\mathrm{E}k(x) = y$ for any given y when k and x are independent variables. Therefore the use of $\mathrm{E}k(x)$ to construct one-way function of a message of indefinite size needs care. A good solution is illustrated in Figure 9.8. It is composed of a number of 'units' of the form

$$y = \mathrm{E}k(x) +_2 x.$$

Given y, this equation cannot be solved for k and x jointly. Since k is a key, y is obviously a one-way function of it. It is a one way function of x for all k because of the way x enters twice into the equation — assuming the $\mathrm{E}k(x)$ is a good cipher.

In the arrangement of Figure 9.8, the message M is split into blocks of 56 bits, the M_i, and I is treated as an initializing value, but I could be part of the message

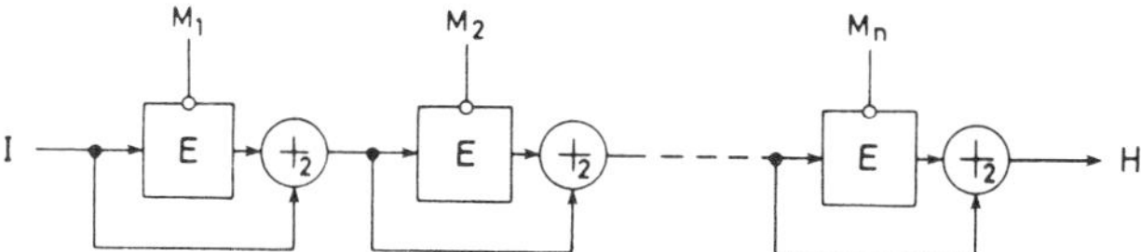

Fig. 9.8 Proposed one-way function of a message

too. The I value can be used as the value chosen randomly at the moment of signature to frustrate the Birthday method of attack described in the previous section.

This same attack could have been used on the $H(M, I)$ function if the feed-forward paths had been omitted in Figure 9.8. Because the calculation could be done in reverse, values of H and I could be matched by half-messages with about 2^{32} variations. A true one-way function of both variables is needed.

There is a need for a standard method of calculating the one-way function and assembling the results in a block for RSA signature. Reference 4 gives an example of the kind of standard that is required. The RSA block should contain some redundancy, such as a set of zeros, so that if the encipherment of the signature fails to restore these values, the fault can be located either in the signature or the public key being used to test it. If the correct redundancy appears after enciphering the signature, there is confidence that the one-way function is correct. If it then fails to match the function derived from the received message, the fault must lie in the message.

9.4. THE FIAT–SHAMIR PROTOCOLS FOR IDENTIFICATION AND SIGNATURE

A new type of signature was introduced by Amos Fiat and Adi Shamir in 1986. This was an extension of an earlier idea, which though it did not provide a signature, allowed one party to authenticate itself to another with the certainty that, during the procedure, no information had been passed between them. Such 'zero-knowledge' protocols are of some theoretical interest because they can provide a proof (assuming that factorization of large numbers is difficult) that there is no hidden channel of communication. For some applications, there are other means of authentication that are satisfactory, given the usual assumptions about secure cipher algorithms. The new protocols have an additional interest because they use mathematical ideas similar to those of public key cryptography yet the complexity of carrying out the calculations is much lower. For this reason they have been adopted for use in smart cards, where the computational power, and sometimes the storage, is limited.

We shall first describe their underlying mathematical basis then describe examples of Fiat–Shamir protocols for identification and signature. These will be the earliest versions of the protocols which were published, suitable for our purposes because of their relative simplicity. Since that date, several important variations have been discovered which reduce the computational load even further and can optimize the protocol by trade-offs between the various computational tasks and the sizes of keys. We shall not attempt to describe all these variations but give a general idea of the performance they can achieve.

Mathematical basis of Fiat–Shamir protocols

All the calculations performed in these protocols are in finite arithmetic with a modulus m which is the product of two primes p and q. In this respect it resembles the arithmetic of the RSA cipher and signature. These protocols are based on the difficulty of finding a square root when the factorization of m is unknown. In what follows we shall assume that p and q and m are large numbers. If we take all numbers between 0 and $m-1$ and square them, modulo m, we will get only approximately $m/4$ different results. It follows that most of the numbers in this range do not have square roots but the one quarter of them which do have square roots can be used for our purpose and these are called *quadratic residues*.

In nearly all cases a quadratic residue has four different square roots. If we know that x is the square root then clearly $m-x$ is also a square root; it can be thought of as the negative of x. The four square roots of the quadratic residue therefore occur in two pairs of the form x, $m-x$.

It is very easy to show that finding a square root in this modulo m arithmetic is a difficult problem. Suppose we could always find at least one square root of x. Starting with a random number t we could form $x = t^2$ then find the square root s such that $s^2 = x$. It would follow that $s^2-t^2 = 0$ hence $(s+t)$ $(s-t)$ is a multiple of m. Since s was chosen at random, there is a probability of one half that $s=t$ or $s+t=m$, which would make the above identity rather trivial. But in 50 per cent of cases it would provide a factorization of m. Therefore if it is always possible to find a square root without too much computation, only a little more effort would enable one to factorize m. Since factorization is known to be a difficult problem we can see that finding a square root in modulo m arithmetic (where $m = pq$ and the factors have been chosen judiciously), is at least as hard as factorization.

On the other hand, if the factorization of m is known, finding a square root is relatively easy. We can use the Chinese remainder theorem, having found the square root of x modulo p and modulo q. This can be made particularly simple if p and q both have the form $4m+3$, where m is an integer.

To summarize, the mathematics on which the Fiat–Shamir scheme is based is simply the great difficulty of finding square roots modulo m and the relative ease of finding square roots if the factorization of m is known. The one technical difficulty presented by this method is that square roots can only be found if the number is a quadratic residue. We can overcome this problem, as we shall see, by making use of the fact that about one quarter of the numbers are quadratic residues and that these residues are distributed in a quasi-random fashion.

The basic identification scheme

An identification scheme concerns two parties, A and B, where A wishes to prove his identity to B. He could do this by giving a password, but that would present B with the means of impersonating A when identifying himself elsewhere. In a true identification scheme we wish A to be able to identify himself to B while, in the process, not giving B any information which might enable him to impersonate A. The beauty of the scheme to be described is that it can be proved that no such knowledge passes to B. By the use of cryptography with symmetric algorithms we can see how the identification can be performed; we described this in section 7.3. These schemes based on symmetric algorithms do not have the zero-knowledge property which is an attraction of the Fiat–Shamir schemes.

The protocol is based on the repetition of a simple, three-message exchange which we shall now describe. We can think of the party A as a smart card which holds a secret value enabling it to be identified to any other party B (typically a terminal) which can operate the protocol. The type of protocol is called *identification based* because A (the card) stores a secret parameter related to its own overt identity. Let I be the identity of the card A, for example the name of its holder, the card number, account number or a combination of these. In order to reduce this information to a usable size we can form a one-way or 'hash' function f and produce $\mathrm{f}(I)$. The card will hold I and make it available to anyone, but it uses $\mathrm{f}(I)$ for its calculation. In fact, we really need a number which is a quadratic residue so we actually use $\mathrm{f}(I, c)$, where c is a small number concatenated with I and some different c values can be tried until $v = \mathrm{f}(I, c)$ is a quadratic residue. The card makes c available along with I for anyone who needs it to verify the protocol.

A central authority, such as a card issuer, knows p and q and can therefore easily calculate the square root of v, namely a solution of $u^2 = v$, modulo m. The u value is the secret held in the card A to enable it to prove that it has been issued by the central authority, who alone knows p and q and can thus find the square roots.

The only assumption about the party B which is to verify the identity of the card is that it knows an authentic value of m and has received from the card A the value of I and c so it also can calculate v, since the hash function f is public. The problem is to find a dialogue between A and B by which A can prove that it knows the square root u without revealing anything about the value of u.

The dialogue which is repeated is as follows and all the calculations are modulo m.

1. A finds a random number r and sends a message to B containing the two values $x = r^2$ and $y = v/x$. B can easily check that $xy = v$.
2. B makes a random decision and sends one bit to A, $e = 0$ or $e = 1$.
3. If $e = 0$, A sends r to B.
   If $e = 1$, A sends $s = u/r$ to B.

B (typically the terminal) can now easily check that either $r^2 = x$ or $s^2 = y$, according to its earlier choice of e.

Clearly the card A must *not* send both r and s to B because it could then construct $u = rs$ and would know the secret of the card. A shows its ability to provide either r or s and if it can do this whenever required, this indicates that it really knows both. It therefore knows u. By leaving the choice between r and s to the 'challenger' B, any possibility of finding a square root by precalculation, for example r as a square root of x because x was calculated as r^2, is eliminated. Either one of the two numbers r and s could be produced in this way, but not both since $xy = v$ is a property which can be checked by B in advance of making its decision.

In a single cycle of this three-message exchange, a card A which does not really know u could be lucky enough to win. Its chance of masquerading as A by good luck is one in a million if the dialogue is repeated twenty times.

All the calculations required in this protocol are relatively easy, multiplications and divisions rather than the exponentials of the RSA method. This makes it, computationally, very much easier than the use of the RSA which requires, on average, 768 multiplications for a full-sized exponential.

In many applications, the one in a million probability of forgery is sufficient to deter cheaters. The most inveterate gambler would be unlikely to offer a forged passport or use a forged credit card with such a small probability of success and a high probability of suffering a penalty. In those cases where the reward for success might be so high as to tempt someone, this probability can be lowered by using more cycles of the three-message dialogue.

The more general Fiat–Shamir identification protocol uses, instead of a single value v derived from I, a vector of k such values. In place of the random bit chosen by B is a random binary vector of k bits. The case we have just described, originally put forward by Shamir, is $k = 1$.

In the original scheme, the probability of 'masquerade by luck' is 2^{-t} where there are t cycles of the three-message dialogue, because each dialogue uses the binary choice. The improved scheme has k binary choices per cycle and achieves a probability 2^{-kt}, so kt can be smaller than t was in the original scheme. For example, four cycles of dialogue with five binary choices give the same low probability (one in a million) as before.

The preparation by the card issuer is more complex because there are k different quantities $v_j = \mathrm{f}(I, c_j)$ for different small integers c_j which generate v_j values that are quadratic residues, modulo m. The authors suggest trying $c = 1, 2, 3 \ldots$ and using the first k values which make v a quadratic residue. The card stores the secret values s_j such that $s_j^2 = v_j^{-1}$ modulo m. The card also holds its identifying information I, the modulus m for its calculations and the values c_j corresponding to each of the s_j. Before the main dialogue begins the card will send the terminal I and the c_j values so that it can generate v_j for $j = 1$ to $j = k$, ready for the identification procedure.

Each cycle of identification has three steps, as before:

1. A finds random r and sends $x = r^2$ to B.
2. B makes its k random binary choices and sends a binary vector $e_1, e_2, \ldots, e_k$ to A.
3. A sends

$$y = r \prod_{j=i}^{j=k} s_j^{e_j}$$

to the terminal.

(Thus the card takes the set of secret s_j values selected by the terminal's random choice, multiplies them together, multiplies by r and sends this quantity y to the terminal.)

4. The terminal's check of these relations is obtained by squaring the equation, giving

$$y^2 = r^2 \prod_{j=i}^{j=k} s_j^{2e_j}$$

Since $r^2 = x$ and $s_j^2 = v_j^{-1}$ this can be written

$$x = y^2 \prod_{j=i}^{j=k} v_j^{e_{ij}}$$

This check can be carried out by the terminal, using the set of v_j it has already calculated.

The Fiat–Shamir identification scheme is very suitable for a smart card application, being compatible with the storage and processing ability of these devices. Terminals or other units which have to recognize the genuineness of the presented identification data I merely need the authentic value of the modulo m.

Fiat–Shamir signature scheme

Dialogues of the kind we have described can convince B that the credentials presented by A are genuine, so they can be a preparation for a valuable transaction. But if the transaction goes wrong in some way and a dispute occurs, how can the 'challenger' B present evidence to an arbitrator that he did indeed identify A by the chosen dialogue? A transcript of this dialogue has no conviction whatever because it could be made up in such a way that the choice of the 'random' bits was always the 'lucky' one. Essentially, a signature which can be verified afterwards by an independent arbitrator must be produced entirely by A, not by means of a dialogue with B. Then the signature can convince not only B but, subsequently, anyone else who has the modulus m.

We shall describe how A generates a signature for a message M. Again, this scheme is identity-based and if the identification data is I, the vector of k values v_j are formed as $v_j = f(I, c_j)$ where a set of small integers c_j are chosen which will make each v_j into a quadratic residue. When presenting its identification data I, the values of c_j must also be given so that whoever verifies the signature can reconstruct the vector v_j. As before, the signer A stores the secret values s_j such that $s_j^2 = v_j^{-1}$, modulo m.

The signature process by A has the following three steps which represents the t cycles of operation we met in the identification process, but here they are all carried out together.

1. Find t random values r_i and calculate $x_i = r_i^2$.
2. Compute a one-way function $f(M, x_1, x_2, \ldots, x_t)$ and use the first kt bits of this result as the values of a matrix e_{ij} for $i = 1, 2, \ldots, t$ and $j = 1, 2, \ldots, k$.
3. Calculate the signature which is a vector of t values as

$$y_i = r_i \prod_{j=i}^{j=k} s_j^{e_{ij}} \qquad \text{for } i = 1, 2, \ldots, t.$$

The product is taken over all the k values of s_j but only those are selected for which $e_{ij} = 1$.

At the end of this process the signature is complete and can be included, along with the message M and sent to any verifier, such as B who will be able to check the signature, knowing only the modulus m. In order to perform this verification

B needs, in addition to the message, the identification data I, the set of small integers c_j, the matrix e_{ij} and the signature vector y_i.

In order to verify the signature, B must first calculate the vector $v_j = f(I, c_j)$ for $j = 1, 2, \ldots, k$; next, the verifier calculates

$$z_i = y_i^2 \prod_{j=1}^{j=k} v_j^{e_{ij}}$$

If the calculations have been correct, these z_i equal the x_i used by the signer A, in the same way that we observed for the identification procedure. The final step is to calculate $f(M, z_1, z_2, \ldots, z_t)$ and compare its first kt bits with the given e_{ij}. If they match, the signature is verified.

When using this procedure, a suitable value of kt must be chosen to prevent forgery by trial and error. In this case, a forger could try a large number of values of signature until eventually he gets agreement for the kt bits. It would therefore be unsafe to use $kt = 20$ because a million trials are quite practicable. Fiat and Shamir suggest $kt = 72$ for example $k = 9$, $t = 8$. This would mean that the stored secret key s_j occupies 576 bytes and the signature data (which are principally e_{ij} and y_i) occupy 521 bytes. The computation uses 44 modular multiplications compared with an average of 768 to generate an RSA signature and, say, 17 to check it with a small exponent such as 65537. Compared with the RSA signature there is a saving in computation by a factor of about 18 and an increase in the size of the secret key by a factor 9 and in the size of signature by about 8.

The development of the Fiat–Shamir protocols has continued and there are further tradeoffs which can reduce the size of the signature at the expense of increasing the size of the secret key, while at the same time reducing the computational task of signature checking. A further tradeoff is possible which reduces the computational task of signature generation at the cost of increasing the task of checking. The characteristics of some of these methods are shown in Table 9.1 and compared with RSA signature. First we give the methods described

Table 9.1

Method	Size in octets			Computing load in full-size multiplication	
	Public* key	Secret key	Signature	Generate	Check
RSA	64	64	64	768	2†
Fiat–Shamir:					
$k = 9$ $t = 8$	6	576	521	44	44
$k = 64$ $t = 1$	40	4096	72	33	33
Small v:					
square root	104	4096	72	33	1
cube root	64	4096	72	34	2
16th root	76	1024	72	36	4
Small s:	1024	128	72	22	66

* Includes the specification of c values or choice of small primes.
† Public exponent is 3 (cube and cube root).

above, then methods in which the vector v_j is reduced in size and in addition those in which the square root is replaced by a cube root and a 16th root. Finally, there is a method with a 16th root and small values of s_j. The precise values of the entries in this table depend on details of how the computation is carried out and how such values as the c_j values are represented. They may therefore be arguable, but the table shows the correct trends. All the methods show a clear improvement over RSA in the matter of complexity of key generation at a cost in key sizes.

9.5. SIGNATURES EMPLOYING A SYMMETRIC CIPHER

There have been many proposals for signature methods which use a symmetric cipher. The principal motivation seems to be a belief that the security of public key ciphers is less well founded than that of a 'certified' symmetric cipher such as the DES. This is a matter of opinion. It might also be thought that the computational load of the RSA cipher is a disadvantage but, as we shall see, the signature methods based on DES have a number of disadvantages. We will describe the principles of DES-based signatures but do not recommend them, since we believe that their inconveneince greatly outweighs any disadvantage of the RSA signature method.

The scheme we describe was invented by Lamport and described in 'New directions in cryptography' by Diffie and Hellman.[5] It is sometimes called the 'Lamport-Diffie' signature. The principle is best described by showing how it could sign a 1-bit message. For both of the possible values, 0 and 1, of this message, random values x and k are chosen and the DES algorithm used to calculate the corresponding values $y = \mathrm{E}k(x)$. Among the six quantities, these relations hold

$$\begin{aligned} y_0 &= \mathrm{E}k_0(x_0) \\ y_1 &= \mathrm{E}k_1(x_1). \end{aligned}$$

The signer makes these calculations and reveals, as his public key, the values x_0, y_0, x_1 and y_1. He keeps the values k_0 and k_1 as his secret key. This completes the computation of the keys. When the message is sent, if the value of the message is 0, the signature is k_0, and if the value of the message is 1, the signature is k_1. The receiver can check the signature by enciphering with k_0 or k_1 as key the appropriate value of x and checking whether the result equals the corresponding y value. It is impossible for an enemy to discover the k value from published x and y values assuming that the DES cipher is immune to known plaintext attack. The signature, consisting either of k_0 or k_1, verifies that the message came from the sender who had previously computed and revealed the x and y values.

This is a big expenditure of data just to sign a 1-bit message. The public key contains 256 bits and the signature 56 bits. Supposing that a one-way function $H(M)$ of 64 bits were to be signed, the signature would contain 3584 bits and the public key 16384 bits.

When the signature has been used by the sender, the corresponding public key values are worthless for any future signature. Half of the previously secret k values have been revealed so their corresponding x, y values cannot be used again. The other k values also cannot be used because if they were revealed they could be used by an enemy to sign a falsified first message. Therefore, each message requires the use of a completely new set of public key material. A frequent sender of signed messages must publish a large quantity of key material and these data are used up as he produces signatures. Before all the key material has been consumed he must publish a new batch and use one of the existing sets of public keys to sign it. In this way the continual publication of new key material can be authenticated by signature. This form of signature is heavy on storage capacity and on the use of communication facilities.

The property being used in the Lamport-Diffie method is that y is a one-way function of k. This enables y to be revealed and k used later to prove that the originator of the message was the person who knew y, since only he could have employed the one-way function to calculate y. The purpose of the x values is to ensure a good supply of one-way functions based on the DES.

Rabin's signature method

A different method of using a symmetric cipher was invented by Rabin.[6] This method requires the sender of the message to take an active part in signature validation. It is illustrated in Figure 9.9. The sender chooses at random an even number of DES keys, for example the 36 keys $k_1, k_2 \ldots k_{36}$. Having computed the one-way function H of the message to be signed, he generates as signature the 36 values

$$S_i = \mathrm{E}k_i(H) \quad \text{for } i = 1, 2, \ldots 36.$$

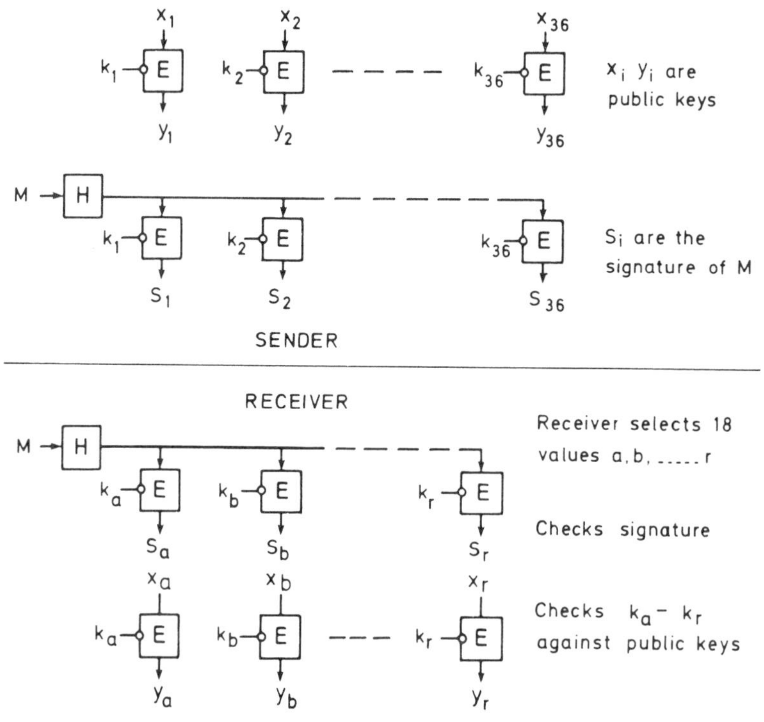

Fig. 9.9 Rabin's signature method

When the receiver wishes to verify the signature he chooses any 18 values of i and asks the signer to reveal this subset of the keys. The receiver can then verify that all 18 of the revealed keys match the value of $H(M)$ according to the equations given above.

If the sender had been obliged to reveal all the keys, the receiver could then forge message corresponding to these. If there is a dispute, the receiver produces the message and the received signature while the sender reveals to the judge all the keys and the validating material. The rule applied by the judge is that if 19 or more of the keys in the signature are correct the receiver wins and the signature is held to be valid but if 18 or less of the keys in the signature are correct, it is assumed to be a forgery by the receiver. A sender who wanted to beat the system could put exactly 18 correct values in the signature and hope that these were the ones chosen by the receiver for validation. The validation would work and the receiver would believe he had a true signature. Later, the sender would challenge it and show that only 18 of the signature values were correct, thus winning his case. The probability of getting the right subset of 18 out of 36 is 10^{-10} and the risk is too great for this trick to be useful. If the sender tried to improve the odds by making slightly more than half of the values correct, the decision after a dispute would go in favour of the receiver. Thus the strategy in which exactly half of the keys are revealed at validation seems to be optimal. Any even number of keys can be used and 36 was chosen because it gave a sufficiently low probability of fraud by the sender. The 36 component signature contains 2304 bits and the data provided by the signer for verification is 1008 bits. But the authentication of the 18 keys supplied at the receiver's request must be based on previously authenticated data, constituting the public key of the sender. For this purpose, x and y values are needed for each of the 36 keys, which occupies 4608 bits. There is some economy compared with Lamport and Diffie's method. The public key material is 'consumed' by using it for a signature and this is the snag in signatures based on a symmetric cipher.

The practical value of signatures based on symmetric ciphers is at present a matter of opinion since there is little practical experience of using any signature method. Our opinion is that the need to renew the public key material constantly, coupled with the large amount of data in each public key, makes these methods very awkward to use. For verifying signatures by the RSA method, most receivers of messages would be content to use a stored value for the sender's public key, referring to a key registry from time to time either to receive new keys or verify old ones. Methods using symmetric ciphers require too much organization and effort to handle the public key material.

Arbitrated signatures

It is possible to use a symmetric cipher for digital signatures without having the data expansion, large key size and 'one-time key' properties of the methods we have just described. In order to do this, the two parties, sender and receiver, must have an arbitration service which they both trust. Figure 9.10 shows the principle. As usual, it is not the whole message but a one-way function, $H(M)$, that is signed, together with the initializing value I, if one is used. The sender

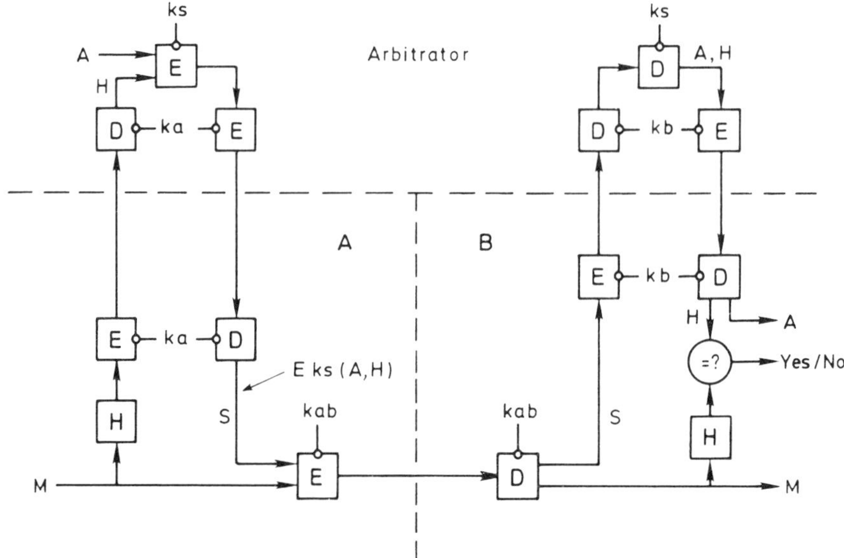

Fig. 9.10 Arbitrated signature

obtains the signature by sending $H(M)$ to the arbitrator, using a protected channel with a key *ka* which is known to the sender and the arbitrator but to no others. Along with the one-way function, enciphered as E*ka*(H), he sends the plaintext value A which identifies him to the arbitrator so that the arbitrator can extract from store his own key *ka*. The arbitrator deciphers the message to obtain H, concatenates it with A and enciphers this pair of values with its own signature key *ks*. The value of *ks* is known only to the arbitrator and is used only in a TRM where it is carefully protected. The signature S = E*ks*(A, H) is returned to the sender, enciphered under the key *ka* with which the sender and arbitrator communicate. At the sender, S is recovered and the signed message (M, S) is sent to the receiver. If necessary, it is enciphered by a key *kab* known to sender and receiver, as shown in the figure.

The checking of the signature also employs the arbitrator. The receiver sends the arbitrator the value of S enciphered under a key *kb* known to himself and the arbitrator, together with the plaintext value, B, so that the arbitrator can extract the correct key for decipherment. The arbitrator deciphers the value S with his secret signature key *ks* and returns the values (A, H) under suitable *kb* encipherment to the receiver. Now the receiver is provided with two quantities, A which verifies the source of the message M and H which can be compared with the one-way function $H(M)$ obtained from the message itself. If these match the message is assumed to have been delivered unchanged and to have come from the sender A.

There are many variants on this procedure. The channels between the arbitrator and the participants A and B could be protected by integrity checking (message authenticator) instead of encipherment if the application allows this. The signature S could be formed by a one-way function instead of encipherment, then the verification process must be given the values of A and H and return the signature for comparison with the S value derived from the message. Replay attacks must be prevented by sequence numbering or time and date stamping. The essential requirement is to detect any active attacks on the communication channels and Figure 9.10 shows just one way to do this.

Both sender and receiver depend on the integrity of the arbitrator and the arbitrator must keep a set of the keys, like *ka*, *kb*, for secure communication with all the users that it serves. The method we have described is substantially that described by Needham and Schroeder[7] where the arbitrator is called the 'Authentication Server' a term which has been taken up by other papers which comment on Needham and Schroeder's methods. Since *ks* is an exceptionally valuable secret, it should probably be restricted to one arbitrator or authentication server, yet the number of users this could effectively serve is limited. Messages can pass between arbitrators using keys which they hold in common, so if the receiver B calls on a different arbitrator to help verify the signature, that one can send the request back to A's arbitrator for servicing. In this way a wider scheme could be organized, at the cost of greater complexity and more communication. Probably, this method is most suitable for a large, but closed, organization rather than universal public service. The dependence on the arbitrator is extreme so there must be provisions for restoring the service after a failure, consequently for back-up values of *ks* and the other keys the authenticator holds.

Other schemes for arbitrated digital signature have been described, for example by Matyas[8] and some of them involve using the arbitrator as part of the communication path between A and B. The one we have described is closest to the usual concept of a signature and, like a public key system, it allows a signed message to be verified independently by a number of receivers or by a judge, assuming that all these parties have established their communication keys with which to communicate securely with the arbitrator.

All signature schemes depend on a trusted service if they are to be used widely. There is a difference between the dependence on the arbitrator and the dependence on a key registry. There can be several key registries, all maintaining the same data and authenticating it with different public keys. Someone who wants to be very sure about a signature can request the public key from several registries and compare the result. Neither failure nor subversion of any one registry will upset this system.

If, in a similar way, several arbitrators are to be provided they must have the same secret key *ks* so that a signature can be verified by any of them. Replicating the secret key weakens its security, so the usefulness of replication is doubtful.

9.6. THE PRACTICAL APPLICATION OF DIGITAL SIGNATURES

The arbitrated signature is easy to implement because of the availability of fast DES devices but it requires a trusted arbitrator, a reliable and always available arbitration service and the distribution of individual secret keys for communication between the arbitration service and those it serves.

Signatures of the Lamport-Diffie or Rabin kind seem very impractical. The need to provide a continual supply of authenticated public key material seems particularly troublesome. But it is of some theoretical interest that such signatures exist.

Signatures by the RSA method have uniquely useful properties which could greatly improve the use of electronic messages in commerce. After a small expenditure in establishing authentic public key values the users can provide signed payments, contracts, receipts and all the other legally binding documents and messages which are part of normal business. A single, signed message can pass through a chain of people or organizations, each of which needs to take some action and can verify the signature.

Signatures can use the same public keys that are used for encipherment. A key registry is the source of authentic public keys and the signature process is used to authenticate the information coming from the public key registry. Therefore, the whole system depends on the public key of the key registry itself and this important datum can be established by multiple methods of publication which would foil any attempt to falsify it. No doubt the registry's public key will be changed from time to time but there is no problem in maintaining an historical record of old keys.

In a large network which uses digital signatures for the authentication of important messages, the administration of the public key registry must be designed very carefully. The most difficult problem is that of registering new public keys, which means giving the owner of the key in question a certificate signed by the registry which will be accepted by other participants as guaranteeing the genuineness of the public key. The most obvious way to do this is to require participants to bring their public keys to the registry with sufficient documentation to authenticate the person presenting these keys on behalf of the participant. But it has been said that, if a representative of the participant must be present at the registry, it will be equally easy to distribute secret keys at that time and build up an authentication scheme similar to an arbitrated signature. A digital signature would still have advantages in reducing the need for holding and carrying secret keys but the advantages would be even greater if public key registration could safely be carried out remotely.

The essential feature of a remote registration procedure is to allow a period during which the owner of the key can verify that it has been correctly registered before the registration becomes effective. Matyas[9] has defined and analysed such a procedure which goes in three phases. Suppose that a person P wishes to register a new key with the registry R. They must first agree the timetable for registration by fixing the three instants at which the phases of registration begin. The first phase allows temporary registration of the key, the second phase provides for its review (and potentially its cancellation) while the third phase is the period during which the key is effective. This may have an end point at the expiry time and date, which can be specified in the certificate.

During the temporary registration period, and before the next phase begins, P must send to the registry the public key, identifying data and a time variant quantity X. R responds by returning all this information together with a digital signature, which forms a temporary certificate. If there is an intruder, R may be presented with more than one such message, in which case it responds normally to the first one but gives a 'failure' response to the second one, the failure response also being signed, and discontinues the registration.

When the second phase begins, registration is no longer allowed but the participant can cancel the registration by sending a message with his identity and a time variant quantity Y. A response is sent by R confirming the cancellation and returning the identity and Y. Cancellations will be accepted until the third phase begins. During this phase, P must check the temporary registration and cancel it if it is false.

As soon as the third phase begins the registration is effective, assuming that only a single registration has been received and no cancellation. It is now possible to enquire for the value of the public key and receive in response the certificate containing the identity and the public key value signed by R. This is the effective certificate and may be used as evidence of the value of the public key.

If the procedure fails at any stage it must be restarted by P, with a new time schedule agreed with R.

An intruder can prevent the registration of a key by putting forward a false registration during the temporary registration period arranged for P. A false key could be registered if P is prevented from sending a cancellation message during the review period. To avoid this, P may have to find other means of communication and the procedure should allow these to be used for cancellation if the normal channels are being blocked. Clearly, no remote registration procedure can be entirely secure, but the Matyas procedure is convenient and effective where large numbers of terminals have to be registered. Where the registration of a public key justifies more effort to ensure its security, multiple channels between P and R and the attendance of representatives to authenticate the key directly will provide all that can be achieved.

Public keys used by individuals will change and it will be difficult to keep an historical record of all of them. There is a requirement to give a more permanent basis to a signed document when it has a long life. For this purpose a signed document could be 'notarized' by applying the signature of a public body (perhaps the key registry). But it is easier to apply to the registry for a copy of the signer's public key which is, according to custom, identified and signed by the registry. The registry's key is a matter of permanent historical record. This 'certificate' from the registry enables the origin and validity of the document to be proved at any future time. There is another form of notarization which checks the signature and gives a time and date stamp in the certificate. If a certificate for the public key is available, it is really only necessary for the notary to sign and time-date stamp the signature value as 'seen at that time'. Many signed messages have a short useful life and their signatures will not need notarization.

The heavy computational requirement of RSA can be counted as a disadvantage for signature, as it was for RSA encipherment. With RSA chips it should be possible to sign a message (i.e. a one-way function $H(M)$ of a message) in a fraction of a second. This is a useful speed for signature where human interaction is involved. Where signatures have to be checked, the special case $e = 3$ enables this to be done fast, as for example at a message switch handling messages from a hundred or more sources. Another convenient value of e for signature checking is $2^{16}+1$.

Revocation of signatures

What can a person do if he discovers that others know the supposedly secret key he uses for digital signatures? Perhaps the first he will know of the loss of secrecy is the appearance of such a falsification, bearing his digital signature. He must then declare the signature invalid and attempt to revoke all signatures made since the stealing of the secret key, until he can examine the messages and check which of them are valid and which false. Unfortunately, there is usually no way to discover exactly when the secret was lost and it may be necessary, in order to be cautious, to revoke all signatures with that secret key. The revocation should take the form of a public notice, together with an urgent message to the key registry to mark the corresponding public key as invalid. It is important that someone who needs to verify a signature, particularly when the signed message is of great significance, should check with the key registry. In practice, this accidental loss should be very rare and the disadvantage to the person who loses the key, who has to examine a number of messages for validity, is a just punishment for carelessness. But there is a more sinister aspect to revocation of keys, which is the possibility of its use for fraud.

Someone who has entered into agreements by digitally signed messages can escape from the agreements by falsely declaring a loss of his secret key and revoking all the documents signed with it. It has been suggested that this can be prevented by having signatures notarized by a trusted authority, with the time of notarization attached. In this way, signatures presented for notarization before the secret key is declared lost will be valid. Unfortunately, the victim of theft of his secret key may not detect the fact for some time and in the meantime false documents could be signed and the signatures notarized. So the time-stamping of signatures is not a complete solution to the problem.

Some have drawn the conclusion that this possibility of fraudulent revocation undermines the usefulness of digital signatures but we do not share this view. Whatever problems it creates are problems for any kind of digital signature, including also the 'true signatures' based on symmetric ciphers and arbitrated signature. All true signatures depend on secret keys which can be lost. Arbitrated signatures also depend on secret keys, namely the keys used for communication with the arbitrator. If one of these keys is lost, an enemy can masquerade and present documents which will receive another's signature. Potentially, digital signatures can greatly simplify the automation of many business procedures, giving greater security than handwritten signatures provide, so there is an incentive to find ways to minimize the damage of fraudulent signature revocation.

When the existence of evident forgeries alerts someone to the loss of his secret key, he must declare this publicly as well as notifying the key registry and notarizing authority. After the public declaration, anyone holding a document which has been signed with this key should be entitled (if the document is still of any current value) to present it for re-signature. Then the original signer must either re-sign it or declare it to be a forgery. For those documents declared forgeries, a dispute can arise, which is no worse than the disputes that result from any kind of fraud. To gain by fraudulent revocation, the perpetrator must win the dispute. If he has made a substantial gain by revocation, this will throw suspicion on him, though it may be unjustified.

Revocation cannot easily be used for continued fraud. A second case of alleged loss of a secret key might suggest extreme carelessness, but would also throw additional suspicion if the person made substantial gains on this second occasion. People would be less willing to do business with someone if they knew that signatures were liable to become invalid, whether from fraud or carelessness. This is similar to bankruptcy, which can either be the result of ill fortune or a method for defrauding creditors. Someone who makes a habit of going bankrupt will find people less willing to do business with him.

It is curious that many theoretical papers about digital signature ignore the simplest method to prevent the revealing of a secret key (to revoke signatures) which is to make it secret from its owner. Practical methods of safeguarding the signing key always have this property. The key is contained in an enclosure which the owner, like any thief, cannot penetrate.

A new signature mechanism that is to be widely used needs a legal structure to support it. The replacement of a manual signature by a digital one is not itself a difficult change because the validity of the digital signature can be agreed between the parties in a contractual agreement with a manual signature. All the law requires in most cases is that the agreed method of signature should be reasonably safe against falsification. The establishment of a public key registry and notarization service needs a legal basis, and the misuse of this system should become a criminal offence with suitable penalties. The laws covering the revoking of signatures and the arbitration of the disputes that arise are aspects of the additional legal support with which digital signatures, as a widely used method, must be provided.

The principal applications for digital signatures seem to be the numerous small transactions which pass between companies for ordering supplies and services and paying for them. These are probably the most frequent source of small-scale fraud, which might be controlled by using digital signatures. Manual signatures are effective because of the strict view of the law towards forgery, but in large numbers signatures on documents are mainly unchecked. On the other hand there is a danger that automated systems generating and checking signatures can be the subject of 'automated fraud'. Digital signatures can greatly improve and simplify the design of automated systems, but, like all data security precautions, they defend against some attacks and leave others unchanged.

9.7. APPENDIX: THE BIRTHDAY PROBLEM

This is a classical problem in probability theory which is easily solved and gives an unexpected result.

The problem is to determine how large a group of people (for example students in a classroom) is needed to make it likely that two or more of them have the same birthday. By 'likely' we mean 'more probable than 0.5'. Assuming their birthdays are chosen at random from 365 possibilities the answer is that 23 students are enough.

The result is sometimes described as a paradox because it is so small. It happens that this result and others like it are often useful in measuring the security of systems, so it is worth knowing the theory. We have applied a similar result in this chapter to show how someone might be tricked into providing a signature for a document he has never seen.

Let us restate the problem in more general terms. Numbers are chosen at random in the range of n values 1, 2, ... n. After r such numbers have been chosen, what is the probability that two or more of them are equal? We can derive a simple approximation, valid when r^2 is small compared with n, by asking how many comparisons can be made among the r numbers. There are $r(r-1)/2$ ways in which two numbers can be selected and compared. Each comparison has the probability $1/n$ of giving equality. So the probability of an equality among the whole set is approximately $r(r-1)/2n$. This shows that we have to compare r^2 with n, and this helps our intuition to dismiss the apparent paradox. We are looking at the result of about r^2 comparisons, not just r.

Applied to cryptographic security problems this r^2 factor can be crucial, because the numbers are large. Suppose we had to do 2^{64} tests, each occupying 1 μs, then this would occupy 584 942 years, which makes it hopeless. If we can devise a method based on finding two samples with equal values, the birthday problem suggests that the number of samples needed from this enormous population of 2^{64} would be about 2^{32}. At 1 μs per test, the time taken would be 4 295 s, which is trivial. The reason for the great economy of effort is that equality in a population of r samples can be discovered by sorting, a process needing about $r\log_2 r$ operations, much smaller than the r^2 comparisons we used in our argument.

The exact solution of the birthday problem is not difficult. Suppose we write the n different possible 'days of the year' on the blackboard and let the r students come into the room one at a time. Each one crosses off his own birthdate on the blackboard and sits down. If the date is already crossed off, the test ends because an equality of birthdate exists. We want to calculate the probability that this does *not* happen, so there are still no matches when the rth student has crossed off his date.

For the first student, the probability is 1. For the second there are $n-1$ dates on the board and his probability of a 'miss' is $(n-1)/n$. For the next, $(n-2)/n$ and the probability that there has been no hit is now

$$\frac{n}{n} \cdot \frac{n-1}{n} \cdot \frac{n-2}{n}$$

After the rth student has entered, the probability of no match is

$$\begin{aligned} p &= n(n-1)(n-2) \ldots (n-r+1)/n^r \\ &= n!/(n-r)!n^r. \end{aligned}$$

For $n = 365$ and $r = 23$, $p = 0.493$, the solution of the original birthday problem. The probability that a match *has* happened is 0.507, with only twenty-three students.

There are many related problems with similar results. One of these is used in this chapter to show how a signature might be forged. In this variant there are two sets of samples, of sizes r and s respectively, taken at random from n values with repetition allowed. The question asked is: 'How big do r and s have to be to make it probable that at least one value appears in both sets?' Matching within a set is ignored. Comparing any of the r samples with any of the s, yields rs comparisons. The probability of a match is approximately rs/n which can become non-trivial for numbers r and s which are very much smaller than n, when n is large. For example, working with sets of size 2^{64} is not feasible but sets of 2^{32} may sometimes be usable. Again, the computational trick is that matching can be discovered by sorting the complete set of $r+s$ objects, in much less than n

operations. Sorting 2^{33} items takes about 2^{38} operations, much less than the 2^{64} operations needed if the entire population had to be searched.

The essence of exploiting the 'birthday paradox' is to look for coincidence within a set or between two sets, not to look for specific values.

REFERENCES

1. Shamir, A. *A fast signature scheme,* MIT Laboratory for Computer Science, Report No. MIT/LCS/TM-107, January 1978.
2. Kohnfelder, L.M. 'On the signature reblocking problem in public key cryptosystems', *Comm. ACM,* **21**, No. 2, 179, February 1978.
3. Ong, H., Schnorr, C.P. and Shamir, A. 'An efficient signature scheme based on quadratic equations', *Proc. 16th Symposium on the Theory of Computing, Washington,* May 1984.
4. Davies, D.W. and Price, W.L. 'The application of digital signatures based on public key cryptosystems', *Proc. 5th ICCC, Atlanta GA.*, 525–530, October 1980.
5. Diffie, W. and Hellman, M.E. 'New directions in cryptography', *Trans. IEEE on Information Theory,* IT-22, No. 6, 644–654, November 1976.
6. Rabin, M.O. 'Signatures and certification by coding', *IBM Technical Disclosure Bulletin,* **20**, No. 8, 3337–3338, January 1978.
7. Needham, R.M. and Schroeder, M.D. 'Using encryption for authentication in large networks of computers', *Comm. ACM,* **21**, No. 12, 993–999, December 1978.
8. Matyas, S.M. 'Digital signatures — an overview', *Computer Networks,* **3**, No. 2, 87–94, April 1979.
9. Matyas, S.M. 'Initialization of cryptographic variables in a network with a large number of terminals', TR21.1000 IBM Cryptographic Competency Centre, Kingston, NY, August 1986.

Chapter 10 ELECTRONIC FUNDS TRANSFER AND THE INTELLIGENT TOKEN

10.1. INTRODUCTION

Payments can be made by cash, cheque, credit card and in many other ways. New electronic methods of payment are being introduced to improve the speed and convenience and to reduce costs by eliminating paper vouchers. In addition to these objectives there is the paramount requirement of security against attempts to defraud banks or their customers. The possibility of electronic payment systems depends on the relationship of trust between banks and their customers.

Each payment begins with an instruction by the payer. It must be possible to identify the source of the instruction so that the authority to make the payment can be checked. Traditionally, payments between bank accounts have been initiated by a signed instruction from the customer to his bank, such as a cheque, but there are now other forms of payment using paper documents such as credit-card vouchers and credit transfers. Since the essence of these documents is in their information, they can be replaced by digital messages if the necessary checking of authority is possible. The taking of cash from a cash dispenser, using a plastic card and a personal number, illustrates how far we have come from the concept of a signed authority for each payment, yet this is a type of transaction that has not had serious security problems. On the other hand, credit cards, depending for their authority on a signature, have been the target for fraud on an increasing scale. Each payment mechanism brings with it new security questions.

Electronic funds transfer (EFT) operates in 'retail banking' to give the customer a more convenient service, like 24-h service from cash dispensers, to save expense by avoiding the need for capturing data manually from paper vouchers and to improve security as much as possible. When EFT operates between banks and from corporate customers to banks, it serves mainly for convenience and speed of transaction.

Experience shows clearly that, with careful design, EFT can be much better protected against fraud than any method which uses paper documents. New technology such as public key ciphers and digital signatures may be able to offer even greater protection. In this chapter we shall describe some representative EFT systems, giving an indication of their security requirements and the available technology to meet them. We will begin with existing and well-understood systems and move towards those that are still in planning or no more than a future possibility.

When cash dispensers and automatic teller machines (ATM) were introduced, the need to protect them against fraud was obvious and, with a few exceptions, these systems have been reasonably secure. Growing evidence of fraud in other systems with poor protection has made it clear that any new payment method should be very secure in its protection against fraud. No system can be completely secure, so the design should be matched to an assumed level of technology in the way the system may be attacked, with a margin of safety against attacks at a higher level still.

The need for greater care in the design of automated systems is not just a reaction to the growing sophistication of fraudsters. When the protection against fraud is based on human vigilance, the enemy cannot be sure how this is being applied and what rules are in operation at a given time. New precautions can be added where they are needed without upsetting the whole system. In this respect, automated systems are less flexible and the security measures have to be designed and built into the system from the start. These precautions can be studied by the enemy over a long period and an elaborate attack can be prepared if the potential gains are large enough. Because the system is automated there is a danger that the fraud may also be automated in the sense that, once a successful attack has been devised, it can be repeated rapidly to produce a substantial gain to the enemy (and loss to the bank or its customers). Furthermore, the essence of the new systems is their convenience to the user, so any bank which had to withdraw a payment system because of mounting fraud would lose many customers. These are reasons why the design of future electronic banking systems requires a much greater attention to data security than was ever the case in the past, and larger safety margins.

The security of major systems is not preserved primarily by secrecy in their basic design. Discussion of the principles of security in EFT is valuable because it underlines the risks and, in the long run, leads to better methods. But it would be wrong to describe in detail the ways in which working systems can be attacked. In this chapter we shall describe the principles governing the design of various payment mechanisms without describing details which are too close to the potential vulnerabilities of any actual system.

From a cryptographic viewpoint, the attacks on payment systems are active attacks which seek to change data for the benefit of the attacker. Therefore the precautions are primarily those of *authentication* and *integrity* which enable unauthorized changes to be detected, so that payment transactions which are fraudulent can be inhibited before any damage is done. It seems that the need for data security has been understood in banking earlier than in other activities where data security may, in fact, be an equally urgent requirement. Banking provides an excellent example for the application of principles described in the rest of this book. The techniques used in this chapter come from earlier parts of the book, and here it is their application which interests us. We rely heavily on the verification of personal identity, the subject of chapter 7, but the predominant means of identification of payments has been the password. In the form used in banking, the password is usually short because the customer is expected to remember it and it is known as a *personal identification number* with the acronym PIN. When the password includes letters as well as numerals it is sometimes known as a *personal identification code* or PIC, but we will use the name PIN for either kind.

To begin this chapter there is a review of payment methods. Then three general types of payment are introduced and studied from a security viewpoint. We begin with payments between banks where, because of the degree of trust, the security requirements are more easily met, but the risks are great because of the high value of payments. We continue with payments between banks and their customers, such as cash dispenser transactions.

Throughout this chapter we shall be describing payment mechanisms that are implemented by many different kinds of financial instruction such as banks, savings and loans and credit-card issuers. To avoid the expression 'financial institution' at every point we shall refer to them conventionally as 'banks'. When their role in a transaction is specific, they will be denoted by that role, for example 'card issuer'.

An increased complexity appears in the security measures for ATMs when these are shared between a number of banks. The problems are similar to those of 'point-of-sale' payment systems which are the subject of the next section and both have the characteristic that the banks which might lose money from security violations are remote from the point of transaction and dependent on telecommunications. Thus shared ATMs and point-of-sale payment systems usually employ messages between central processors and remote terminals. The possibility of off-line systems for point-of-sale payments is discussed, and this gives a possible role for *intelligent tokens*, sometimes called 'smart cards'.

Having discussed payments from bank to bank and transactions between customer and bank, the final section of this chapter looks at payments between customers in which the bank need not be on-line at the time of payment. These methods depend essentially on the public key signature and there is no current proposal for their widespread introduction, but they are worth studying as a future possibility. They also point to the future value of digital signatures in almost all kinds of payment system.

10.2. ESTABLISHED PAYMENT MECHANISMS

The three main payment systems in operation today are cash, cheques and credit cards. Cash payments are the greatest in number and cheque payments the greatest in value. For example in USA, in the mid 1970s, cash payments formed about 88 per cent of the 300 billion transactions per year.[1] (We use the convention that 1 billion = 10^9.) Of the \$7 000 billion transacted in these payments approximately 96 per cent was by cheque even though the survey excluded payments over \$10 000. There is some uncertainty about the cost per transaction of these payment methods, but it is clear that cash payments were the least expensive at about 1.5 cents per transaction whereas cheques and credit-card payments each cost about 50 cents at that time. The cost of cash payment lies in the manufacture and replacement of coins and notes and there is a difficult trade-off between the cost of manufacturing notes and their resistance to counterfeiting. Like all anti-forgery measures, this resolves into a technical battle, in this case between the forgers and the legitimate bank-note printers and paper makers.

The bank cheque

The cheque payment mechanism is based on a signed instruction from an account holder to his bank to make the payment. This instruction or cheque is given to the payee as evidence of the payment so that, if the payer can be trusted, the payment can be regarded as complete, enabling the goods or services to be supplied. Figure 10.1 shows the outline of this payment mechanism which is surprisingly complicated and at first sight not a good candidate for automation. The payee or beneficiary, Bill, presents the cheque for payment to his own bank which sends the cheque to the payer's bank since the instructions contained in the cheque are destined for them. The payer's bank debits Ann's account and the payee's bank credits Bill's. The settlement is made between banks in the form of a payment covering all the transactions in a given period, such as those in one day.

Since there are many banks, bilateral arrangements between banks would be clumsy, so both the handling of the cheques and the inter-bank payment is usually provided by a 'clearing system'. As the volume of cheques increases (at about 7 per cent per year) an efficient clearing system has become essential. The big practical disadvantage of the cheque system is its slowness. Even with automation, several days must elapse before Bill's bank can be sure that the cheque has been cleared and allow Bill to use the funds that have been paid to him. The period of uncertainty prevents either Ann or Bill from using the funds for several days. Where very large payments are concerned, this is a serious cost to the customers and it favours the banks, which can use the amounts in transit to gain interest.

The physical sorting of cheques to return them to their payers' banks once threatened to undermine the cheque payment system and it was saved only by the introduction of cheque sorting using magnetic ink characters. In principle, the whole system could work without paper as soon as the essential data have been captured, the cheque itself remaining at the branch of Bill's bank where it was presented for payment. The use of 'cheque truncation' depends on the legal system under which the banks are operating.

The security of the cheque mechanism depends on the manual signature on

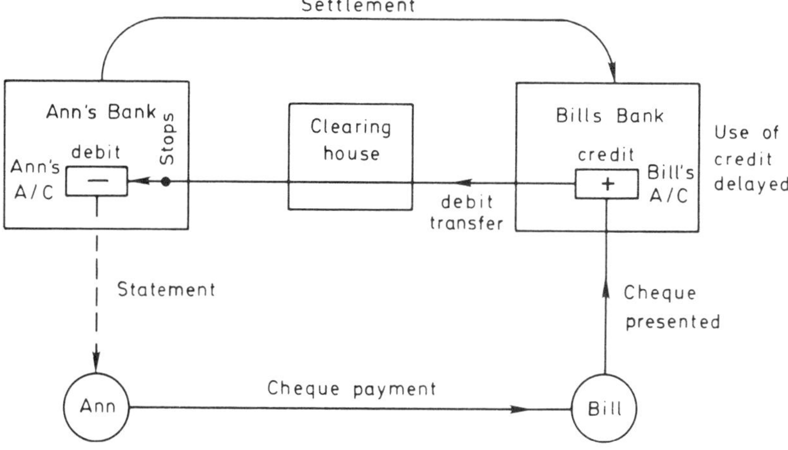

Fig. 10.1 Cheque payment mechanism

the cheque. The weakness of the signature as a security measure is that the enormous volume of cheques which are handled prevents the verification of individual signatures in most cases. Dynamic signature methods which were described in chapter 7 are of no value here, but the automatic verification of signature patterns on paper (the static signature) is showing some promise. Nevertheless, cheque fraud is a considerable problem. Cheque guarantee cards give some protection to shop-keepers, up to a limit set by the banks, in return for checking the signature and copying a number on to the cheque from the cheque guarantee card, but the cost of fraud then falls on the banks.

The paper cheque with its manual signature is not an ideal subject for automation yet it embodies an important principle, which is the authority given by the payer and made apparent to the beneficiary before it threads its way through the clearing system. Changed into a modern form as a message with a digital signature this can be regarded as an 'electronic cheque'. At the end of this chapter we show that this kind of message can be a basis of many different payment systems.

Credit transfer

The cheque is a *debit transfer* when it passes from Bill's bank to Ann's bank because it asks the latter to debit Ann's account. Other transactions are *credit transfers* because they carry the payment with them, for example in the payments between banks which make the settlements for cheques. When an account holder at a bank makes a 'standing order' for regular payments, these payments are carried out by credit transfers. For example, a magnetic tape from the bank of payer Ann may contain among its many transactions a request to credit Bill with a stated sum. The magnetic tape containing this payment also entitles Bill's bank to debit the account it holds for Ann's bank so that the total of the transactions on the tape cancel out. The transport of these magnetic tapes can be replaced by file transfers using data communication.

Figure 10.2 shows the effect of a typical credit transfer from Ann's account in

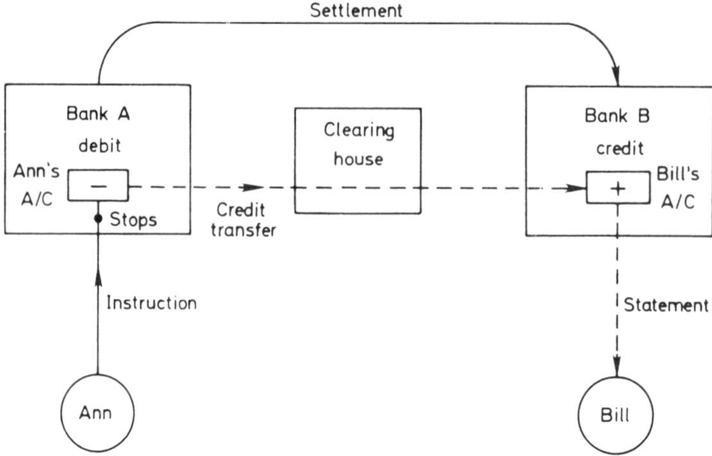

Fig. 10.2 Credit transfer

bank A to Bill's account in bank B. If it starts with a paper document, this can end its journey at A, while the information flows on. The settlement between the banks covers a multitude of individual transactions. Sometimes the credit transfers are initiated by a magnetic tape containing a large number of transactions. In the UK, a bank customer can make an arrangement to send the information directly to the 'Bankers' Automated Clearing Service' where the sorting of entries takes place. For this purpose, tapes or diskettes carry the transactions, or they are sent over a public data network.

A form of banking, named a Giro, which favours credit transfers, is common in European countries. This is a single 'bank' which can very easily make credit transfers between accounts held by its own customers — the two banks and the clearing house of Figure 10.2 are merged into one organization. Because of the need to let Bill know he has been paid, a short statement of account is made whenever a credit arrives and (since the paper is all in one place) the credit transfer form itself goes with that statement as confirmation of the payment and its source.

The Giro has a long history in Europe where Post Offices operate the systems using the postal service to carry the paper work. Typically, Ann sends a credit transfer to the Giro centre by post, which then makes the transaction by adjusting the accounts of both Ann and Bill, then sends an advice to Bill that the payment has been made. Since the two accounts are adjusted simultaneously, there is no 'float' on which the Giro obtains interest. The first notice to Bill of the completion of the payment comes with the advice, so the completion of the deal in goods or services may have to wait on the two transits through the postal system, and posts are not getting any faster. The Postal Giro has an extra convenience that the document can also contain the order for goods or services, making one letter complete the deal. Since Bill receives the payment with the order, there must be some trust, so this simple and convenient device is used for small payments. Giro payments and credit transfers generally are particularly useful for regular payments that are not constant in value such as for telephone service, utilities, tax and insurance.

The simplicity of the postal Giro is due to the monopoly position of national Giro services, which means that the whole transaction between customers' accounts takes place in one centre. There is no problem of paper-sorting or settlement between clearing banks. It follows that Giro services can be automated easily, except for the postal part. Eventually the cost of the post will make this less economic than on-line electronic payments. Bank credit transfers enable many payments to be covered by one cheque and the credit transfers, once their data have been captured, are easily automated because the instruction on paper stays at the customer's own bank. At present, the credit transfer seems better able to fit in with information processing than the debit transfer or cheque, but if paper documents do not have to be moved around (for example if they stay at the point-of-sale in an on-line system) the distinction is less clear. The route taken by messages in point-of-sale EFT is, in fact, closer to that of the cheque.

Summary of the properties of payment methods

Returning to the three principal methods of payment used today, their advantages and disadvantages can be summarized. Cash is the cheapest to operate and loses very little by fraud. The payment is quick and there is no 'clearing delay'. It is not surprising that cash is the most frequent payment method. The main objection is that cash can be stolen. Its disadvantages are found in collecting large quantities of small payments (public transport, payphones and parking meters) and in large payments when violent crime is a danger. Cheques are best for large amounts when the cheque can be cleared in time or the payer can be trusted (or brought to court successfully). For small payments in shops they are too easily used for fraud, too expensive and payment is slow. For very large payments, the clearing delay loses interest to the customers and favours the bank. Credit transfers are best for multiple and large payments. Credit cards have characteristics similar to cheques, except those concerning payment delay, where the customer has more choice. Considering their disadvantages, cheques and credit cards should be overtaken by more automated payment mechanisms as soon as the economics of on-line operation are favourable and the security of electronic methods is assured. Cash will remain the most frequent payment method.

10.3. INTER-BANK PAYMENTS

A large customer of a bank may present a magnetic tape containing thousands of credit transfers. In the UK these can be handled directly by Bankers Automated Clearing Services (BACS) where the items are sorted according to the destination banks and passed on to them also in batch form on magnetic tape. In the future, most of these batches will be handled by file transfer through a data network. The security requirements are principally the authentication of the files to prevent unauthorized changes and it was for this purpose that MAA was designed. Electronic funds transfer based on the batch handling of credit transactions has been in operation for a long time. This service is provided by BACS for large- and medium-sized bank customers and is the world's largest clearing house in terms of transaction volume.

When very large sums are paid between bank customers a cheque is not used because it puts the funds out of use for too long. In its place, telephone or Telex messages can be sent by one bank to give an immediate request to another bank to make a payment on its behalf. The payer makes the request to his bank. That bank validates the request, making sure that it is genuine and that the individual has sufficient funds, then passes on the request to the beneficiary's bank which makes the payment, debiting the account which the payer's bank holds with it. In this way the payment is completed within minutes or hours instead of days. Because telephone or Telex messages are vulnerable to fraud by impersonation it has become normal to add to each message an authenticator using secret keys established between correspondent banks for this purpose. Traditionally, the name given to this authenticator is a 'test key'. The secret keys employed were often in a form of tables, typically two sets of tables held by different individuals. The calculation of test keys was necessarily rather simple because it was carried out by a clerk.

The volume of these payments increased and so did their complexity. Errors due to misunderstanding of the complex transactions had to be resolved by telephone calls and their resolution depended on a degree of trust betwen correspondent banks. These semi-automatic payments continue but are increasingly being replaced by use of custom-built, message-handling systems of which the prime example is the international Society for Worldwide Interbank Financial Telecommunications (S.W.I.F.T.) system. Following on the introduction of S.W.I.F.T. national systems have been introduced such as clearing houses interbank payment system (CHIPS) in USA and clearing houses automated payments system (CHAPS) in UK. After describing the S.W.I.F.T. system, we shall compare its service, and the method of providing it, with CHAPS.

The Society for Worldwide Inter-bank Financial Telecommunication S.C.

The society is a bank-owned, non-profit cooperative society, directed by more than 1400 shareholding member banks which are located in more than 60 member countries. It began as a result of a study by European banks in the late 1960s which led to the formation of the company in 1973 and the first operation of its network in 1977. It has members in North, Central and South America, Europe, Africa, Australasia and the Far East. The costs of its message transfer service are paid for by members using a tariff designed to recover the costs according to numbers of connections, addresses and volume of message traffic. In addition to providing the service and the associated information and training it acts as the forum for agreement on standards.

All messages are handled by store and forward procedures through one or both of the 'operating centres' located in Belgium and USA. The society installs, in member countries, its 'regional processors' which are connected by leased lines to the operating centres with some extra connections so that single line failures do not disconnect any country. Countries with heavy traffic such as USA, UK, Germany and Italy have multiple regional processors and each regional processor has duplicate CPUs for reliability. All connections to the network go through the regional processors which act as concentrators. The operating centres contain duplicate equipment and customers can fall back to an alternative regional processor so that the worst effect of an equipment failure is a short delay. Figure 10.3 shows part of the network as it was in 1984. Since then, the communications system has been redesigned for SWIFTII.

A S.W.I.F.T. terminal is generally a small computer which can 'stand alone' or be used as a front end to the bank's payment system computer. The majority of connections use a 'S.W.I.F.T. Interface Device' (SID) with software supported by S.W.I.F.T. but about 30 per cent use other methods, principally the 'Direct S.W.I.F.T. Link' (DSL) software running on International Business Machines (IBM) main frames. For small users, and as a fall back for others, Telex was used but this has not proved convenient and a microprocessor-based terminal offered by a S.W.I.F.T. subsidiary company has taken over this function.

The S.W.I.F.T. system can report the number of messages sent and received and stores the messages it has transmitted so that they can be retrieved until 14 days have elapsed. This helps banks to resolve any difficulties about the content of messages or the completeness of their traffic.

Message format standards

Strict standards for message format can be imposed because the composition of messages is assisted by the SID in nearly all cases. The agreement on standards has been one of the major advances brought by the S.W.I.F.T. system because

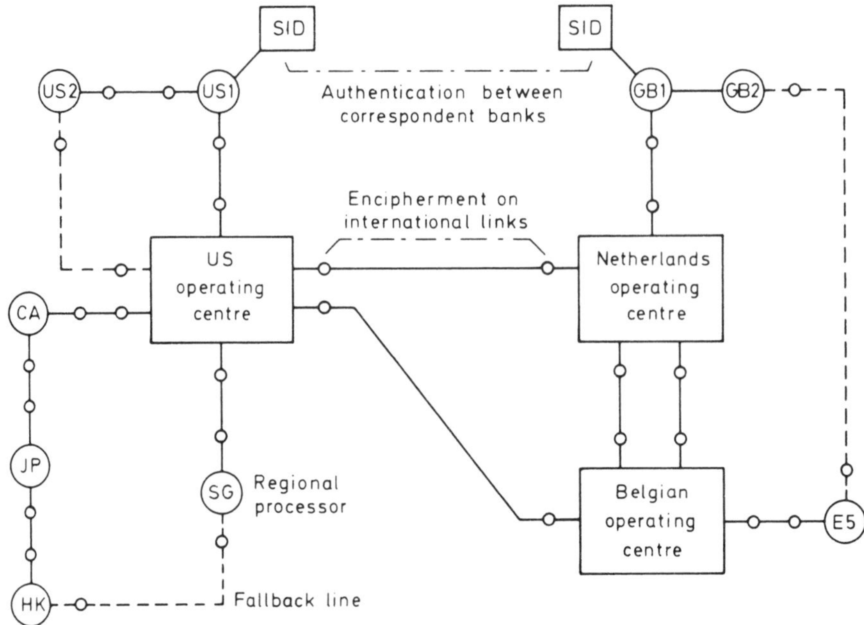

Fig. 10.3 Part of the S.W.I.F.T. network (1984)

its uniform terminology and notation prevent some of the misunderstandings that might otherwise occur in complex interactions. The nature of this complexity will appear later. Working groups of S.W.I.F.T. members derive these standards by discussion and agreement and continually review and update the standards for new applications of the system.

Each message format is denoted by a message type (MT), for example MT 100 is the type known as *customer transfer* which is an inter-bank payment order originated by a bank's customer in favour of another bank's customer. Message types are divided into categories according to their first digit and there are at present eight categories (number 6 is not defined) which are listed in Figure 10.4. In each category n, the groups numberd n90 to n99 are for purposes common to all categories, for example group MT 499 is a free format message concerning documentary collection. The names of categories are self evident except perhaps for categories 4 and 7.

'Documentary collection' is a service provided to an exporter of goods to enable him to receive payment at the place of delivery in return for the shipping documents which give title to the goods. The exporter's bank instructs a bank at the place of delivery, which collects the payment for him. This avoids payment in advance, where the buyer is exposed, or billing the buyer after delivery (open account) where the exporter is exposed. An alternative way to handle the situation is 'documentary credit', for which category 7 messages are designed. The buyer instructs his bank which provides credit (up to a certain sum and time limit) enabling a bank in touch with the exporter to make payment on receipt of stipulated documents. The buyer receives the documents in return for the money, as before, the difference being that the exporter has already been paid.

A message contains a number of fields some of which are mandatory and others

Category	Example groups
1. Customer transfers	100 Customer transfer
2. Bank transfers	202 Bank transfer in favour of third bank
3. Foreign exchange, loans/deposits	300 Foreign exchange confirmation 350 Advice of loan/deposit interest payment
4. Documentary collections	400 Advice of payment
5. Securities	50n Orders and offers
6. Reserved for future use	
7. Documentary credits	71n Issue and amendment
8. Special payment mechanisms	88n Bank card authorization
9. Special messages	900 Confirmation of debit

Some of the message groups have standards that are agreed but are not yet implemented.

Others exist for interim use while awaiting finalization of standards.

Fig. 10.4 Examples of S.W.I.F.T. message groups and categories

optional. The demarcation of each field is by a message tag, for example the tag '15:' denotes the test key. Taking as an example the customer transfer (MT 100), the mandatory fields are

20: transaction reference number
32A: value date, currency code, amount
50: ordering customer
59: beneficiary customer

The fields are sufficient for the simplest kind of transaction where the ordering customer — the person making the payment — has an account with the bank which sends the message and the beneficiary customer has an account with the bank which receives the message. The sending and receiving bank do not appear as fields in the message text because they are contained in the header of the message which directs it through the system.

For technical and accounting reasons, not all pairs of banks can usefully exchange messages. Full connectivity would imply a million or more possible links each with an authentication key and account relationship, which is not practical. Therefore, smaller banks specialize in transfers with certain countries, offering this service in competition with other banks. Consequentrly the accounts of the ordering customer and the beneficiary customer may not be in the banks sending and receiving the messages.

As many as four other banks in addition to the sending and receiving bank may be involved in the transaction and it is this kind of complexity which led to errors when payment messages were less formal, before S.W.I.F.T. existed. Figure 10.5 shows the messages and payment in the most complex case. These messages have fields denoting

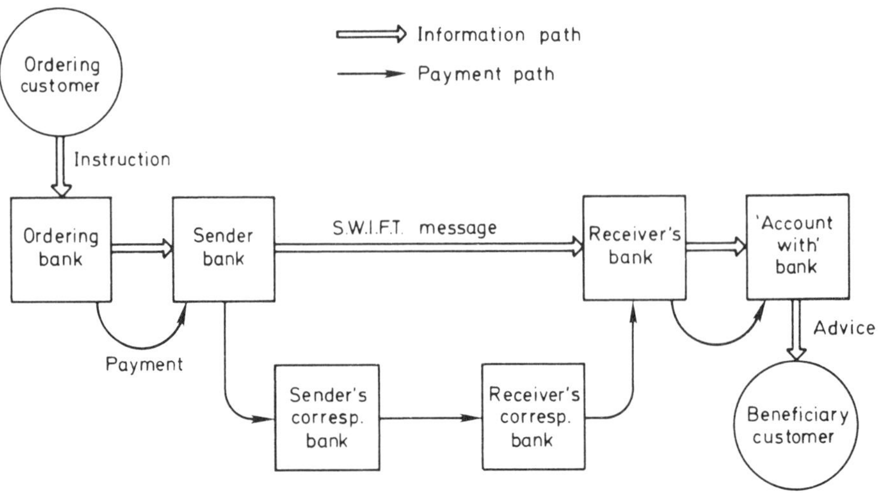

Fig. 10.5 A complex customer transfer by S.W.I.F.T.

52: ordering bank
53: sender's correspondent bank
54: receiver's correspondent bank
57: 'account with' bank.

The ordering bank sends a message to the sending bank (which originates the S.W.I.F.T. message) and also transmits the funds to this bank. The receiving bank makes the payment to the beneficiary via the bank which holds his account, that is the 'account with' bank.

The payment between sending and receiving bank is often made by an account held by one of these banks for the other. Whether this is the sender's account with the receiver bank or the receiver's account with the sender bank depends on the currency in which the transaction is made. If the currency is that of the country of the receiver bank then the receiver debits an account held by the sender in this bank. Alternatively, if the currency is that of the country of the sending bank, the sending bank credits the account of the receiver which it holds for him. Credits and debits for these accounts and statements of account can be sent as special messages in category 9.

In some cases it is necessary for the payment to go through intermediate banks working either on behalf of the sender or receiver. The message says that the payment will be made this way by quoting the correspondent banks in the fields with tags 53: or 54:. Any of the additional four banks referred to in these optional fields can be present or absent in a customer transfer message. Messages to the sender's and receiver's correspondent banks to effect these payments are quite independent of the S.W.I.F.T. message which goes direct from sending to receiving bank.

Formats are defined in a similar way for a large number of message types. New message types are being developed. In each message group there is the possibility to send free format messages for purposes not covered in the standards, using the digits 99 to signify free format, for example 199 in the customer-transfer group.

There is no doubt that the careful definition of these message types and their formats has been an important byproduct of the introduction of the S.W.I.F.T. message service.

Security in the S.W.I.F.T. system

The most important security features are authenticity and integrity. There have been a number of spectacular frauds by generating bogus payment messages between banks under manual systems of operation. In each case the authentication of messages was at fault.

The general properties of an authenticator algorithm were described in chapter 5 with some examples. The details of the S.W.I.F.T. authenticator are not public knowledge. As in any authenticator, the sharing of this key between sender and receiver bank and its secrecy from all others is the basic requirement for secure authentication.

The responsibilities for authentication are clearly defined. The algorithm is provided by S.W.I.F.T. (and in some cases the software in a terminal within which the algorithm is implemented). Keys are exchanged bilaterally between banks and are not known to S.W.I.F.T. or its personnel, but S.W.I.F.T. has issued guidelines about the exchange of authenticated keys suggesting the procedures to be used and the timing of key changes. Members can regard S.W.I.F.T. as a forum to assist their co-operation in the security matters but the responsibility for the safe exchange of keys lies entirely with the members.

An authenticator does not prevent messages being replicated, deleted or stored and retransmitted later. These requirements are handled by the sequence numbering of messages. Sequence numbering within the S.W.I.F.T. network ensures that messages are not lost or duplicated. The connection between S.W.I.F.T. and its users is safeguarded by input and output sequence numbers. The S.W.I.F.T. operating centre checks the format and input sequence number of messages offered to it and refuses those which have format errors or wrong sequence numbers. The output from the S.W.I.F.T. operating centre contains an output sequence number which must be checked by the receiver. The transaction reference number provides an end-to-end sequence control for each pair of banks and is included in the part of the message to which the authenticator is applied.

The interface between S.W.I.F.T. and a member bank must be protected against the introduction of false messages which have correct sequence numbers and authenticators. This is a matter for the individual banks who carry the ultimate responsibility for declaring that a message is valid and should be passed into the S.W.I.F.T. system. For this reason, SIDs and other connection arrangements normally provide for the assembly of a message by one person and the checking of the message and its release for transmission by a second person. The software and hardware of the connection must be protected against unauthorized changes because they administer these precautions and also store the authenticator algorithm and its keys.

Terminals coming on line to the system must verify their identity by a password. The passwords are issued by S.W.I.F.T. in the form of tables which are sent in two parts, A and B. By sending them separately the interception of a complete password set is made less likely. Password tables must be acknowledged to S.W.I.F.T. by an authorized person before they are activated on the system. Each table contains a sequence of passwords listed against a sequence number. Each login employs the next password in the sequence so that interception of passwords by line tapping gives no clue to the next password that will be needed. To avoid the trick of extracting passwords by impersonating the S.W.I.F.T. system to the terminal, each password is given a response number which the user must receive from S.W.I.F.T. and check before beginning transactions. The formalities and procedures for introducing new password tables and the method of fall back when problems occur are carefully defined.

In the central part of the S.W.I.F.T. system which consists of the operating centres, regional processors and connecting lines, the security is the responsibility of the S.W.I.F.T. organization. Access to the system, its software and the users' messages is strictly controlled and so is physical access to the areas containing the computer systems. The international lines which connect operating centres together and join them to regional processors are protected by encryption to preserve the privacy of the banks' messages. It is a matter for the individual bank to decide whether to use encryption on the line between its SID and the regional processor but S.W.I.F.T. will offer help. Privacy, though important to some, is not such a cardinal security factor as authentication. This is why end-to-end encryption is not provided, whereas authentication is end-to-end. The authentication allows the use of keys which are not known to S.W.I.F.T. but end-to-end encryption would prevent the useful role of S.W.I.F.T. in storing, validating and retrieving messages.

The chief inspector's office has overall supervision of security and privacy and performs security audits. Internal and external auditing are carried out at random intervals to make sure that the security of data and of all S.W.I.F.T. operations is being maintained.

The S.W.I.F.T. system provides an excellent lesson in the careful design of system security. Similar systems like CHIPS and CHAPS, though they differ in detail, employ similar principles for authentication, sequence numbering, login and other security features.

The Clearing Houses Automated Payments System

The Clearing Houses Automated Payments System (CHAPS) network carries payment messages between banks, like S.W.I.F.T. The basic security requirements of the two systems are similar, but when they are examined in more detail, differences appear, due to their different institutional framework, geographical coverage and settlement methods. A comparison of the two systems is instructive.

S.W.I.F.T. provides the facility for carrying messages between any of its banks. Any pair of banks which use it make their own arrangements for keys to authenticate messages travelling between them and arrange a route for settlement of debts between them. CHAPS is operated by a small group of 'settlement banks', who all do substantial business with each other, so it is a closer-knit community. These banks offer a means for large payments between about 300 other banks in the UK, most of them branches of foreign banks which use London as a business centre. The attraction of the settlement banks as intermediaries lies in their facility for *same-day* settlement using accounts they hold at the Bank of England. Cheque-clearing settlement in the UK takes place via CHAPS, in addition to its main function of payments between bank customers. The facility of same-day settlement for individual payments is called 'town clearing'. It has been done in the past by carrying paper documents within the compact business centre of 'the City' — a square mile (2.5 km^2) containing the offices of most of the 300 banks involved. In order to let the Bank of England arrange its funds before close of business, the individual payments stop at 15.00 hrs and the settlement between the thirteen banks follows immediately. About 19 000 payments took place per day with a total of £17 000 million. The movement of paper documents in the city is by bank messengers on foot.

The method worked well but had two limitations due to its old-established transport method. It was limited to the city. Banks elsewhere had to correspond with settlement banks by telephone or telex messages which was inconvenient and required careful discipline to keep it secure. The service was expensive per message and, though this cost was small compared with the interest saved by immediate access to funds, it led to a lower limit of £10 000 on individual transactions allowed to use the system. The CHAPS payment system is an automation of town clearing, using a computer-based message system between the settlement banks (of which the Bank of England is one). Some of these banks share gateways into the system, for economy. There is no limitation on where the point of entry to CHAPS is placed, which makes it possible for Scottish banks to have access points placed near to the banks they will serve. Additions to the group of settlement banks are not often made but further access points are feasible whenever they are needed.

The message format has been chosen with the aim of easy conversion to or from S.W.I.F.T. standards because instructions for payments via S.W.I.F.T. will often initiate payments through CHAPS. When a payment has been sent into CHAPS, settlement that day is part of the implied agreement, therefore a bank acting on a S.W.I.F.T. payment message must be sure of the source of the funds. Members of CHAPS compete in providing settlement facilities, balancing the value of the business against the risks they take. If a CHAPS payment has to be cancelled, this can only be done by a contrary payment, which needs the agreement of the payee.

Figure 10.6 shows the physical structure of CHAPS. Unlike S.W.I.F.T. it has no central operation, but uses messages transferred through the packet-switched public data network called Packet Switchstream (PSS). The integrity of the system depends on 'gateways' implemented in Tandem non-stop computers with software that has been jointly developed. Each message receives a logical acknowledgement via PSS back to the originating gateway, so that the performance of the system is continually monitored. The high availability of PSS and the Tandem computers is fundamental to the design. The gateways administer time stamps, sequence numbers and running totals for each bank pair. With each message from A to B, the total 'paid' by A to B since the last settlement is reported,

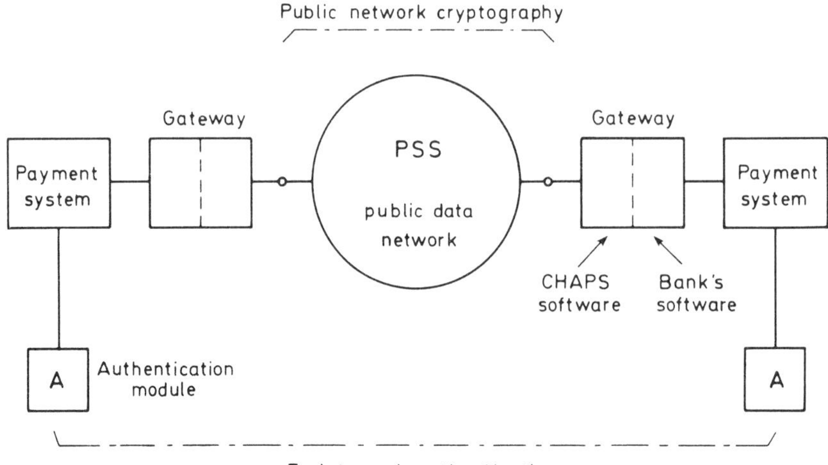

Fig. 10.6 Structure of the CHAPS system

like a bank statement that is always up-to-date. Consequently, at the end of business for the day, agreement on the accounts can be immediate and settlement follows by a message to the Bank of England.

Each bank's arrangements for message processing outside the CHAPS interface are its own concern. The large banks run a payment process in their own main-frames. Smaller banks (those with less CHAPS activity) can run their payment process in the Tandem that holds the gateway.

Clearing houses automated payments system is different from S.W.I.F.T. because same-day settlement through the central bank is an integral feature of its operation. The physical and organizational structure is also different. It is operated by a more centralized group of banks but the network has a more distributed structure. The security requirements are very similar to those of S.W.I.F.T. and are primarily concerned with authentication of messages and less critically with the confidentiality of messages.

The authenticator used in CHAPS is the Data Encryption Standard (DES) in its cipher-block chaining mode. It is implemented in the tamper-resistant hardware modules which hold the keys and perform the tasks of generating and checking authenticator values. The key management reflects the structure of CHAPS in which each pair of banks has an individual relationship. It employs master keys used to protect session keys which are transmitted over the network. Though the method of authentication has been agreed by the settlement banks, authentication is an end-to-end matter and is the responsibility of the two banks concerned. The payment system has been designed to be transparent to the message content, except for its procedure of handling the payment fields.

Confidentiality is provided by encipherment at the point of entry to PSS and decipherment at the exit from PSS. The main security feature is the authentication procedure, implemented in the bank's payment process with the aid of the special hardware module. Because the two processes of generating and checking authenticators are carried out in the physically secure modules, the ability to generate authenticators can be confined to the sender. In this way, some of the useful properties of a signature are obtained in a system which uses a symmetric cipher, the DES.

10.4. AUTOMATIC TELLER MACHINES

Early automatic teller machines (ATMs) were essentially cash dispensers which had only one function, to deliver cash in the form of bank notes and debit a corresponding bank account. To identify the user, tokens were used in the form of cards of various kinds. Some early machines used punched cards that had only one life — they authorized one payment only and were not returned to the user, who had to obtain a supply of cards from his bank. There were also magnetic cards with a limited life, for example twenty withdrawals of cash. These early machines functioned for one bank, dispensing cash only to customers of that bank. To reduce the risk of lost cards being misused it was usual, right from the beginning of cash dispensers, to issue a personal identification number (PIN) to the customer which is entered on a keyboard to complete the identification. Many of these early cash dispensers were off-line; they worked on their own without connection to a central data base. They were located at first inside the bank where they provided an alternative to the human tellers. After some experience had been gained, cash dispensers were installed in the 'through the wall' position in which they were part of the bank building but available from outside it when the bank was closed. In this way banks could provide 24-h cash service — a factor which is becoming even more important with the growing competitiveness of banking organizations.

A third kind of location which was entirely away from the bank developed later. This can provide greater convenience but the cost of installation makes it suitable only in places with a large access to users, for example shopping centres and very large work areas which are remote from banks. The real value of these remotely sited cash dispensers or ATMs arrives when a single ATM can be used by the customers holding 'cash cards' of many different financial institutions.

The increase in remoteness from the bank increases the problems of physical security. One of the earliest crimes against 'through the wall' cash dispensers was to pull them completely out of the building with a heavy machine. Now these dispensers are built like a strong safe and fixed to the concrete foundation.

By adding further facilities, cash dispensers began to deserve the name of 'automatic tellers'. These can be used for enquiries about the status of accounts, for transfers between the customer's own accounts (such as deposit and checking accounts) and possibly for credit transfers from a customer's account to some other person's account. The machines can deliver travellers cheques instead of money and they can accept deposits, though in doing the latter they add little to the facilities of a bank deposit box. The most important function of automatic teller machines is still that of dispensing cash (or travellers cheques in countries where these are needed) and this is the operation requiring the greatest attention to security because any weakness which allowed cash to be withdrawn without authority would quickly render the service untenable to the banks. One of the primary requirements of a cash dispenser or ATM transaction is, therefore, to identify the customer as reliably as possible.

Automatic teller machines raised a host of legal problems concerning the liabilities of the bank and the customers when things go wrong. In the USA this has been covered by legislation, both state and federal. In other countries banks have been able to make their own agreements with customers and there has been comparatively little problem with disputes.

Generally we can say that the ATM facilities are greatly liked by bank customers, to the extent that short queues form at ATMs even when human tellers are free. Where banking hours are short, the use of ATMs outside hours has enabled banks to compete with other financial institutions having longer opening hours and these have, in turn, been persuaded to install ATMs to give 24-h service.

On-line and off-line operation

There are several advantages in connecting an ATM to a central data base so that transactions can be checked before money is paid out. The account balance can be checked and the card identity compared with a central file of cards against which payments will not be made — the so-called 'hot card list', 'negative file' or 'blacklist'. But on-line connection is expensive and has in the past been unreliable in some countries because of communication problems. So there is a strong incentive to use off-line operation if the security risks are not too high.

Checking the PIN against the card identity can be done in the terminal if the PIN is related to the account number and other card data by an algorithm, for example the use of a cipher with a secret key. Some details of these methods are described later. Having the relationship dependent on a secret key which is the same for all ATMs is a risk, because the key is held at so many places. Compromise of the key makes the whole system unsafe because an enemy is able to discover the PIN of any stolen cards or make his own cards and determine the PIN which will match them. The physical security surrounding an ATM makes it very unlikely that the key will be discovered by intrusion. The weakness, if any, is in the way the key is loaded into the ATM. The key can be the result of a calculation employing data from several sources or the algorithm can be a complex one requiring several keys. It can then be arranged that no one person has access to all the data necessary to forge cards.

The keys are used in two places in the system, at the ATMs for checking the relationship of the PIN with the account number and other card data and at the bank headquarters where PINs are calculated for distribution to customers. It is probably in the central area that the security of the keys is most at risk and this type of risk appears in some form in an on-line system, whatever method is used centrally for checking PINs. Therefore the additional risk in an off-line ATM system due to the multiplicity of places at which PIN checking is carried out can be small if the system is designed well. The physical security of the ATMs is an essential factor in this evaluation.

An off-line terminal can be equipped with a blacklist and this can be updated periodically. Since the bank's liability typically begins when the stolen card is reported, there is a strong incentive to update the blacklist quickly, within a day at most. The organization needed to do this increases the effective cost of off-line systems.

The greatest danger in off-line systems lies in the lack of effective communication between one ATM and another through the central data base. For example, if a card is stolen and the PIN is known or can be discovered and if the enemy can make several copies of this card, these can be used rapidly in many ATMs and an off-line system will not react by blacklisting the card until the next updating cycle. In an on-line system, the unusual activity on a particular account is easily detected and stopped. An on-line system can enforce complex rules limiting the usage of a card such as: total amount withdrawn per day or week or month; or number of withdrawals per day. With an off-line system such rules are unenforceable except when the token itself can carry secure information about its usage.

In some early ATM systems, data about the card's usage were stored in an area of the magnetic stripe that was rewritten each time the card was used. The data were enciphered, but this protection is illusory because a card can be returned to an earlier state by restoring its data. An on-line system can protect against this 'replay' threat with a sequence number but the storage on the card is not needed. To make storage on the card effective it must be physically and logically protected. This is the 'intelligent token' which will be discussed later.

In an off-line system using magnetic-striped cards there is a great need to prevent the copying of cards or the manufacture of new ones. This is the purpose of a 'security feature' incorporated in the card. An example of a security feature is the EMI watermark (see page 179) which stores a number on the magnetic-striped card in a way which cannot be altered or forged. Someone who copies a card, even onto another card with magnetic watermark, will not be able to copy the watermark number. Because a mathematical relationship exists between the watermark data and other data on the card, any copied or forged card without this relationship can be detected. The simplest relationship to use is a cryptographic one with a secret key k. For example if W is the watermark number and I the other identifying information on the card, such as the account and card serial numbers, a reference number R is calculated where

$$R = \mathrm{E}k(I, W).$$

This is stored on the card. Only a card with the correct reference number will be accepted and the reference number cannot be calculated without a knowledge of k. The notation (I, W) indicates that the data in these fields are enciphered as a cryptographic chain, thus if either I or W is changed it is necessary to recompute R in order to generate a valid card. This, we assume, the enemy cannot do because the key k is stored securely. In essence, the reference number R is another kind of password.

It is possible to refine this method by making R a digital signature of the data contained in I and W. This signature is computed by the bank using the secret key when the card is generated but any enemy, using old or stolen cards (which have their own W values) to copy a card for which the PIN is known, will not be able to compute the new signature. The advantage is that a digital signature can be checked by using a public key, therefore the secret key k is used only centrally for card encoding. The disadvantage of this method is the large amount of data employed in a signature, typically 512 bits, which does not fit easily in typical magnetic-striped cards.

To summarize the comparison between off-line and on-line operation, there is no limit to the methods which can be applied to checking card validity in the on-line system but off-line systems are limited to checking the PIN against the card data, comparing with a local blacklist and at most preventing the repeated use of one card at a particular ATM. The security advantage lies very strongly with on-line operation. With the growing tendency towards shared ATM systems capable of carrying out transactions for many separate financial institutions, we shall find that there is an additional motive for on-line operation.

Some operators of ATMs require their terminals to continue giving service when the central data base is out of action, for example during routine maintenance. This involves an extra risk, which is reduced if the periods of off-line operation are not at regular times and are undetectable. But the off-line state of an ATM is revealed by its inability to give the account balance. The safest policy is to equip an ATM system in such a way that off-line periods are rarely necessary.

PIN management

Personal identification number (PIN) management is the subject of a US National Standard[2]. Three types of PIN are distinguished: assigned derived PINs, assigned random PINs and customer-selected PINs. To the customer the only detectable difference is between a PIN he is assigned or a PIN he can choose himself.

Assigned PINs are usually numeric and typically have between four and six digits. In the UK, for example, PINs of four decimal digits are widely used. In practice, PINs are not usually memorized, as they should be, but written down, sometimes on the card itself. It is not so unsafe to write the PIN in an obscure way in some part of a diary but, if the bank were to suggest this and give examples, a resourceful crook would know which schemes were widely used. Ultimately, the bank has to depend on the good sense of its customers and recognize that the security which derives from separating the PIN from its card is at best only partial.

Since many people now find that they have been assigned many PINs for different purposes, memorizing these PINs becomes difficult. One of the advantages of customer-selected PINs is that the customer can choose one value and use it for all purposes. There seems to be an advantage in alphabetic PINs to make a memorable word and thus avoid the need for writing it down. Alphabetic PINs do not imply an alphabetic keyboard because a ten-key layout of the form shown in Figure 10.7 is used — this is the proposed American National Standards Institute (ANSI) standard and is based on the telephone keypad. Unfortunately there is support elsewhere for a layout based on the calculator keyboard, where the top row of keys is 7, 8 and 9.

There is clearly a need for careful guidance to customers. Where the customer selects his PIN he should be given time to think about it, not required to choose it when the account is opened. This can be done by assigning a temporary PIN which makes the ATMs usable at once and then allowing him to change this to his own chosen value when he wishes. The card issuer can accept the customer's chosen value by post or telephone if he is given a serial number unrelated to his account (or temporary PIN) with which he can validate his new choice. It can

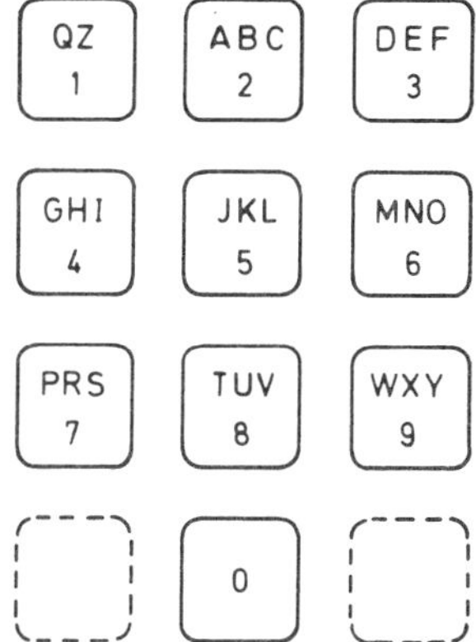

Fig. 10.7 Alphanumerical PIN pad layout

easily be arranged that no employee need see the association between the customer's chosen PIN and the account number or name and address.

Any system employing such a short password as a four decimal digit PIN must be careful to prevent methods of discovering the PIN by searching through all of the 10 000 possibilities. On-line systems make this restriction easy. Typically, ATMs allow three attempts to give the correct PIN, then refuse payment. Early systems retained the card after the third incorrect try but if an ATM card also functions as the credit card this card retention is very unpopular. In an off-line system it would seem dangerous to give the card back because it allows PIN searching to continue.

An excellent source of advice on PIN management is the *PIN Manual* published by MasterCard International[3] which is reproduced as chapter 10 in *Cryptography* by Meyer and Matyas.[4]

Algorithmic PIN checking

When off-line operation is used or when an on-line system must be capable of checking PINs in a temporary off-line state, the method of checking must depend on an algorithm, typically a cipher with a secret key. Such methods of PIN checking are described as examples in the ANSI Standard. Figure 10.8 shows the method by which a (small) PIN of N decimal digits can be derived from the customer's account number. First the account number is padded to 16 decimal characters, using zeros or some other constant value, then the resulting 64-bit number is enciphered using a secret DES key. From the 64-bit result a decimal number of N digits is derived.

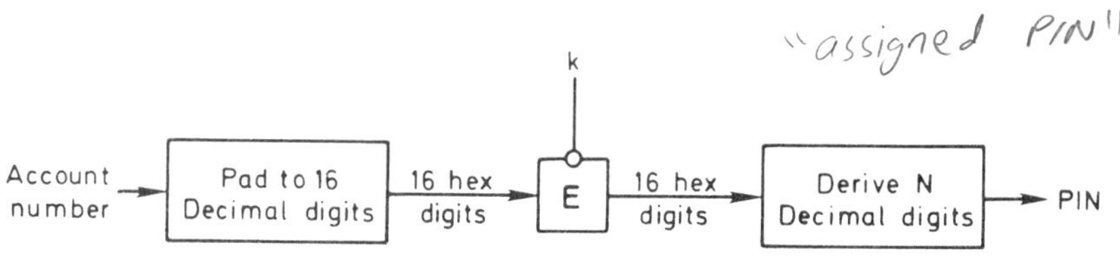

Fig. 10.8 Derivation of PIN from account number

The method of deriving the decimal digits is to examine the 64-bit result 4 bits at a time, starting from the least significant end and accept only those 4-bit groups whose binary number is less than 10. The accepted binary numbers form the decimal digits. This is similar to the method by which decimal digits are derived for cipher feedback (see page 92).

The PIN management standard anticipates PINs of between 4 and 12 decimal digits, but typical PINs have 4, 5 or 6 decimals. In most cases these will be derived from the 16 hexadecimal digits comprising the 64-bit output of the DES. If, by bad luck, insufficient digits are generated, the remaining hexadecimal characters can be used, subtracting 10 from their binary values to get a digit value lying in the range 0 to 5. The very small bias introduced in this way is generally acceptable.

This method generates an assigned PIN. If a customer-chosen decimal PIN is needed, the same procedure is used to derive an *N* decimal number, then the chosen PIN is added to the derived PIN, by decimal arithmetic without carries (individual digits are added modulo 10). The resulting decimal number is called an 'offset' and this is stored on the card. Because of the approximate randomness of the derived PIN, it is not possible to deduce the chosen PIN from the value of the offset.

When a card encoded in this way is used with an on-line ATM, PIN checking can be done in one of two ways. The offset can be read from the card with the account number and the PIN checked algorithmically at the terminal, or the PIN can be transmitted to the centre and checked against a file of chosen PINs.

Instead of storing an offset on the card an alternative is to store a number derived cryptographically from the PIN, which is called a *PIN verification value* or PVV. Because the range of PIN values is so small, an attacker could make a table showing the relationship of the PIN and PVV, so in practice the PVV is made a cryptographic function of both the PIN and the account number. This is illustrated in Figure 10.9 where the derivation is by encipherment, followed by producing *N* decimal digits in the way described earlier. In practice the encipherment uses a double length key because this key has a long life and might be subject to prolonged attack. The PVV can be recorded on the card and used for local PIN checking, but since a compromise of the key would be disastrous, this kind of checking should only be performed in a very secure environment. Alternatively, if the PIN is sent, under cryptographic protection, to a central place, it can be checked against a file of PVV values, which is safer than holding the PINs themselves.

Typically the PVV has four decimal digits, but this has a disadvantage. Suppose that an attacker is making a search of PIN values to find the correct one. With a four-digit PIN he will obtain the correct value with probability 1/10 000. However, the PVV is a random function of the PIN hence there may be other

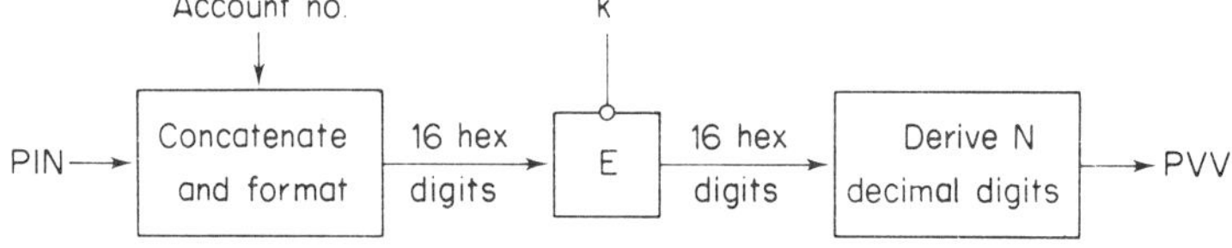

Fig. 10.9 Derivation of a PIN verification value

PIN values which give the correct PVV. The average number of PINs associated with a valid PVV value is $e/(e-1)$ which is approximately 1.58. This increased probability of finding an acceptable PIN (if not the correct one) makes the PIN searching attack somewhat easier. This effect can be reduced to a negligible one by employing a PVV field of five digits, when the number of PINs corresponding to a valid PVV is on average 1.05.

The dialogue for an on-line ATM

The essential transaction between the Automatic Teller Machine (ATM) and its central data base contains two messages, a request from the ATM to the centre for authorization of the transaction and the response from the centre which is either an authorization or a denial.

Typical contents of the request message are

ATM identity,
Account number and other identifying information from the card, such as a card serial number,
Watermark or other security number read from the card,
PIN provided by customer, enciphered,
Amount of cash requested,
Sequence number of transaction,
Authenticator for all the other data.

The response could contain the following

ATM identity,
An operation code signifying payment approval or denial, or denial with card retention,
Sequence number of transaction,
Authenticator for all the other data.

Both messages need authentication otherwise the request could be modified to reduce the amount debited to the account or the response could be altered to change a denial into an authorizaton. Authentication could consist of enciphering the entire message since the serial number, if managed correctly, prevents the substitution of a previous message. The essential cryptographic requirement is to encipher the PIN so that a line tapper cannot discover which PIN value is associated with a given account.

In spite of all precautions the response message might not get through, then the customer would suffer because his account would be debited but the money would not be paid at the ATM. To help resolve these problems, a log is kept at the ATM, sometimes in the form of a printed tally-roll. Such a log is also needed to satisfy the internal accounting of the bank when the ATM's payments are reconciled with the cash loaded into it.

It might be thought that greater certainty could be obtained by a third message, from the ATM to the centre, confirming that the payment had been made. The lack of this message would signal to the bank a possible discrepancy, without waiting for the customer's complaint. This would enable the rare examples of system failure to be detected earlier. But no sequence of messages in the dialogue can provide certainty, because the final message could be lost and there would still be doubt about the outcome. This is an example of the impossibiity of final agreement when messages are exchanged by a channel which is not perfect. This is known as the 'Two Generals Problem' and has been studied carefully, but there is no effective solution. Whether a two-message or three-message dialogue is used affects the state of knowledge at the centre but does not solve the basic problem. A permanent record at the ATM is essential and this should if possible resist forgery attempts, in order to reduce the possibility of employee fraud. If a magnetic recording is used, authentication or digital signature on the record can provide this additional security.

This dialogue of two or three messages is only one possibility. There is another scheme in which the PIN is not sent from the ATM in the request but, instead, an 'authentication parameter' is returned with the response. This is described later on page 309 in the context of shared ATM systems.

The essential requirements of message authentication and PIN encipherment can be provided by any of the standard methods described in earlier chapters. A less conventional method for PIN encipherment and approval authentication is described as an example in the PIN Management Standard. This is illustrated in Figure 10.10. It employs a decimal counter in the ATM which is reset only when the keys are changed. The output of this decimal counter is enciphered twice to generate 32 hexadecimal digits from which are derived, by the method described earlier, sufficient decimals to encipher the PIN and a further 5 decimal digits which become the approval authentication code. The PIN is enciphered by modulo 10 addition of the generated digits.

At the other end of the communication line a similar algorithm is employed with its counter in synchronism and, to keep the counters synchronized, two or three digits of the counter are transmitted with each request message. At this central place, the same $N+5$ decimal digits are derived, where N is the number

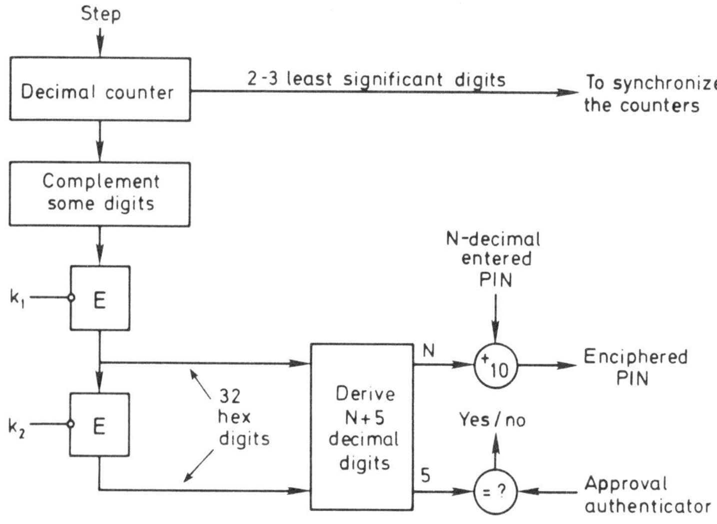

Fig. 10.10 Method for PIN encipherment and approval authentication

of digits in the PIN. By subtracting these values, the PIN is deciphered and checked against the local records. If the payment is authorized, the 5-digit approval authentication code is returned in the response message. The method does not handle the authentication of the request message, which is an essential security requirement.

We should consider the ATM and the central system as 'communicating entities' and consider how they authenticate each other in one of the ways described in chapter 5 (see page 126). The sequence number from the ATM is returned in the response by the centre and both messages have authenticators; this authenticates the centre to the ATM and prevents intrusion by a recorded earlier message. The sequence number must not repeat during the life of the authentication key. The centre authenticates the terminal by holding the sequence number last received from each ATM and requiring the next one to be incremented correctly.

The security of this sytem depends on the secrecy of keys — a different key for each terminal — the integrity of the ATM software and hardware and the integrity of the software and database at the centre. Since the ATM is physically secure, the biggest dangers at the terminal come from those who are allowed physical access. Preferably, the electronics should not be accessible to bank employees and critical parts of the electronics should be repaired only by replacement under strict control. At the centre, PIN files should be secured by encryption. The control of card and PIN issuing, and the recovery of lost PINs, must be carefully designed so that no single employee (or small group of employees working closely together) has access to enough information to reconstruct corresponding PINs and card data.

Shared ATM systems

When a group of banks join together to make all their ATMs available to the customers of the whole group, the task of approving payments takes on a new dimension. It is unlikely that the methods the various banks use for verifying PINs against card identity will be the same, therefore no single algorithm operating in an ATM can verify the PINs of all the customers it serves. Banks could not be expected to trust other banks with their secret keys for verifying PINs and it would not be practicable for lists of all the PINs against card identity to be stored in all the banks. From this it follows that PIN verification must be done centrally by each card issuer, that all the ATMs must be on-line when they serve customers of other banks, and that the ATM processors of all the banks must be connected by a communication network.

Within this network the banks must exchange authenticated messages, they must be able to carry PINs securely from one bank's processor to another and they must maintain accounts of the transactions carried out which are satisfactory to all the banks, so that settlement can take place for payments that have been made by one bank to another's customer. All this implies the exchange of secret keys between the banks for authentication.

Many schemes for PIN encipherment and message authentication in shared networks have been proposed. In the context of point-of-sale systems (treated later in this chapter) methods using session keys have been described[5,18] and also methods using personal keys stored on the customer's card.[6] Some shared ATM networks use a system which makes extensive use of system keys, known as 'zone keys' and is similar in concept to 'node-by-node' encipherment, described in chapter 4 (see page 106). All methods have some advantages and disadvantages. The method we shall describe is called 'zone-encryption' and it is resistant to customer fraud and line tapping. It requires very careful handling of the system keys so that bank employees cannot discover keys of their own or other banks. Later, we shall show how the intelligent token (or smart card) and public key encipherment and signature can, separately or together, improve the security of shared systems against employee fraud.

Typically, in the zone-encryption method, a group of banks appoint an authority to operate on their behalf for the interchange of messages for ATM payment approvals. This 'interchange' is represented by a processor which acts as a node to which all the processors of the banks involved in the ATM system are connected. In general a bank is both an issuer of cards and an operator of ATMs. When a customer goes to a bank which is not his own, that bank becomes the *payer* of money to the customer. The other bank involved in the transaction is the *card issuer*. The issuer holds the account that is to be debited as a result of the transaction and must credit the payer bank during the settlement process. Figure 10.11 shows the links involved. The ATM belongs to bank A which is the payer and the ATM is connected to the processor HA which is in turn connected through the interchange I_1 to the processor HB of the bank B which is the card issuer. The request message from the ATM takes the path ATM-HA-I-HB to the card issuer. The PIN is verified at HB against the card identity, the status of the card and the account are checked and a response is returned giving or denying approval for the payment. The approval response has two functions, it allows the ATM to issue the cash and it also allows the payer bank A to present an item in its account for settlement by bank B. In the figure we also show a second interchange I_2 because we assume that shared ATM systems will grow by the joining together of interchanges to generate, eventually, very large systems. The principles which are adopted for shared ATM systems should be capable of extension to a hierarchy of sharing.

The system must keep to a minimum the opportunities for fraud by employees of one bank against another. Clearly the banks must share secret keys if symmetric ciphers are used for authentication and encipherment. These keys must be handled in a way which reduces the dependence of one bank on another's

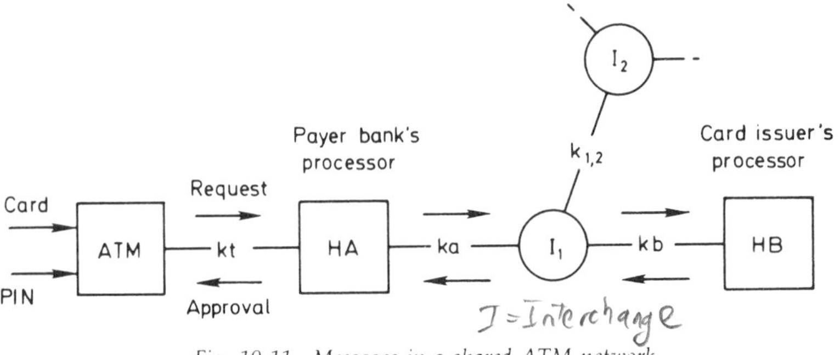

Fig. 10.11 Messages in a shared ATM network

security. In the figure, each link is associated with one secret key. The various ATMs attached to processor HA can have individual keys of which *kt* is an example. For connection to the interchange, bank A uses key *ka* and bank B uses key *kb*. Between the interchanges I_1 and I_2 the key $k_{1,2}$ is used.

The PIN entered by the customer at the ATM is carried in the request message to the processor HB and the approval is returned via the payer bank's processor HA to the ATM. The encipherment of the PIN in the request message, the authentication of the request and authentication of the response employ the keys shown on each link, therefore a transformation of the request message to a new key takes place at processor HA and at the interchange I_1.

A transformation at the interchange is necessary if a large number of banks is involved and it is not convenient to establish a key for each bank pair. Dividing the network into 'zones' in this way is essential if it is to be capable of unrestricted growth by combination with other banks or groups. The transformation at HA is not strictly essential, because bank A might decide to use a single zone-key *ka* for all its ATMs and for interchange. It could be argued that this increases the risk because a discovery of one key then exposes the entire network of bank A. In principle, discovery of *ka* exposes the network even when each ATM has it own key, *kt*. Probably the storage of *ka* is better protected than that of the keys *kt* in the terminals, but this is arguable, since ATMs are physically strong. For generality we shall assume separate terminal keys.

Whenever a message is transformed between one zone-key and another, as in the interchange, there are potential security risks. Keys of two different banks are present and might be exposed, the PIN is deciphered before re-enciphering and thus appears in plaintext form and the authentication of the message is renewed — a process which is vulnerable to active attack. Such a procedure cannot be safe in a mainframe processor and it is, therefore, recommended strongly that all the sensitive data and functions are within a physically secured module. Figure 10.12 shows what happens to a message inside such a module.

Keys for all the links entering the node are held inside the secure module, but if there are too many keys they can be called in from storage outside, on condition that the keys are enciphered by a 'storage master key' while they are outside the module. In the example of Figure 10.12, the key k_1 is associated with the incoming message M_1 and the key k_2 is associated with outgoing message M_2. First the authenticator A_1 of M_1 is checked using k_1. If this fails, the message sent forward is not the request or response message but a diagnostic message to state the nature of the error. It is sometimes suggested that a failed message be sent forward with a wrong value of authenticator, to propagate the failure, but this loses useful fault-tracing information. If the authentication process checks, a new message is assembled using the PIN value which has been deciphered with k_1 and re-enciphered with k_2. There seems no good reason to use different keys for the encipherment and authenticator procedures. The new message is equipped with a new authenticator A_2, assembled and sent forward.

Instead of separate encipherment and authentication, the whole message can be enciphered. If precautions are taken against reply or resequencing of messages this is equivalent to authentication. It has been argued that encipherment makes treatment of the request message at the card issuer more difficult because the whole message must then be deciphered for authentication, and the PIN is

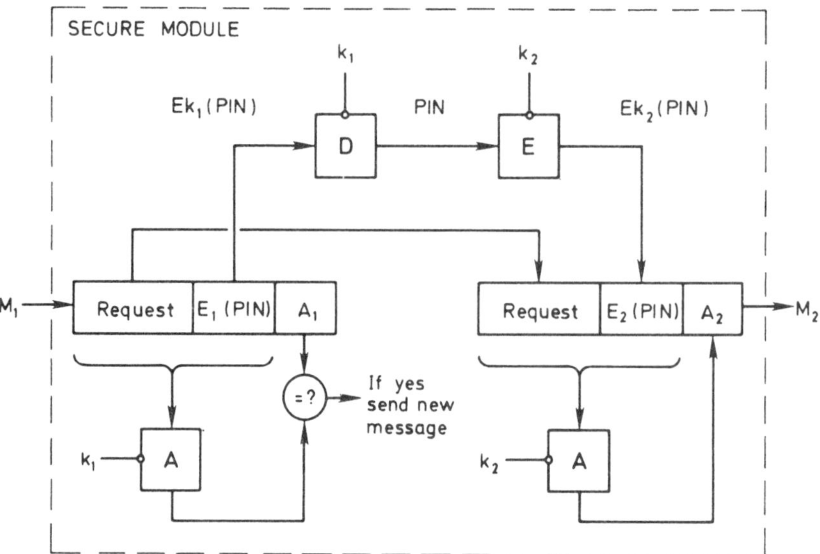

Fig. 10.12 Message conversion between zone-keys

revealed. But if the decipherment takes place in a secure module where PIN checking is done this does not reduce security.

The precautions against replay or resequencing require that message sequence can be checked at the receiver. The ATM's request messages can easily maintain their own sequence counters so that all messages they authenticate are different, but the checking process at the issuer will not be able to keep a note of the sequence numbers of all sources if there are very many banks in the interchange. In this case a date and time field is the best solution, with the precautions already noted in chapter 5. Authentication between a bank and its own ATMs should be able to use sequence numbers.

There is a different way to combine authentication and encipherment using the same key which is to compute an authenticator in the usual way over the message (excluding the PIN) and add this to the PIN before incorporating it in the message. The addition could be modulo 2 addition of binary digits, or modulo 10 addition of decimal digits, according to the way the authenticator is coded. At the receiver, the authenticator is recalculated and, by subtraction from the authenticator in the message, the PIN is obtained. If PIN checking fails the cause may be either a wrong PIN value at the ATM or an authentication failure. Perhaps it does not matter, since the outcome is to refuse payment in either case. Figure 10.13 illustrates this method and Figure 10.14 shows how easy it is to recompute the authentictor/PIN at the interchange or wherever a new key is needed.

Each bank must trust the other banks in the interchange system to the extent that their handling of keys is safe and the physical protection of the secure module is good. Each bank has within its grasp sufficient information to derive the PINs of transactions it handles. But this is inherent in the use of PINs, since they are in any case exposed at the point of entry into an ATM.

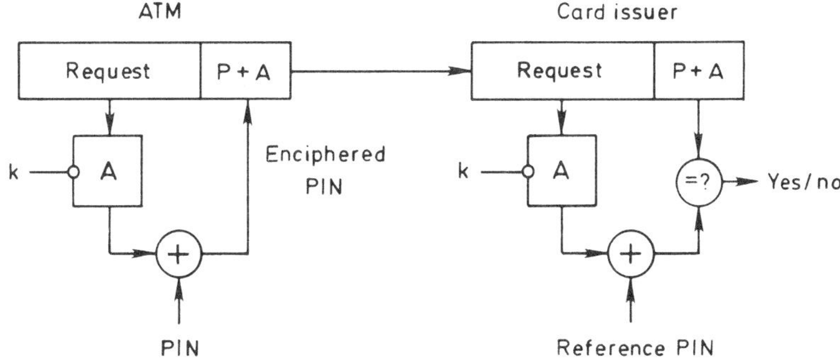

Fig. 10.13 Combination of authentication and PIN encipherment

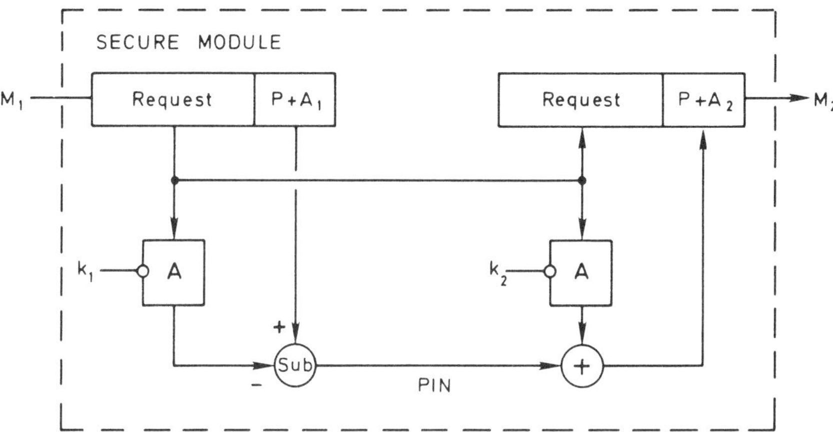

Fig. 10.14 Message conversion for P + A method

Checking the PIN with an authentication parameter

An alternative method of checking the PIN is to transmit from the card issuer to the ATM a function of the expected PIN sufficient to make the check. This quantity is called an 'authentication parameter' (AP). Suppose the pin value is P and there are data D on the card's magnetic stripe, including the account number and other things. Some of these extra data may deliberately be made 'random' or unpredictable. The authentication parameter A is a complex function $A = \mathrm{f}(P, D)$. For example, the function might be derived from data X related to the account number and other card related data Y. A suitable function f would be $X +_2 \mathrm{E}k(X)$ where $k = Y +_2 P$, since this is a one-way function of both X and $Y +_2 P$. The reference value of A is held at the card issuer's processor. When it receives a request containing the account details for this card it first verifies the account status. If this is satisfactory it authorizes the payment in its response, subject to the correct PIN being given and returns the response value A for the check to be made. At the ATM, when the PIN is received, with X and Y derived from the card, a local value A' is calculated. If $A' = A$ the payment is made and a confirmation message is returned to the card issuer, which records the transaction and updates its accounts. This third message is essential if the AP method is used, since the outcome is still uncertain when the card issuer gives its response.

The message in which the AP goes to the ATM can state other PIN criteria, such as the number of tries the customer has, to get the correct PIN, and whether the card is to be retained if all fail. The trials can be conducted locally, whereas the other method, in which the PIN goes to the centre, requires network messages for each trial. The number of trials must be a parameter in shared ATM systems because there are several issuers and each sets his own rules. Some may not even use a PIN, and in that case the card issuer does not want the ATM to ask for one. So the AP method is more economical in terms of network messages if there are in practice enough re-tries of the PIN.

The AP method of PIN checking does not remove the need to transmit data in secret. Knowledge of A does not allow $Y +_2 P$ to be determined directly, even if X can be deduced from the account number in the message. But if Y is also known it is usually easy to find P because of the restricted range of passwords in bank systems. Many use only 4 decimal digits, giving a task of only 10 000 trials to find P. The card formats used by different banks vary and some have little data Y that are not deducible from the account number. A general purpose network able to handle messages for many banks should regard the secrecy of the AP just as seriously as that of the PIN. The obtaining of the PIN value by a search is not a result of the particular form of AP function employed. It is due to the small range of PIN values, since any 'one-way function' can be inverted if its range is small.

Even allowing that there is little security advantage, the AP method can have advantages in limiting the worst-case number of message per transaction. The 'confirmation message' which goes from ATM to card issuer is an overhead which some consider necessary to help resolve the cases where messages are lost. Some packet-switched networks allow three messages for a unit fee in the 'fast select' mode, so the confirmation message may be 'free'.

Public key cryptography in a shared ATM system

The security requirements of enciphering PINs and authentication messages can be met by public key ciphers and signatures.

Taking first the case of a single ATM network, it is only necessary to have one set of keys, a secret key at the centre and a public key which is stored in each ATM. This allows encipherment of the request and confirmation messages and signature of the response from the centre. Secrecy of the response is not needed unless the AP method of PIN checking is used. A further requirement is to authenticate the request message, since otherwise an active attack could modify the request or introduce bogus requests. To avoid the need for the ATMs to hold their own secret signature keys, they can use a password to identify themselves, since this is concealed in the request message. A random number in the request, which is returned in the response, prevents the ATM acting on a replayed response. If a second random number is used by the centre to authenticate a confirmation of payment, an active attack can be detected even if the password were discovered.

In the context of shared ATM systems, public key methods are even more valuable because they avoid the need for zone keys and the corresponding key distribution problems. But the simplifying feature of using the ATM's password for authentication is lost. The ATM must be able to sign its messages using its own secret key. A group of banks in a shared system can easily exchange their own public keys but it would be difficult, even impracticable, for them to exchange the public keys of all their ATMs. This problem can be solved by having each ATM send with its request a 'certificate' of its public key signed by its own bank.

If the shared system becomes very large and expands rapidly by combining with other systems, keeping in each ATM the public key of each issuer becomes difficult. Then a central authority can be established as the key registry and each card contains the public key of its issuer, signed by the registry. The ATMs hold the registry's public key and check the issuer's public key before using it to encipher the (signed) request message.

There is a more radical approach which embraces shared ATMs, point-of-sale EFT and other payment systems, employing a 'smart card' to originate a transaction with a digital signature. This is described in section 10.6 starting on page 329.

10.5. POINT-OF-SALE PAYMENTS

The majority of payments in shops may be cash but it is the cheque and credit-card payments that concern banks. These present two problems, the expense of operating them and the increasing extent of fraud. The operating cost can be reduced by improved data capture, avoiding paper documents. The fraud can be reduced by on-line operation and improving the identification of the cardholder. The PIN, though it has limitations, has proved reliable enough for ATMs so its extension to point-of-sale payments seems inevitable.

From the customer's viewpoint he has not much to gain. Having a different PIN for each of several payment cards will be a nuisance. Moving from a credit scheme to one in which his account is debited at once is a disadvantage. Neither of these things is inherent in point-of-sale EFT but they are possible consequences of the change. The shopkeeper can gain from the more rapid payment and because the discount to the credit-card company is no longer needed. Hopefully the salesperson will find the procedure easier and the responsibility for checking the card signature and validity is ended. The operation should be faster, which helps at busy situations such as supermarket checkouts. The balance of advantage is in favour of extending the use of EFT in shops and phasing out the paper-based systems.

This is the motivation for point-of-sale EFT. The basic principle is illustrated in Figure 10.15. The main parties in the transaction are the customer who holds a card (or other token) and remembers his PIN and the merchant who is to receive the payment for goods or services he supplies. In general these parties have accounts at different banks, the customer relating with the 'card issuer' and the merchant with the 'acquirer'. The acquirer will usually have installed the terminal and connected it to its own payment system. In the figure, the terminal is shown connected to the acquirer's system HA and the acquirer's system exchanges messages with the card issuer's system HB through an interchange. The figure shows only one of several message patterns which are discussed below.

Typical operators of point-of-sale terminals will be those who need to receive payments of modest amounts from the public, such as shops, travel agents, ticket offices and professional services. No one word is adequate to denote this wide range of activity so we will employ the conventional term 'merchant' to designate the person or institution which uses a point-of-sale terminal to receive payments. Whereas an ATM system for a single bank can flourish in the countries where there are nationwide banks, point-of-sale systems are not attractive unless the merchant can use his terminal to accept cards from a large group of card issuers. They are essentially shared systems.

The sequence of operation is that a sale is agreed between the customer and merchant and the payment amount is displayed for the customer's approval. The customer then passes his card through the card reader on the point-of-sale terminal (usually a simple swipe reader) and enters his PIN on a keyboard attached to the terminal.

Before the transaction can be completed a payment-request message must go to the card issuer who will verify that the card is valid, the PIN correct and the account in good order. A response then travels from the card issuer HB to the acquirer HA. This 'approval' response is effectively a payment message which authorizes the acquirer bank to credit the merchant's account with the amount requested. From the acquirer's system an advice passes to the point-of-sale terminal telling the merchant that the transaction is complete and causing it to print a receipt for the customer.

The natural progression of messages is from terminal to card issuer to acquirer to terminal — a triangular route. The first requirement is to validate the card and customer, the second to make a payment from HB to HA and the third for HA to tell the merchant that the payment is made. But a triangular path makes it very difficult to recover if the transaction fails in the middle. Recovery must at any one time be the responsibility of one unit in the system. We can assume this is not the terminal because of its relative insecurity. The scheme of Figure 10.15 places the acquirer's processor HA in the strategic position to handle recovery by re-tries. For this purpose, the outgoing message to the card issuer goes through HA, where the new transaction can be logged for checking its safe return as a payment or a refusal. This pattern of messages also matches the bidirectional nature of most data calls in a switched network.

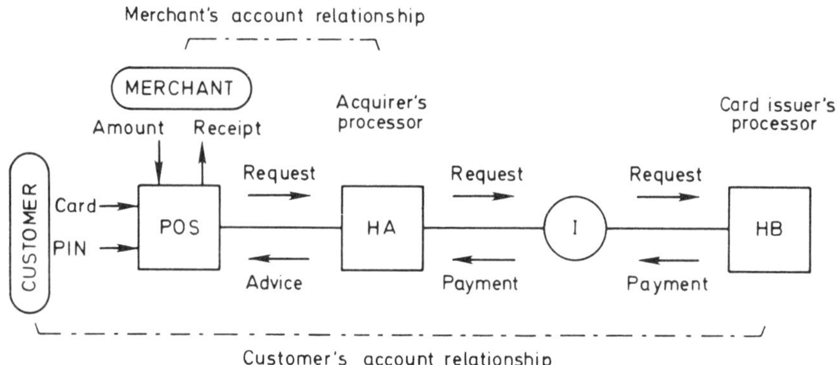

Fig. 10.15 Point-of-sale electronic funds transfer

This scheme requires both the card issuer and the acquirer to be on-line for a transaction to happen. Central checking of the account certainly requires the card issuer, but the actual payment can wait, provided that the merchant can quote the message from HB stating that a payment has been recorded. Where the number of issuers is small and if there is good authentication in the messages, the card issuer can take the central role and the acquirer's participation can be delayed. HA and HB change position in Figure 10.15. Then the onus of recovery falls on the card issuer's processor HB. The message path is terminal — card issuer — acquirer and back to the terminal. In yet another adaption of the system, the responsibility for recovery can fall on a separate processor working for the whole network, much as I does in the figure. There are many ways to organize the responsibilities and the message paths, but the basic security requirements are similar in each case.

In addition to these normal payments there are other types of transaction which the terminal must generate. Since the use of the terminal must be controlled, it is logged on and off by the merchant using a magnetic-striped card and PIN for the purpose. This transaction is verified at the acquirer's system and it may, in some systems, result in the distribution of session keys to the terminal. Another, optional, transaction type is the *refund* which requires special authorization, by the same merchant's card and PIN.

The payment messages can cause instant debiting of the customer's account and crediting of the merchant's account. These immediate transactions are balanced by equal amounts in accounts maintained between the issuer and acquirer. For example, the issuer may maintain an account for the acquirer which he credits with the amount paid. These accounts are settled periodically by an inter-bank transfer.

Point-of-sale EFT provides a better method of capturing data and helps to reduce fraud, but it changes the nature of the transaction by debiting the customer's account at once. Many customers will find this unwelcome and prefer to continue with cheques and credit-card transactions where the delay in payment works in their favour. It is not essential that these systems debit accounts at once. They capture the data and check the card validity in a way which could equally function like a cheque, with a clearing delay, or a credit transaction.

The requirements for security in point-of-sale systems are essentially those we have already examined for shared ATM networks. For on-line systems they can be summarized in two statements.

1. The verification of the PIN entered by the customer must be carried out in association with the issuer's processing system. The PIN value or AP value carried through the network for this purpose must be enciphered to protect it against line tapping or disclosure in any of the processing systems through which it passes.
2. The messages which carry the request and the approval/payment response must be authenticated to prevent their undetected alteration whether by active line tap or in any of the processors through which they pass. Among other attacks, the possibility of a replay must be guarded against.

The principles we have described earlier show us that both the encipherment and the authentication must employ variable elements and the receiver of messages must be able to check the sequence in which they were sent. Any of the authentication methods described in chapter 5 will serve.

If the 'zone key' principle is used there is a different key for each of the three links shown in the figure. The presence in one processor (for example the interchange) of keys shared with a number of other institutions presents risks which can be minimized by using tamper-resistant modules (as we described for the shared ATM network), and by paying careful attention to key distribution. Employees of the interchange organization should have no means of access to the keys installed for communication with their associated banks.

Figure 10.15 shows the messages carried for a normal payment in the chosen message scheme. The request message travels from the terminal to the issuer's processor HB. The intervention of HA and the interchange I in this message transport is needed for the change of keys when the zone encryption scheme is employed. The response message travels from HB to HA as a payment and continues on to the terminal as an advice to the merchant that the payment has been made.

A refund transaction needs special authorization at the terminal, employing a card and PIN belonging to the merchant. In this case the request message carries the PIN, enciphered, to HA where the PIN is verified and the message authenticated. If these are correct, the payment message is sent from HA to HB. The advice then returns from HB via HA to the terminal. The requirements for PIN encipherment and message authentication are similar to those for normal payments.

Point-of-sale payment terminals are part of an interchange system shared between many card issuers and acquirers so that a wide choice of cards may be used. Since it would not be practicable for the card issuers to exchange their blacklists of lost, stolen and otherwise exceptional cards, there is a strong argument for on-line connection of terminals. Nevertheless, the cost of on-line connection and the communication difficulties in some countries will strongly favour off-line operation. Off-line terminals collect together the records of payments and periodically the complete record of payments is sent to the acquirer bank, either as physical record or by a telephone call.

A compromise between the cost of on-line connection and the need for security is to use off-line transactions for payments below a certain value and within a participating group of banks. For payments of high value and those involving other banks, a validation call is made to a service operating on behalf of the card issuer.

The use of public key signatures promises to improve security and increase convenience in payment systems generally. A comprehensive payment system using public keys can be devised which is suitable for large ATM networks, large point-of-sale networks and off-line transactions between bank customers analogous to today's bank cheques. In the final section of this chapter we shall introduce this form of payment using signed messages and show how it can be adapted to a point-of-sale network.

The transaction key method

In Figure 10.15, the handling of keys for point-of-sale EFT is like the method customarily used for ATMs but there is a world of difference between the physical security of an ATM and of a point-of-sale terminal. Because of the need to protect the money it holds, an ATM is made strong and can physically prevent anyone from reading the secret keys it holds. On the other hand, when point-of-sale terminals are widely distributed in shops a criminal gang can easily obtain a few, take them apart and discover, after some effort, how to read the keys contained inside. They might, in the course of reading the key, virtually destroy the terminal so that it cannot be used again. Even so, if they had recorded the transactions for a long period before breaking into the terminal, possession of the keys could give them all the PIN values that have been exchanged. Terminals can be made more secure by further tamper-resistance features but, in the end, their keys can be read out.

To make point-of-sale EFT systems more secure, special forms of key management have been developed, specifically to prevent the kind of attack we have described, called *back tracking*. There is another kind of attack called *forward tracking* in which the terminal is put back into use after the keys have been discovered and the subsequent messages it handles betray the PIN values. But anyone who is skilled enough to do this could probably arrange for the terminal to record for him all the PIN values as they are entered. Therefore, forward tracking is such an advanced attack that, even if only back tracking is prevented, the system is safer.

A scheme of key management which, if the circumstances are right, prevents both forward and backward tracking is *transaction key*. The principle was first explained by Beker, Friend and Halliden[7] in a form in which each terminal maintains a special key in common with each card issuer and this key, the transaction key, is changed at every transaction in a way in which an enemy would find it hard to follow. This is an excellent principle, but when there are very large numbers of terminals and a fairly large number of card issuers, the multiplicity of keys makes it difficult to administer. In particular, as terminals come on line or are replaced after maintenance, all the card issuers have to be updated with their transaction key data. For this reason, most practical applications of the transaction key principle employ the transaction key principle between the terminal and an *acquirer*, with each acquirer having to handle only its own set of terminals. The scheme has been extended to protect the PIN while it continues in transit to the card issuer, as we shall describe below.

There are two special features of the transaction key method, the way in which it handles the authentication and integrity checking of messages and the way in which it maintains the transaction key and modifies it after each transaction.

Authentication and integrity checking are provided by the MAC based on the data encryption standard. When a MAC is produced the results of the cryptography are 64 bits from the final DES operation. Normally, only the left-hand 32 bits are used as the MAC and the right-hand 32 bits are discarded. In the transaction key method the right-hand 32 bits (called the *MAC residue*) form an important part of the scheme. We shall abbreviate MAC residue as MAR. When the terminal sends a request to the acquirer, it includes the MAC in the request but retains the MAC residue (MAR1) for use in authenticating the response. When the response is returned, a MAC is included with it and the MAC residue (MAR2) is retained by the acquirer. Each checks the MAC value on the messages it receives. The special feature of the transaction key method is that the MAC attached to the response is formed cryptographically not just from the message but from the message preceded by MAR1. The acquirer has obtained MAR1 as a byproduct of checking the authenticator on the request. But MAR1 is not sent in the message, it is, so to speak, a 'ghostly' part of the message and the terminal will not be able to check the authenticator unless it precedes it by MAR1 before applying the cryptographic process of MAC calculation. As we shall see later, MAR1 is not the only ghostly part of the response message, there may also be an authentication parameter (AP) which we describe later. This idea of including in each message a ghostly residue of the previous message before authenticating it effectively chains the messages together and assures the terminal that the acquirer has properly authenticated the request. If more than two messages are used in the dialogue, for example there may be a confirmation from terminal to acquirer, the same chaining principle can be used, the response having its MAC formed with a ghostly value of MAR2 preceding it.

Perhaps more important than this authentication scheme is the way that the transaction key is formed and used. For this purpose the terminal maintains a 56-bit *key register* in the DES format for each acquirer it must work with and the acquirer holds the same values, having one stored for each terminal with which it communicates. After initialization to the same values, the key registers (KR) must synchronize their values after each transaction so that the terminal and acquirer can continue to communicate.

Figure 10.16 shows how the key register is used at the terminal and how it is updated. The function shown as O is a one-way function $O(k, x)$ derived from the DES operation as

$$O(k, x) = Ek(x) +_2 x$$

The symbol we use for 'O' shows which is the key input by the small circle.

From KR and data read from the card tracks, two transaction keys are formed tk_A and tk_E. The first of these is used for authentication and the second for enciphering the PIN between the terminal and acquirer. The input DC1, DC2 and DC3 are all derived from card data. DC1 and DC2 come from the account number and are therefore known to the acquirer but DC3 uses some of the discretionary

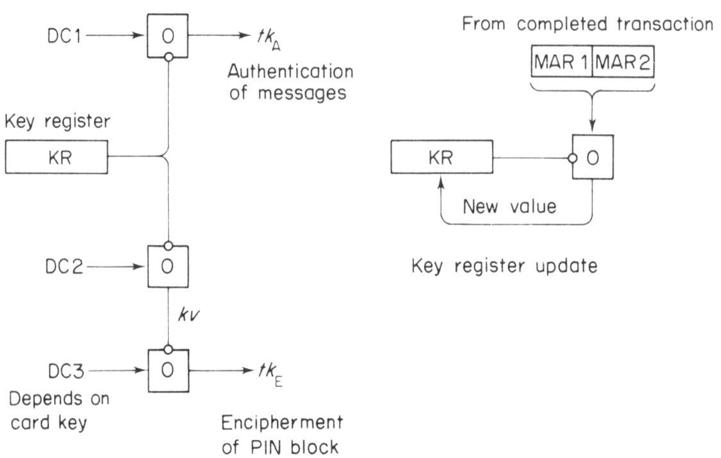

Fig. 10.16 Transaction key production and key register update

data field together with the account number for its calculation. Data from the discretionary field on the card is not sent with the transaction and is not known to the acquirer. The purpose of forming tk_E in this way is to encipher the PIN using a key which can normally only be calculated by the card issuer from its knowledge of the contents of the discretionary data field. Note that the quantity *kv* is known to the acquirer. This preparation of tk_A and tk_E occurs at the beginning of the transaction and these values are used throughout. At the end of the transaction the KR value is updated in the way shown in Figure 10.16, using as input a concatenation of the two MAC residues which are known to both terminal and acquirer at the end of the transaction. When there are more than two messages in the dialogue, there are special rules determining which MAC residues shall be used. After this update, KR depends on the entire progress and outcome of the transaction. If an attacker knew how it was initialized and wanted to track its value he would have to record every message and also discover the discretionary data from which DC3 was derived. The value DC3 is sometimes known as the 'card key' because it has a degree of confidentiality to the terminal and card issuer. It is not difficult to read it from the card but the need to maintain an unbroken chain of complete data including the card key makes forward tracking very difficult. Backward tracking is impossible because of the one-way function.

We have seen how tk_A is used to authenticate the request and response. The enciphered PIN must reach the card issuer who does not have the means for maintaining a KR value. To enable the card issuer to reconstruct tk_E in order to decipher the PIN, the acquirer calculates *kv* and sends it to him in the request message as shown in Figure 10.17.

There is an implication in the way that the scheme is officially described that *kv* should be protected by encipherment for transit between acquirer and card issuer, though this is not mandatory. Certainly, if *kv* is known and the card key, the PIN is not safe from decipherment. Depending on the level of security desired users may wish to encipher *kv* and this encipherment uses a conventional key which is changed from time to time but not per transaction.

In order to authenticate that the response came from the card issuer, an authentication parameter (AP) is returned from the card issuer to the acquirer when it

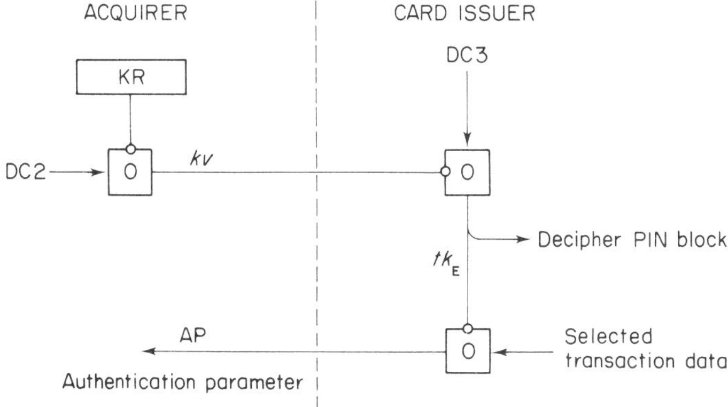

Fig. 10.17 Authentication of card issuer response

approves (or not) the transaction. This is formed by operating with tk_E using the one way function on certain critical data taken from the request message. If the AP is correctly returned, the transaction is assumed to be approved. In order to confirm this from acquirer to terminal, AP is concatenated with the response message when its MAC and MAR are formed, though it is not returned in the response message sent. Together with MAR1 it forms the 'ghostly' part of the message for authentication purposes. When authenticating the response the terminal can also form AP and concatenate it with the received message to apply the cryptographic process of authentication.

The transaction key method has a number of variants, one of which is incorporated in an Australian national standard.[8]

Initialization of the key register can be done in a number of ways. If the problem is sidestepped by using a known constant for the initial value, assuming that all the cards have unpredictable data in their discretionary field, it will soon become unlikely that anyone has tracked the value of KR. But it is better to use an unpredictable starting value and this can be provided by having the first transaction (or a few transactions) performed with cards that are carefully preserved from illicit use. This could be done by an installation engineer or a representative of the retailer. If synchronism between KR values at the terminal and acquirer is ever lost and must be restored, a known constant must be placed in KR and the initial transaction repeated, so the best plan is that the retailer should perform this ritual. There are other more elaborate initialization methods but the ones described will usually be adequate.

Suppose that the transaction is approved by the card issuer and the response is correctly produced by the acquirer but due to a fault, perhaps in communication, the terminal cannot complete the authentication and regards the transaction as invalid. In this case the terminal will not update KR but the acquirer will and there will be a loss of synchronism. To avoid this, whenever the acquirer receives a request which does not authenticate, it tries the authentication again but with the KR value it would have used if the previous transaction had not completed. Therefore the acquirer must maintain not only the fully updated value of KR but also the previous value. With this protocol, successful receipt of a request implicitly confirms that the previous transaction was completed at the terminal. This is not essential and transaction key systems may use a confirmation message instead.

The transaction key method is an effective way to avoid the risks of tracking backwards or forwards. The most important protection is against backwards tracking because this is technically much easier for the attacker.

Derived unique key per transaction method

An alternative type of transaction key has been developed in the USA with the ungainly title 'derived unique key per transaction'. As in the transaction key method, the keys employed for enciphering the PIN and authenticating the messages change at each transaction but they change in a fixed way which is independent of the transaction data. By careful design this can provide protection against tracking backwards but not, of course, against forward tracking. For many purposes protection against the greater risk of backwards tracking may be considered adequate. The method is conceptually simpler than transaction key and simplifies the problem of synchronization, but does not provide such good protection.

The derived key method, as we shall call it, provides a unique key per transaction for encipherment of the PIN between the terminal and the acquirer. A key derived from this one can also be used for authenticating the messages. We shall simply describe how the single key is calculated. For PIN encipherment and message authentication between the acquirer and card issuer, a normal kind of session key, often called a *zone key* is used. The cryptographic processes in the acquirer will re-encipher the PIN and must have a good level of logical and physical protection.

When the terminal is initialized, it is loaded with an initial key I and the same key is registered at the acquirer. There is an identifier to enable the acquirer to retrieve this key whenever a transaction arrives from the terminal. The value of I determines the whole sequence of subsequent keys. There are procedures in both the terminal and the acquirer for generating the derived key at any stage. The procedure can generate only a finite number of keys but this number, which is $2^{20}-1$ (equal to 1 048 575), is considered large enough for the probable life of most terminals. If necessary, the terminal can be reinitialized with the new I value.

At the terminal, the key derivation process takes place in a PIN pad and is sequential in nature. It uses one-way functions in a manner which does not allow previous values to be deduced, even if the secure environment is broken into and all the data extracted. It cannot therefore derive the key each time from I but must be able to produce the next key from stored data and always destroy sufficient previous information to avoid backtracking.

At the acquirer a different process takes place in a security module and here the starting value is always I and the derived key must be produced with a reasonable amount of computation given only the sequence number of the key. To avoid synchronizing problems, this sequence number is sent with every request message.

We shall first describe the rule by which all the keys are generated and it will then be easy to see how the security module at the acquirer can produce any one of the keys on demand with a modest amount of computation. Next we shall describe the different procedure used at the PIN pad which stores just sufficient intermediate values to enable the next key to be deduced.

If there had been no requirement for rapid production at the acquirer, a very simple process could have been used, such as iterating a suitable one-way function. This easily produces a sequence which cannot be tracked backwards but generating the thousandth key at the acquirer would need one thousand steps, which is not practical. Instead of using the one-way functions in a linear sequence, they are used in a tree structure so that the number of steps required to reach the end of the tree is much less.

Figure 10.18 shows the tree structure with the initial key at the top. From each node there are two downward paths of which the left one has no processing, it simply produces a copy of the key above. The right path in each case has a one-way function and this function is different at each level of the tree so that the results at the lowest level form a sequence of distinct keys which can be reconstructed quickly, given I. In practice, these keys are taken from level 21 but not all of them are used.

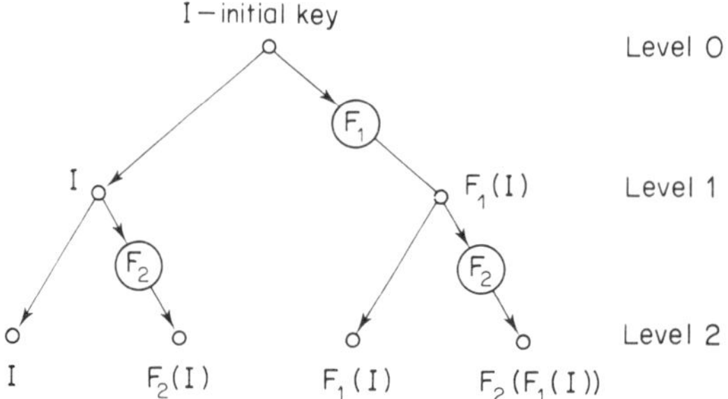

Fig. 10.18 Generation of sequence of keys

To generate a key with sequence number S it is sufficient to express S in binary form and, starting at the most significant end, use the binary pattern to decide whether or not to apply a one-way function. If the digit is 1, the function is applied. In this way, a key can be generated with at most 21 operations of the one-way function. The one-way function used is derived from the Data Encryption Standard and a fast chip can be employed in the security module at the acquirer.

To get even more rapid operation, the sequence of values S is limited to those which have no more than ten 1s. This more than halves the number of operations required and the number of keys is approximately halved except that, since I is not used, there are actually $2^{20}-1$ keys in the sequence.

A special rule is needed to increment S in this way. We shall introduce two functions which will be useful in describing later the program at the terminal. Let COUNT(S) be the number of '1s' in the binary representation of S and LOWBIT(S) be the position of the least significant '1' bit in S where the most significant bit is counted as 1 and the least significant bit as 21. By convention, if S = 0 then LOWBIT(S) = 0.

The rule for counting such that S never exceeds COUNT(S) = 10 is:

IF COUNT(S) IS LESS THAN 10 THEN S = S + 1.
ELSE S = S + $2^{21-\text{LOWBIT(S)}}$

Using this rule, if the weight of S has already reached 10, 1 is added into the place of its least significant 1 bit and therefore the weight will either stay constant or decrease, otherwise 1 is added to the least significant place in the usual way.

The method by which the current key of the sequence is derived at the acquirer is obvious from the structure of the tree. At the PIN pad, the procedure depends on a number of stored intermediate values each corresponding to a node in the 21-level tree. There is at most one stored intermediate value for each level of the tree and these form an array K(X) where X is the level ranging between 0 and 21. There are two such pointers in the algorithm, CP which points to the current key and SP which is a moving pointer used in preparing the stored array for the next key.

The algorithm is shown in Figure 10.19 where the key is derived as K(L) in the second line. Briefly it can be described as follows:

If S is odd, the required key is already stored in the 21st level as K(21) and, after retrieving it, little more has to be done except deleting the key from the array.

```
    L = LOWBIT(S)
    K = K(L) Remark: K is the current key
    IF L = 21 THEN GOTO 'X' Remark: add value of S
    IF COUNT(S) = 10 THEN S = S + 2^(21-L): GOTO 'Y'
    FOR P = L + 1 TO 21
    K(P) = F_P (K(L))
    NEXT
X:  S = S + 1
Y:  CLEAR STORED VALUE K(L)
    IF S = 2^21 THEN DISABLE PAD
    END
```

Fig. 10.19 Generation of the key at the PIN pad

If S is even, then L = LOWBIT(S) gives the level at which the key is stored. This corresponds to tracing the key from the twig of the tree at level 21 upwards, using only paths with no one-way function in them, until the stored value is found at a higher level. After retrieving this key and deleting the stored values, the stored values at all the lower levels must be recalculated using the pointer P.

An improvement to the derived key method

We know that forward tracking is possible if the stored information can be obtained from the PIN pad. With the derived key method as described, knowledge of one key of the sequence reveals some of the subsequent keys, all those corresponding to S values obtained by inserting extra 1s in the binary representation of S. The worst case is the key for $S = 2^{20}$ from which all subsequent keys can be derived. In practice, the value of the key can only be obtained by cryptanalysis but the value of some of these keys to the cryptanalyst is such that this attack might become tempting. It can easily be avoided by a small change, introducing a one-way function into the left-hand branch at the lowest level of the tree. Fortunately, this never increases the number of steps in the calculation at the acquirer. In place of line 2 of the program the following two lines are introduced:

```
IF (L IS LESS THAN 21 AND COUNT(S) IS LESS THAN 10)
THEN K = F_X(K(L)) ELSE K = K(L)
```

The extra one-way function at the bottom of the tree is shown as F_X. With this improvement it is possible to employ K(0) as the first key since it equals $F_X(I)$ and does not reveal I. Then the initialization of the PIN pad is simply to set $S = 0$ and $K(0) = I$. Without this improvement, when the terminal is initialized, the updating shown in Figure 10.19 must be performed before any keys are used.

EFT-POS with public key cryptography

The central problem of EFT at the point of sale is the vulnerability of the terminal. The two attacks we have discussed are those in which the data contained in the terminal have been extracted and the attacker either tracks backwards to reveal PINs for earlier transactions he has recorded or puts the terminal back into service and obtains PINs for future transactions which will be recorded. In favourable circumstances, both kinds of attack are prevented by the transaction key method and backward tracking is prevented by the derived key method. If public key cryptography is used to protect the PIN, this inherently prevents back tracking and forward tracking. When the terminal has been broken into, only the public key used for encipherment is revealed and this gives no help with decipherment. Therefore, public key cryptography is a neat solution to the problem of terminal vulnerability.

A detailed design for using public key cryptography in EFT-POS has been worked out for the UK national system. We shall describe this briefly though not with details of key management at higher levels. Figure 10.20 shows the secret and public keys held by the main entities which take part and the relationship between them. For example, sk_T is the secret key held in the terminal and pk_T is the corresponding public key held by the acquirer. These keys are generated by the terminal during its initialization. Similarly the acquirer, card issuer and key registry each generate a pair of keys and make the public key known to others, as shown in the figure. The arrows point from secret key to public key and indicate how a signature can usefully pass from one entity to another. Such signatures actually accompany the request messages from terminal to acquirer to card issuer and the response messages in the opposite direction as the diagram shows.

Each entity, when it generates its keys, must register the public key with the key registry — a body which functions for the whole system. The key registry has its own secret key sk_R and the public key pk_R is made known, with careful precautions for its authenticity, to the terminal, acquirer and card issuer. When any of these entities requires an authentic public key it is received in a message from the card registry containing the key, the identifying data and a digital signature using the registry's key. Before using any public key for the first time, each entity checks its signature in this *certificate* which it has received. Terminals receive their certificates from the acquirer when they need them.

Until about 1987, it was difficult to find equipment to carry out the public key cryptography fast enough. By using signal processors, such as the Texas Instruments TMS320 with suitable software, a low cost solution for operation at reasonable speed in the terminal has become available. At the same time a specialized RSA chip has appeared which can perform these functions fast enough to meet the requirements of the acquirer, card issuer and key registry. With a short public exponent (65537 was used) the operations of encipherment and

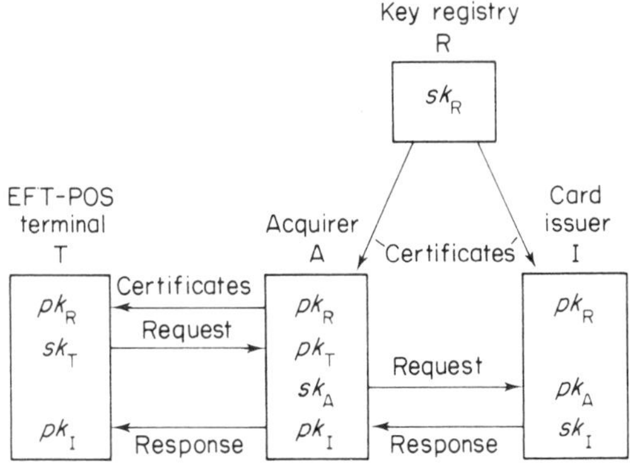

Fig. 10.20 Public key relationships for EFT-POS

signature checking are made very fast. Signature employs a one-way function using an RSA operation with a public key distributed by the key registry for which no secret key has been retained when it was generated.

With these keys in place, the security services for the request and response messages are as follows.

The PIN block is enciphered using the public key of the card issuer. There are many card issuers, but a small number of these issue the majority of cards and the terminal holds a cache of authenticated public key values for this purpose. If a card arrived for which no public key is held, the terminal requests this public key from the acquirer, which holds a complete set and provides them as certificates. Checking the certificate at the terminal is a rapid process. PIN block encipherment is provided from end to end, terminal to card issuer and this meets one of the main security requirements in a simple way, without depending on human operations to handle secret keys at the acquirer.

Authentication of messages is performed by the digital signature technique which simplifies key distribution since secret keys remain where they have been generated. The request from the terminal is signed with sk_T and this signature is checked at the acquirer which holds the corresponding public key. It would be nice to check the same signature at the card issuer, but that would complicate key management because card issuers would have to hold public keys for all terminals. Instead, the message is re-signed by the acquirer using its own secret key for which the card issuer holds the corresponding public key pk_A. Re-signature at the acquirer places some trust on the acquirer's software but authenticity of the request can, to some extent, be covered by the authenticity of the response. The response is signed by the card issuer with its secret key and this response can be checked by the acquirer and also passed to the terminal for checking. Provided the essential information on which the transaction depends is repeated in the response, the necessary trust in the acquirer re-signing the request message can be largely eliminated. The advantage is that both PIN block encipherment and transaction authenticity are virtually end to end.

Digital signature has another bonus. If an event is recorded in the audit trail at each entity as a signed message, and assuming these have sequence numbers to prevent insertion, deletion and copying, they can subsequently be checked by an auditor using the public key and they provide non-repudiation in its purest sense.

In practice, an EFT-POS scheme using public key cryptography may have to differ from this ideal in a few respects, but the description of the principle shows that all the requirements can be met in a very complete and simple way using public key cryptography. It is possible to arrange for all the key pairs to be replaced from time to time, if the users require it, except that the terminal's secret key would remain unchanged during the life of the terminal or, if it has to be changed, the terminal would be re-registered and treated as a new terminal for purposes of initialization. Terminal initialization involves registering the public key and storing in the terminal the public key of the key registry. The easiest method is to bring the terminals to a secure area under the control of the key registry for this purpose, though registration schemes based on the principles described by Matyas can be used.

In order to perform the signature reliably and handle the full level of traffic, acquirers and card issuers need more than one security module capable of performing all the cryptographic functions required. These security modules must share the secret key and an exchange of secret keys between these modules requires careful design to minimize its dependence on the operators and provide the necessary dual control. In a similar way, the registration of public keys requires careful operational design and similar considerations of dual control. Standardization of key management for EFT-POS will probably cover all the details of the messages exchanged and some aspects of initialization, but not the higher levels of key management since a degree of confidentiality here helps to keep systems secure.

As soon as it is freed from the constraints of processing power, public key cryptography in this context provides a simpler solution than symmetric algorithms. There is some additional message overhead but, with ingenuity, this is reduced to little more than is needed to make up the units of 64 bytes employed in RSA cryptography.

Off-line point-of-sale terminals and smart cards

Since an off-line terminal cannot contain a complete list of PIN values against card identities, it must apparently employ an algorithmic relationship between the PIN and data contained on the card. Since card issuers do not wish to share PIN information they must use different keys for their PIN algorithm. Off-line point-of-sale terminals which can accept cards from many issuers would need to hold the keys of all those issuers. Since point-of-sale terminals are physically insecure it is unlikely that card issuers would share their secret keys in this way.

Public key systems cannot help, though there have been suggestions that this is so. We can devise an effective way to use a public key system to check the PIN against data contained on a card. For example, if P is the PIN, I the card identity (account number, issuer, card's serial number) and R a reference value stored on the card, the value of R can be generated by using the secret key sk of an RSA cipher as

$$R + P = (P + I)^{sk}.$$

Any method of addition can be used. For checking the PIN, using the public key pk the relationship used is

$$P + I = (R + P)^{pk}.$$

This does not allow P to be obtained from card information because P is on both sides of the equation.

Unfortunately, this scheme is vulnerable to a PIN search because all the information needed to check a PIN is available and the searching can be carried out by a fast process without using a terminal. No process which uses only public information to check a PIN can be secure against PIN searching, unless the range of the PIN is sufficiently large, which probably means ten digits or more and is not acceptable for payment systems used by the public.

The remedy seems to lie in the use of tokens which contain processing and storage ability, the so-called 'smart cards' which we call *intelligent tokens*. These contain a microprocessor and a data store which does not need electrical power to maintain it. When the token makes connection with its terminal, the microprocessor and store access circuits are provided with power and the token is able to apply its intelligence in order to protect data in its store and control access according to pre-arranged rules.

Physical security requirements of the intelligent token

If it were possible to open up the token, read all the data within it and then make a similar token containing these data, the token would have very little advantage over a magnetic-striped card. It gains its advantage either by preventing the reading of its data or making it exceptionally difficult for the enemy to manufacture cards with the required properties. This difficulty of manufacture is a consequence of making the token in the form of a thin plastic card, so that readily available electronic components cannot be used to counterfeit it. It would seem a very difficult, though perhaps not impossible, task to read the data from a card. The effort required would only be worthwhile if it gave a large pay-off in the form of duplicate cards which could be exploited for fraud.

PIN checking in an intelligent token

By checking the PIN within the token itself, it is possible to use an arbitrary PIN value. There is no key for checking an algorithmic relationship and the card issuer need share no secrets with the acquirer's terminal. The principle is simply that the PIN value is entered on a keyboard in the terminal, transferred directly to the card through its electrical interface and the checking of the PIN is carried out within the token. The token then signals to the terminal if the PIN is correct. The token itself can enforce rules to prevent PIN searching. It can also apply rules on behalf of the issuer to limit the amount of payment the card may make within a certain period. In doing this, the card depends on a current date given it by the terminal. Falsifying this date in the terminal produces transaction records which are unacceptable to the issuing bank and is unprofitable to the merchant.

For example, the issuer could impose a limit on the total payment provided by the token so that the customer requires a new token to continue making payments. The issuer could limit the amount of payment in any single transaction or series of transactions to the same account, or limit the total number or amount of transactions in one day. Any of these rules can be applied within the token just as well as by an on-line system. Only the absence of a comprehensive and up-to-date blacklist makes off-line operation less secure in this respect.

The matching of the PIN within the token could use a reference PIN value. Alternatively, a one-way function could be used, which does not require the PIN to be stored and this might seem to be more secure, for the PIN cannot be read from the token (even if the reading technology is available to the enemy). However, this approach is not really more secure because the small range of the PIN allows it to be found by searching whenever information is available about how the checking takes place.

The result of a successful PIN check may be to generate a properly authenticated request message for sending to the card issuer. In an on-line system this results in an authenticated approval message telling the merchant that the payment has been made. In an off-line system there is a further difficult requirement, which is to assure the merchant that the PIN checking has been properly carried out and that the match is genuine. Even without a micro-processor it would not be difficult to construct simple logic which always gave the 'PIN match' response to the terminal, whatever PIN was inserted. Verifying the PIN within the token is easy but conveying this message convincingly to the terminal can be difficult.

The use of a one-way function would, it seems, allow the terminal to check the PIN it has been given with the PIN held in the token without the ability to read the PIN from the token. For example, the terminal generates a random number x which it sends to the token and receives in reply the response $EP(x)$, where x has been enciphered using for the key a quantity derived from the reference PIN, P. By performing the same calculation with the PIN keys in by the customer, the terminal is able to verify the PIN without being able to discover the reference value in the token. In this way the terminal verifies that the token actually contains the pin and is 'intelligent'.

Unfortunately, this scheme fails because of the small range of PIN values, which allows PIN searching. Figure 10.21 shows an alternative procedure which partially solves this problem and allows both the terminal and the card to verify that the entered PIN is correct. In place of the encipherment function E we have used here a one-way function O such that, if

$$y = O(k, x),$$

a knowledge of values of y and x does not allow k to be determined. We can think of k as having at least 56 bits to prevent searching. The function O could be replaced by encipherment with the DES algorithm, but an encipherment function is not essential and the one-way function $O(k, x)$ can serve just as well.

In the procedure of Figure 10.21, the PIN, P, which can be thought of as a 16-bit number, is concatenated with a 40-bit number C to form a 56-bit 'key' for the one-way function. The terminal sends the value x to the token which returns the response

$$y = O(P \| C, x).$$

TERMINAL		INTELLIGENT TOKEN
Offered PIN P′		Holder's PIN P Constant C
find random x	→	x
y	←	y = O(P‖C,x)
send P′	→	Is P′ = P?
is y = O(P‖C,x)?	←	If true, send C
If true, accept PIN and token		

Fig. 10.21 Protocol for intelligent token authentication

There is no way in which the reference value P could be obtained by searching unless C is known. After this response, the terminal transfers to the token the trial value, P', which has been entered by the customer. The token's processor checks this against its reference value P. If the two match, the token returns the 40-bit number C and the terminal then computes $O(P \| C, x)$ and checks whether this agrees with the value y received earlier.

A token made by an enemy without a knowledge of P would be unable to give the response y. Because y and x have 64 bits there is probably only one value of $P \| C$ which matches the x, y pair and this contains the 16 bits of P in their correct places. There is no way in which the token could find a value of C to satisfy the final check in the terminal unless it contained the value P at the time of its last response.

Thus the token is able to check that the correct value of P was entered at the terminal and the terminal is able to check that the token contained this value, assuming that the PIN was correctly entered.

Having once obtained the value C, the token is an 'open book' to the terminal but this is no loss since the terminal could equally well store the value of P associated with a successful match. The value C is of no greater value to it than P.

Following the correct match of the PIN, the card would be prepared to generate an authenticator so that the transaction record sent by the acquirer to the card issuer will be accepted as a valid authority for payment. In effect, this functions like a bank cheque and is an instruction to the issuer to pay a specified amount into the account of the acquirer. No PIN need be carried, because the valid authenticator is a sign that a correct PIN was entered.

The authenticator for this message employs a key which is stored in the token and is known only to the card issuer. This key is specific to the customer and when the payment is processed by the issuer, the customer identity in the message will allow his authentication key to be extracted from a table and used to check the authenticity of the entry.

We can now see clearly the advantage of the intelligent token. Its intelligence is used on behalf of the customer and the card issuer in the potentially 'alien' environment of the point-of-sale terminal. The token checks the PIN to verify who is presenting it, then authenticates a message to its 'master' which is the card issuer's computer system. From the issuer's viewpoint this is secure. The authentication key is customer specific, so there is no large gain from extracting the key from a token.

The acquirer is in a less favourable position because his terminal cannot verify the authenticator value. The method we descibed above (see Figure 10.21) goes some way to authenticating the token but cannot verify the end result — a message which is the authority for the card issuer to make the payment. In the off-line situation the merchant may be vulnerable, depending on what kind of indemnity the card issuer gives.

To help the acquirer it has been proposed that the token add a second authenticator using a key known to all terminals, which his terminal can check. But such a key contained in all terminals and all tokens is really too exposed.

Used in on-line systems, smart cards greatly improve security compared with the easily copied magnetic stripe. Off-line, they also improve security, but leave the residual weakness we have described. Public key cryptography can help.

10.6. PAYMENTS BY SIGNED MESSAGES

Our discussion of payment systems has emphasized the value of on-line connection to bank processors which enables card validity to be checked, the use of stolen cards detected and the account of the customer examined before the payment is authorized. Today, many payments are made without reference to a bank, for example by means of a cheque. Either a bank has been willing to guarantee the payment up to a certain limit or the beneficiary must trust the drawer of the cheque.

There will in future be many occasions when payment through a communication network is useful but reference to a bank is not convenient. For example, an on-line information service charges a fee but a caller does not have to be a known subscriber to the service; he can be anyone who 'presents a bank card'. The need for this method of payment, an 'open shop in information services' was pointed out in a note by Raubold.[9] It is the electronic equivalent of a bank cheque, which requires some trust between the parties. In this kind of transaction, the communication path betweeen the two parties is already established as the means by which the information service was provided. Payment for tickets for transport and entertainment can follow the same pattern.

The implementation of an electronic cheque requires technically little more than a digital signature facility with a key registry to authenticate public keys. A hierarchy of keys is proposed in which a registry authenticates the bank's public key and the bank authenticates the customer's public key. In order to allow the content of the cheque to be validated with a minimum of data beyond the cheque itself it should contain all the items listed in Figure 10.22. These items fall into three sections. The first is, in effect, a certificate by the key registry which authenticates the bank's public key. It also includes an expiry date so that the bank need not worry that an old signature is being used after it has expired. Anyone can verify the bank's public key using the signature and the public key of the key registry, which is available in many different places to avoid falsification. The key registry could be the central bank of a country, for example.

The second section of the cheque contains the customer identity and his public key, signed by the bank and therefore verifiable using the public key stated in the first section. This again has an expiry date to prevent its indefinite use. These first two sections of the cheque are constants which appear on every cheque drawn

1 Bank identity	2 Bank public key
3 Expiry date	4 Signature of 1–3 by key registry
5 Customer identity	6 Customer public key
7 Expiry date	8 Signature of 5–7 by Bank
9 Cheque sequence number	10 Transaction type
11 Amount of payment	12 Currency
13 Payee identity	14 Description of payment
15 Date and time	16 Signature of 9–15 by customer

Fig. 10.22 Electronic cheque

by this customer during their period of currency. They are followed by the payment information which forms the third section of the cheque data.

A sequence number ensures that each cheque is different from any other and provides a method for detecting duplicates. The transaction type is provided in order to use this same format for other purposes, described later. The amount and currency of the payment and identity of payee describe the payment to be made and there is a field for 'description of payment' which helps the parties to associate the payment with its purpose in their records.

Because electronic cheques can easily be copied, the drawer of cheques or his bank should examine the account to look for possible duplicates. Items 9 and 15 of the cheque allow them to be sorted into a sequence in which duplicates would immediately show up. It seems more practical for the bank to eliminate duplicates, and this is simplified if there is a time limit for the presentation of cheques for payment.

The final signature, by the customer, covers all the variable information in the cheque. In order to form this signature the customer needs a secret key which he must protect against disclosure.

A corporate customer of the bank will protect his secret key in a physically strong enclosure. Access to this signature can be controlled by physical keys, passwords and tokens, with strict limits on the extent to which it may be used by each individual. Private customers of the bank can carry their keys more conveniently, for example in an intelligent token or 'smart card'. A necessary protection against misuse of a lost token is the conventional PIN but perhaps the amount at risk justifies using a dynamic signature or some other personal characteristic to protect the digital signature against misuse. A terminal is needed in order to generate a cheque, sign it with the aid of the token and send it to the beneficiary. This can be any intelligent terminal which has the interface for connecting the token and a PIN keyboard or a pad for a dynamic (manual) signature, whichever of these is used to protect the token. It will often be the same terminal to which the service has been received for which the payment is being made, for example a theatre ticket issuing machine.

Functioning as an electronic chequebook, the private customer's token can record the transactions it makes and list them for its holders at any convenient terminal. The smart card used for point-of-sale payment has been called an 'electronic wallet'. The cheque is more versatile because it can be sent to anyone who has a means of recording the data and presenting them at a bank. No secrets are present in the terminal. This makes possible 'home banking' with reasonable security. If an account balance is requested, for example, the request is signed (verifying that a correct PIN was used) and the response is enciphered by the bank with the holder's public key. Only the correct token with its secret key can decipher the response.

Having personalized the token for its customer, the bank need keep no record of the secret key. New tokens use new keys and the bank's record of public keys is updated. PIN storage is part of the personalization of the token at the start of its life. The bank needs a record of the PIN only if it wants to remind the customer who has forgotten it. Customers can choose their own PINs but it is not a good idea to allow the PIN to be changed because this introduces an extra risk.

The date and time item in the cheque format depends on obtaining a reliable time value from the terminal. In principle, a date and time service could be provided on the public network with its values signed by the key registry, but this spoils the simplicity of off-line usage. The time stamp may therefore be optional. For corporate cheques, time might be used regularly, to provide timed evidence of payment and to help in logging the use of the cheque-writing facility.

The first motivation for the electronic cheque was an off-line payment for an information service. The service provider collects the cheques and presents them in batches to its bank for payment. They can be sorted and cleared in a conventional cheque-clearing operation. Cheques necessarily involve some trust that the account on which they are drawn is capable of payment. The same electronic cheque could be presented for payment as soon as it is received, using an on-line system, and thus cleared. This is an on-line payment mechanism which verifies the payment immediately. In its appearances to the customers it is no longer a cheque but a point-of-sale payment.

Point-of-sale payments by electronic cheque

The mechanism that we have proposed for a cheque enables an off-line payment to be made to a merchant's terminal. The terminal collects the cheques, comparing them if necessary with a local blacklist, and presents these cheques to its bank in a convenient batch. The bank sorts the entries and sends them to the respective card issuers for payment. The merchant's terminal can verify the signatures and the merchant is sure that the cheques will be accepted as genuine.

For on-line operation, the electronic cheque is transmitted at once to the card issuer bank where the signature is checked and the blacklist and drawer's account examined. If all is well, a payment message, signed by the issuer bank, is sent to the acquirer bank. When this has been checked, a signed advice goes to the point-of-sale terminal. If necessary, the accounts of the customer and merchant can be updated at once and the acquirer bank can issue a signed receipt. These are matters which depend on the precise payment rules, which must be decided when banks introduce new payment services.

By using public key signatures in this way, the security of each stage of the payment process is made dependent on the care exercised by the person or organization that would lose by fraud. Nobody is dependent on another's security. Thus each party is responsible for the security of his secret key and controlling its use. At the end of the payment process, each of the parties holds signed messages which record the commitments of the others. The merchant holds a copy of the cheque and the acquirer bank holds a copy of the payment.

If digital signatures have a weakness, it is the fact that signed documents can be copied and this places on the issuer bank the requirement to detect and eliminate duplicate cheques. The cheque sequence number is administered by the intelligent token and is therefore as secure as the customer's secret key.

The same intelligent token which provides an electronic cheque between individuals can therefore function for off-line and on-line point-of-sale payments. It can also 'cash a cheque' at an ATM with on-line verification.

Figure 10.23 shows how this works in a shared ATM network. The customer's request is formed into a message and presented to the token. Here it is signed

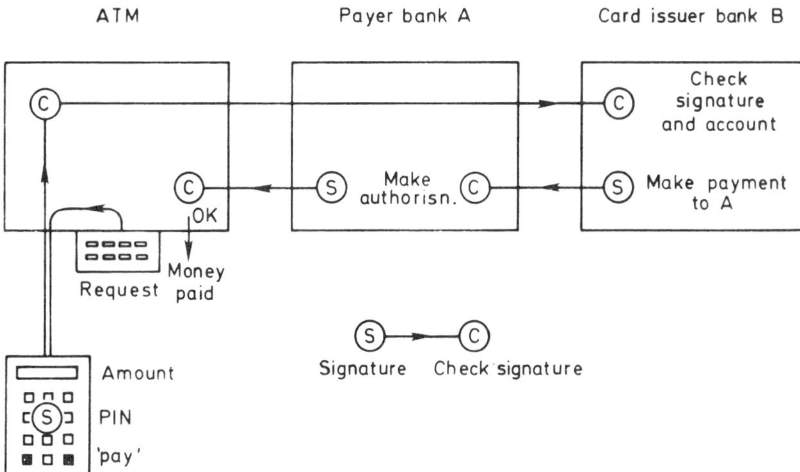

Fig. 10.23 Shared ATM network using digital signatures

and returned to the ATM. The ATM checks the signature, to avoid passing ineffective messages into the system. If it is correct, the 'cheque' passes via the payer bank, A, to the card issuer bank, B. Here the signature is checked and the customer's account examined and, if everything is in order, debited. A payment message signed by B is sent to A. The message and its signature are checked and if all is well an authorization goes to the ATM to release the money. The signed messages form an audit trail. All the messages carry similar information but their purposes are different. To differentiate these messages and provide for even wider use, the format of Figure 10.22 includes the *transaction type* which denotes a customer cheque, ATM request, interbank payment, payment advice and so forth.

If public key signatures come into use first in payment systems, the financial institutions will establish key registries. It would be natural for a verified public key issued by a bank to be accepted as a kind of bank reference. In this way the operation of key registries may become a function of financial institutions.

A development of the intelligent token

The 'smart card' in its current form depends on a separate terminal for its input and output. This is a security risk in several ways. The PIN enters the card from a keyboard, which could be used by an enemy to capture PIN values. The 'cheque' enters the card from the terminal and the card's owner cannot be entirely sure what is being signed. In an on-line payment, the message coming back from the centre is displayed by the terminal and could be falsified. For example it could indicate that a transaction has been cancelled, a refund has been made or a payment refused, when this is not true.

To avoid these serious risks, almost the whole of the signature operation should be under the control of the owner of the secret key. The PIN keyboard should be part of the token, so that the PIN is not entrusted to a 'foreign' device. When a payment is made, the most critical data, which are the amount and currency, should be shown to the payer by a display which he can trust. For this purpose a display is provided on the token. The display also shows any messages the token receives that have been signed by its own bank, such as an advice of a cancelled payment or refund. It is only by placing these vital functions in the token that their security is maximized. Figure 10.23 shows an ATM network using a token of this kind.

There are other functions of the token that are less critical but convenient. It can maintain a local balance of its account which errs on the side of caution, since received amounts will not be credited until a suitable on-line transaction is made. Automatic teller machines can provide an update, but to provide this for 'foreign' cards needs a little more organization and, possibly, a small charge. The account information is easily protected cryptographically, using the token's secret key for decipherment. Since this message concerns nobody else, its integrity can be determined by a secret password held in the token and known to the bank.

Such a token becomes a pocket terminal, capable of coupling to a point-of-sale terminal, an ATM or to a home computer or videotex terminal for home banking. It seems unlikely that home banking can be secure without such a 'sealed' device holding the key to authenticate messages. The coupling can be optical or magnetic, to avoid the nuisance of contact problems. It should have a battery to make the secret key more secure by overwriting it if a physical attack or a PIN search is detected. The extent of physical security is necessarily limited in a small device. This makes it useful to add a time and date stamp to transaction messages so that misuse can be tracked after a token has been reported lost by its owner.

There is no inherent reason why a signature token with its keyboard and display need be thicker than a standard bank card. Calculators of bank card thickness are on the market. But for some time the tokens may have to be thicker than a card to enclose the RSA chip, the battery and perhaps some extra physical protection for the secret key. The customer will expect an advantage to compensate for the increased size. If the token, while limited to a single account, works for point-of-sale debit and credit transactions, ATMs and home banking, this is already a help. In principle a signature token could operate with more than one account, holding the corresponding secret keys and taking its cue for selecting the account either from the terminal it connects with or from its keyboard. Therefore, one token could replace many or all of the cards we now carry. The different signatures can have their own expiry dates. Assuming that its secret key does not have to be altered, a signature facility can be renewed in several ways, either by a number through the post, at an ATM or at a bank branch. The big question is whether financial institutions that are in competition would be willing to cooperate to this extent. The cost of a signature token that is well used can easily be met from the saving in operating the payment mechanisms.

Almost all the technology needed for the 'signature token' is available but it must be put together into one well-engineered packaged. In addition to the technological development, signature and message format standards must be agreed.

10.7. ACCESS CONTROL BY INTELLIGENT TOKENS

We have proposed an intelligent 'signature token' in which the authenticity of the public key associated with the signature is certified by a bank. Other organizations could provide the certification and the key registries but the banking community is probably the best choice for the system to be widely accepted in most countries. When such a system is in place and working it can be used as a means of personal identification for purposes additional to banking. We will look at two possibilities, access control for buildings and access control for distributed information systems. Both applications will need little more than the ability of the token to sign a message. To the banking system, this kind of message has no particular meaning but it must be distinguished from a payment so that the signer cannot be fooled into making a payment. If the 'transaction type' signifies that a payment is being signed, the token's display must show this and display the amount.

A person who is to be granted access to a building must go to an official for his token to be recognized. This procedure inevitably requires an independent method of verifying the person's identity and his right to access. For introduction to the access system, the token is placed in a reader and is given a short test message to be signed. The access control system has then copied the token's public key and can check the signature to verify that the applicant really owns the token (more accurately, that he knows the PIN). The public key is stored, together with the associated data such as the rights and limitations of access and with a local identifier sufficient to allow the access rights to be cancelled when necessary.

To enter the building the token is presented to a terminal, which gives it a message to be signed. If the signature agrees with the *stored* public key, the token and its holder are assumed to possess the stored access rights. The access message must take a standard form so that it cannot be misused for other purposes. In principle, the transaction type 'access' together with a date and time are sufficient. An access control system could fraudulently present the token with a message timed for later, then use that signed message in a bogus token to gain access somewhere else by impersonation. The way to avoid this is to present on the display the name of the building or the authority controlling the building. The signer should always check what is displayed before giving the token his PIN.

By making one token take on the many roles which are presently given to several cards, passes, badges etc., the risk due to a lost token is increased, but cancellation of lost tokens can be made easier. Banks could provide as a service the daily listing of cancelled public keys.

Access control for centralized and distributed information services

For a centralized information service the method of access control can use the same principles that we described for access to a building. The introduction of a new user requires that his right of access be known. If the service is for sale, that right is conferred as a result of a payment. An electronic cheque (cleared if necessary) not only provides the money but also gives the service provider its reference data in the form of a public key.

Subsequent access control uses a different message on each occasion, because of the date and time field, so it is protected against the replay attack. The signed message can be held by the service provider as evidence if there is a dispute about illegal access. But the 'signing on' message does not protect against an enemy who takes over the channel subsequently. At the present time, public key ciphers are too slow for use to protect the channel. If the DES (or a similar cipher) is used, the server can provide a key enciphered under the user's public key and the token can decipher it, using the secret key.

The user's access rights are stored in association with his public key and are put into operation whenever a correctly timed message is received with a signature which matches. For very sensitive or valuable information it is possible to check the public key with the bank which issued it — the fee for this service would be similar to the fee for rapid clearance of a cheque. This scheme of access control is appropriate for a centralized system in which the various users' access rights tend to be independently controlled. If there is a lot of sharing of access with frequent granting of access from one user to another, a more flexible method is needed.

For a distributed information system, access control by *capabilities* is more flexible[10] and it allows sharing of data and procedures to be managed by the users themselves.

A static model of the access rights in an information system is provided by the 'access matrix' in which are stored the access rights of each 'subject' relative to each 'object'. A *subject* can be a person (system user) or a process created by a person or created by another process. By subdividing in this way, the rights of individual processes can be very limited indeed and this prevents the 'Trojan Horse' attack. An *object* can be data to be read or written or a procedure to be invoked and entered. In some systems there are other objects like input/output channels or queues. The entries of the access matrix have a correspondingly wide choice of rights such as read, write, append, enter (subroutine), join (queue) etc.

The method of access control implied by the centralized access control we described earlier amounts to storing, for each object, a table of subjects and their relevant entries in the matrix. On the other hand the *capability* method provides the matrix entries themselves as 'tickets' to be held by the subject and presented when they are needed. Checking their validity is easy. Like a doorman in a theatre the controller looks only at the ticket and is not concerned with identifying the subject. But this carries the requirement that tickets or 'capabilities' must not be forged or copied.

In monolithic systems capabilities are handled in a special way, sometimes with special hardware features such as a 'type' bit in every word, and the user may store and retrieve his capabilities but not create new ones or copy them except with the help of the operating system. In a distributed system the user needs to keep his capabilities entirely under his control, for example in an intelligent work station, in order to present them at any suitable server.

In a distributed information system we would like files of data to be mobile, able to transfer to any convenient file server, and also that procedures can be found in alternative places. Users should be able to share with others their access to data which they 'own' or to their own procedures. But if user Ann gives Bill access to a file she may want to prevent Bill giving the access to Charles without her consent.

These requirements are reflected in the way that capabilities are formed and protected. Each object can have its own scheme for coding and checking the capabilities which give access to it but others should not be able to construct these capabilities. For this purpose the object (or a group of objects) has a secret key which is used to encipher all capabilities for distribution. In this way the capabilities can be assembled and marshalled by users and, to help this, the nature of the capability (the object and means of access) can be associated in clear form, but the 'doorman' looks at the part that was enciphered.

This encipherment does not prevent copying, it only prevents forging new values. We do not want to prevent Ann copying some capabilities to Bill but when Bill presents them we need to know of Ann's intention and permission. It is clear that Ann's capabilities must contain Ann's identifier and a flag to signify ownership. Bill's copy can be coded to prevent further copying.

If these capabilities are to be valid at all those servers where the files or procedures are available, using conventional methods each server would need to recognize Ann or Bill by their passwords, but it is unwise to replicate secret data in this way. Public key signatures provide the appropriate mechanism. A service request from Ann must contain the capabilities needed to carry it out. Her own work station can assemble these. The request should also be signed, timed and dated by Ann. The server will act on this request if the signature matches the reference value of her public key. If necessary one of the capabilities can contain the public key. When she wants to share a capability (one that is marked for sharing) with Bill she sends it to him in a signed message meaning 'I share this with Bill but not for him to own' and giving Bill's public key. The server can check the signature and provide Bill with his copy of the capability, marked for his use but not for copying.

This method of access control allows objects to be replicated and distributed, carrying with them their 'doorman' in the form of a secret key and a method of coding capabilities. They hold no secret information about the users, only their public keys, which can be obtained from other servers. In this application, the intelligent token is required to give a signature to an access request. It may also be required to decipher a message from a server which contains a secret key to protect the communication channel. Such a message should be date and time stamped.

If the application requires the server to be authenticated, the user must know the server's public key and be able to check its signature. Potentially the intelligent token is able to do this but its ability to hold other public keys and perform the interaction needed for checking them raises the question of how much its function should be expanded. In all other matters it has been required only to perform RSA operations with its secret key.

By describing the use of the token for access control we have shown that the use of digital signatures goes beyond banking and that the signature token can serve in these other applications without complicating its function very much. The use of the system is widespread rather than specific to one organization, therefore the economic incentive to develop and install such a system will need many organizations to get together.

10.8. NEGOTIABLE DOCUMENTS

A negotiable document can be given by one person to another and confers some right or value on the person who holds it. The word 'person' includes legal persons, like companies or corporations. Thus it can be owned or sold like any other valuable thing. A theatre ticket is negotiable when it admits not a named person but whosoever presents it at the entrance. A banknote is another example. These documents are often worth stealing so they must be kept in safe places.

There are some payment tokens that are negotiable. These are bought from a shop and used to pay for transport, telephone calls, parking and so forth — usually small payments. Some payment tokens can be used several times, until all their value is consumed. These are called *pre-payment* tokens. An example is the 'Hologyr' card which has one or two tracks of hologram impressed on a plastic surface (and covered for protection). These can be read by infra-red light and small pieces of them successively destroyed by heat. The reading machine reads the next piece of hologram, then destroys it while examining it optically during destruction to verify that it is the genuine thing. Similar tokens can be made with the Drexler material, where very small holes are fused into a metallic film embedded in the card. The security of these cards rests on the difficulty of making or obtaining the materials and constructing something which functions like the real thing.

The intelligent token can also function as a pre-payment card. It must hold some secret data which are recognizable by the reading station. The secret could be a cipher or one-way function or a secret key which takes part in a cipher, authenticator or one-way function. If the enemy can make an intelligent token or something which functions like it, he must be prevented from finding the secret, therefore each challenge to the token and its response should differ from previous ones. The secret is held in each token and in the reading station, so the physical security of both must match this level of exposure. Using public key ciphers, the reading station need not hold a secret, but since the reader can use the same physical protection as the token, or more, this is not a practical advantage. The essence of a pre-payment token's security is the difficulty of forgery, like any other negotiable document.

A general-purpose negotiable document

A signed paper document can have three types of significance, informational, evidential and symbolic. The easiest to provide is its information content, since the whole evolution of information technology has been for the purpose of handling information. But compared with a signed paper document, the ordinary electronic one has lost its *evidential* value. Changes to a paper document show, whereas a digitally stored document usually gives no reliable evidence of its origin or the alterations that may have taken place. The digital signature (chapter 9) restores this evidential significance. With the aid of key registries, digital documents can then be used as evidence of their origin and accuracy, provided that the secret keys have been handled with care.

The *symbolic* function is the one which gives significance to a document's ownership. A theatre ticket symbolizes an access right to a seat at a performance. Negotiable documents are the ones that employ this symbolic significance. Digital signatures are not enough, because a digitally signed document can be copied and nobody can distinguish the original from the copy.

When a situation seems to need a negotiable document it is sometimes possible to change the whole method of operation and avoid that need. For example, some kinds of bank cheque are negotiable, but the 'electronic cheque' we described earlier relates only to one beneficiary account. If duplicates reach that account they are detected and only one credit is made. Negotiable documents can be handled electronically if there is a central register for all documents of a certain category, where their ownership is recorded. All changes of ownership are reported to the register and the changes acknowledged, as part of the procedure for the transfer. A central registry has been established for shares and bonds quoted on the Copenhagen Stock Exchange[11] and there is a project called INTERTANKO[12] to register the ownership during the shipping of bulk cargoes. These registries depend on trust in the central organization and on secure communication between the document holders and the centre. A general scheme to enable (original) documents to be distinguished from copies has been proposed.[13]

If a registry can be established for each category of electronic negotiable document, the symbolic aspect will never be needed. This may be the way things will go, because the infrastructure for cheap and efficient data communication is spreading rapidly. It is not necessary to store the whole document centrally since a suitable hash function or signature can represent it, but it is necessary for each registry to recognize, communicate securely with and authenticate its clients, the document holders. A registry may prove expensive to run because of these security requirements.

We will, therefore, look at an alternative, which is to establish electronic negotiable documents in which the evidential and symbolic properties are managed without registries of ownership. This scheme may prove useful, for example bills of lading are an important form of negotiable document and some countries want to have these documents retained. But even if 'electronic negotiable documents' are not found to have practical value, they make an interesting exercise in security technique.

The information document can take any form but is usually in natural language, which states the purpose of the document. If certain forms are widely used, there will possibly be standards for the wording and format, but these are not essential. There can be any number of copies of this part. The evidence is provided by a digital signature. The value of the document depends on the verification of the signature, and therefore on the availability of an authentic public key. We described earlier a way in which the banking community can provide an authentication hierarchy for cheques and this hierarchy seems entirely appropriate for all other documents used in commerce.

The symbolic part of the document must be unforgeable and contain a secure secret, the existence of which can be verified. The choice we shall make is to use the same device as the 'signature token', with its keyboard and display, except that in this application the secret key refers to the document, not to a person. The token may also need some special, application-dependent features that are described later. To link the token (the symbolic part) to the information and evidential document we use the public key. For example, the signed document could contain these words: 'This document is owned by the possessor of the token containing a secret key to match the following public key: (quote value here).' The signature on the document prevents the quoted public key being falsified and the validity of the symbol token can be checked by obtaining its signature.

A random number is sent to the token and signed. The signature is checked using the public key quoted in the document. If they agree, the token is genuine and is the one relating to the information document.

To create a negotiable document, the originator uses a secure system to generate a pair of matching keys for a signature method, such as RSA. The secret key is installed in the token and 'sealed in' by an irreversible change (like blowing a fuse inside it) so that the secret key can never be altered. At the same time the token is given enough identification to link it with the information document. There is no reason why the whole of the document's information should not also be contained in the token. The public key is quoted in the appropriate part of the information document, such as the statement quoted above, and the whole is signed. The document signature can also be loaded into the token and sealed in, so that no other document can be associated with it. After these steps have been taken, there are both information and symbolic parts to the document, each matching the other. For valuable documents the signature method should be very secure, because an enemy can use a lot of time and computing power attempting to match a fraudulent document to a symbolic token he happens to hold and for which the public key is known. Additional security is given by using only formal documents with a precisely specified wording.

We have described the creation of a document. Checking a document entails these steps:

1. Read the information to verify that it states what it should, that the originator had the authority to issue it and there are no unexplained random parts which might indicate a forgery by the 'Birthday method' (see page 279).
2. Obtain an authentic public key for the originator. This can be obtained from certified public keys in the document itself if they trace back to a well-known key, such as the public key of a national bank.
3. Verify the signature on the document, using this public key.
4. Check this signature against the one contained in the token, if there is one.
5. Send a random message to the token and receive back a signature.
6. Verify this signature with the aid of the public key quoted in the document.

If all the checks are passed, the token is assumed to be the genuine 'symbol' for the information document in question. The value of the combined information and symbol depends on what is stated in the document and on the trustworthiness of the originator.

Destruction of the symbol can be physical destruction, or it could be brought about by a command at the keyboard of the token. Some negotiable documents are not normally destroyed, others come back to their originator for destruction (like a banknote). In the latter case, the destruction can be 'logical' using a password entered by the originator at the time of creation; in this way, inadvertent or malicious logical destruction is avoided. The end result of logical destruction is to destroy the secret key held in the token.

Protection of negotiable documents against theft

A negotiable paper document must be guarded against theft. Electronic 'symbol tokens' must also be kept safely because of the inconvenience of losing them.

Protection against theft or misuse can be achieved logically, removing the incentive to break into a safe in order to steal them.

Each owner can give the token a password and then lock it against further use until the password is presented again. While the token is locked, it will not respond to a request to sign a message with its secret key, so it is useless to a thief. When a transfer is made, the vendor unlocks the token and the buyer locks it by giving it a new password of his choosing.

Since a transfer employs the intelligence of the token it is possible at the same time to enter a record of ownership and dates of transfer, so that each symbol token holds a log. This could be useful if there are suspicions of theft or misuse. Each type of application for these tokens will have its own special requirements and there is a good chance that the software of the token will allow it to help the owners in other ways. For example, supposing an owner loses the password, the document becomes inaccessible. To recover from this there can be an 'emergency password' which unlocks it. Usually, the originator is trusted by all subsequent owners. If the originator keeps the emergency password, any subsequent owner who loses the current password can bring it back to the originator to be unlocked.

The need for electronic negotiable documents is still uncertain, but the possibility of making one (in the way we have described) shows that the signature token can have a range of useful properties in commercial transactions.

REFERENCES

1. Lipis, A.H. 'Costs of the current U.S. Payments System', *Comm. ACM*, **22**, No. 12, 644–647, December 1979.
2. *Personal identification number management and security*, X9.8, ANSI X9 Secretariat, American Bankers Association, Washington D.C., 1982.
3. *PIN Manual; A Guide to the use of Personal Identification Numbers in Interchange*, Master Card International Inc. New York, September 1980.
4. Meyer, C.H. and Matyas, S.M. *Cryptography: a new dimension in computer data security*, John Wiley and Sons, New York, 1982.
5. Houghton, M.R. *An introduction to Electronic Fund Transfer Techniques*, Communications International, August 1980.
6. Meyer, C.H. and Matyas, S.M. 'Some cryptographic principles of authentication in Electronic Funds Transfer', *Proc. 7th ACM/IEEE Data Communications Symposium*, IEEE Computer Society Press, 1981.
7. Beker, H.J., Friend, J.M.K. and Halliden, P.W. 'Simplifying key management in electronic funds transfer point of sale systems', *Electronics Letters*, **19**, No. 2, 442–444, June 1983.
8. Australian Standard 2805 Electronic Funds Transfer — requirements for interfaces, Part 6.2.
9. Raubold, E. *Project proposal 'open shops for information services'*, (unpublished) GMD, Darmstadt, February 1982.
10. Needham, R.M. 'Adding capability access to conventional file servers', *SIGOPS Review* **13**, No. 1, 3–4.
11. *The Danish Securities Centre Act*, Act No. 179, of May 14th 1980.
12. *Legal aspects of automatic trade data interchange*, ISO/TC97/SC16/WG1, document N92, International Organization for Standardization, March 1983.
13. Brüer, J.-O. 'Original copy in the electronic world', *Information security in a computerized office* Part II Linkoping Studies in Science and Technology, Dissertations, No. 100, Linkoping University, Sweden, 1981.

Chapter 11 DATA SECURITY STANDARDS

11.1. INTRODUCTION

A characteristic of advanced technologies is the use of standards, which are needed for many reasons; safety in the case of lifting gear, durability in the case of drive belts and inter-changeability in the case of nuts and bolts. Because the market for the products of an advanced technology is worldwide, the process of standardization is international.

To clarify our terms it must be understood that the English word *standard* can mean two things which are distinguished in other languages. It can refer to standards of measurement such as the units of mass, time, radiation, illumination etc. The National Bureau of Standards (NBS) illustrates this meaning of the word. The other kind of standards which we are discussing in this chapter are known more specifically as *norms* and in many European languages 'normalization' is what we are calling 'standardization'. We shall, of course, continue to use the English word.

In the field of information technology the most frequent purpose of a standard is *interworking*. By this we mean that equipments designed by different teams and in different countries should be able to work together, either by direct connection using plugs and sockets or at a distance through communication networks. Interworking may seem very simple to the user, who would expect a visual display unit to plug into a computer and work. In fact, this level of interworking is very complex and has only partly been achieved. It requires a whole series of standards covering the physical properties of the plug and socket, the electrical signals which pass between them, the representation of binary digits, the procedure for exchanging bits, the representation of messages, their format, syntax and semantics, and higher-level procedures which enable one equipment to negotiate with another the type of service it is capable of providing or wishes to obtain. These complex requirements form a hierarchy and resemble the way that language is built up from a hierarchy of conventions, at the lowest level phonemes and words, then language, with its own very complex structure, and above this the protocol by which we communicate. The analogy between the interworking of computer systems and the complexity of language is very striking and it underlines the inevitably complexity of the standards needed in information technology. Standards of safety, quality and performance are also needed but the interworking requirement is the most characteristic feature of computers and communications.

Standards can completely change the nature of the market in computer components and equipment. Without standards, large manufacturers can tie their customers into a system which is vertically integrated. This benefits large manufacturers in the short term but it splits up the market into small units so that the specialist manufacturer (such as the maker of peripheral devices) cannot grow.

With the aid of standards, specialization among manufacturers can develop and a market which might otherwise seem small, for example the magnetic media, can become big enough to give economy of scale to many competing firms. The competition brings prices down. In this way both manufacturers and users have an interest in standards.

In communications, the need for standards became obvious as soon as the telephone network became fully international. At first the signalling systems of national networks did not work together without a struggle to achieve compatibility at each boundary, but now everything that needs to be decided for interworking is thrashed out by the International Telephone and Telegraph Consultative Committee (CCITT). Instead of trying to resolve interworking problems after they have happened, CCITT is moving towards prospective standardization — developing the international standards in advance of the start of national services. This has its dangers, but is probably better than leaving standards too late.

In computer technology, standards were not welcomed for at least two decades. The big manufacturers discouraged them because they saw their hold on the peripheral market being loosened. Researchers disparaged them because they felt their freedom to innovate would be restricted. There was some merit in these arguments while the technology was immature. Resistance to standards on both counts has now largely ceased.

There are many interests at work, consequently there are standards bodies at both national and international levels and some are working primarily for government users. In this chapter we shall first describe some of the actors in this drama — the standards bodies working in the field of data security — and then compare their different approaches to various data security standards.

To discuss all the detail in each standard or draft standard would merely be to rewrite the standard or reproduce it verbatim. The standards documents must speak for themselves and there are dangers in an attempt to interpret them. What we can do is to guide the reader and perhaps help him to read the definitive documents with better understanding.

The standards authorities

The first moves towards standards for data security were made by NBS (later renamed National Institute of Standards and Technology (NIST)), which is part of the US Department of Commerce. Under the Brooks Act (Public Law 89-306) the Secretary of Commerce is charged with improving the use of automatic data processing in the Federal Government. For this purpose, NIST provides *Federal Information Processing Standards* (FIPS) and technical guidelines, among other things. This work is done by the Institute for Computer Sciences and Technology of the NIST whose headquarters and laboratories are at Gaithersburg, Maryland. The address for enquires about FIPS is

> The Director
> Institute for Computer Sciences and Technology
> National Institute of Standards and Technology
> Washington, DC 20234
> USA.

Because the main work on DES and related subjects was done by NBS, rather than NIST, we retain the designation 'NBS' for this aspect of their work. The standards and technical guidelines are available from

US Department of Commerce
National Technical Information Service
5285 Port Royal Road
Springfield, Virginia 22161
USA.

These standards are maintained by the Institute for Computer Sciences and Technology in the sense that they must be reviewed from time to time to take account of improving technology. Normally they are applicable to all Federal departments and agencies but the FIPS publications relating to cryptography do not apply where classified data is being protected. A number of FIPS publications and some guidelines are described later in this chapter.

In the field of telecommunications, Federal standards are produced under the control of the *Federal Telecommunications Standards Committee* (FTSC). These standards are issued by the General Services Administration and are sold by

General Services Administration
Specification Unit (WFSIS)
Room 6039
7th and D Street SW
Washington, DC 20407
USA.

Both FIPS and the Federal Standards (and Draft Standards) produced by FTSC have had a major influence on work towards international standards. The international standards work has not slavishly followed the US Government examples but both the Data Encryption Standard (DES) and many features of other standards have been incorporated in international draft standards.

National Institute of Standards and Technology and FTSC represent government interests. The forum for US standards generally is the *American National Standards Institute* (ANSI) whose history goes back to an earlier body established in 1918. In common with other non-government standards bodies, the standards it produces are widely respected but not mandatory. The address is

American National Standards Institute Inc.
1430 Broadway,
New York, NY 10018
USA.

Like other standards organizations, ANSI operates through a large set of technical committees, subcommittees and working groups. A special characteristic of ANSI is the support of the secretariats of committees by other organizations representing manufacturers' and users' interests. For example the secretariat for ANSI work on the Data Encryption Standard has been

Computer and Business Equipment Manufacturers Association
(CBEMA)
Suite 500
311, First Street NW
Washington, DC 2000
USA.

For the development of ANSI standards relating to security in financial institutions, the secretariat has been provided by

American Bankers Association (ABA)
1120 Connecticut Avenue NW
Washington, DC 20036
USA.

Major countries have official bodies for the production of national standards, in the sense of norms. Examples in Europe are

The British Standards Institution (BSI)
2 Park Street
London W1A 2BS
UK

Association Francais de Normalisation (AFNOR)
Deutsches Institut für Normung (DIN)

The British Standards Institution was founded in 1901 and its work on data security is entrusted to a committee IST/20 under the general heading of Information Systems Technology. Where possible BSI tries to align its national standards with international standards and sometimes it exerts its effort towards the production of an international standard before beginning the production of a British one. The main technical contributions to international work on data security standards have come from the bodies mentioned above, particularly ANSI, ABA and BSI.

International standards in data processing are produced under the joint aegis of the International Organization for Standardization (ISO) and the International Electrotechnical Commission (IEC). The technical committee responsible for data processing standards is Joint Technical Committee 1 (JTC1), for which the secretariat is

ISO/IEC/JTC1 Secretariat
American National Standards Institute Inc.
1430 Broadway
New York, NY 10018
USA

The work of ISO/IEC in data encryption and its application falls under technical committee JTC1. The work began under the committee predecessor of JTC1, ISO/TC97, which set up its working group 1 in 1980. This was replaced in 1984 by subcommittee 20 (SC20). Therefore the designation of the ISO/IEC committee currently working on data security standards is ISO/IEC/JTC1/SC20; the secretariat for this subcommittee is

ISO/IEC/JTC1/SC20 Secretariat
GMD/Z2.P
Schloss Birlinghoven
Postfach 1240
D-5205 Sankt Augustin 1
West Germany

There are other international bodies, such as the International Telephone and Telegraph Consultative Committee (CCITT), producing standards in various fields of technology. Where the interests of data processing and data communication merge, there is good co-operation between ISO and CCITT.

CCITT is the official body at which telecommunication authorities and operating agencies produce the standards which enable their various systems to interwork. The Committee, together with CCIR, the corresponding body for radio, forms the International Telecommunications Union (UIT) which is an organ of the United Nations. In the most recent CCITT study period work has begun on various data security standards, for example a Directory Authentication Framework. CCITT is particularly concerned with open systems interconnection standards up to layer 4, whilst ISO/IEC committees deal with all layers, 1 to 7.

In addition to the official bodies so far described, there are special interest groups which can have a big influence on standardization by being represented in CCITT or ISO/IEC and presenting views which are known to be a consensus of a large group.

In CCITT there is an interest group which is the European Committee for Posts and Telecommunications (CEPT). Telecommunication authorities and operating agencies in Europe meet in CEPT to generate drafts which, when presented to CCITT, are understood to represent a consensus, in spite of the unofficial status of CEPT. In a similar way, the European Computer Manufacturers Association (ECMA) has had an influence on the work of ISO/IEC by the work it has done to produce standards.

The number of interests involved and their complex interaction makes it impossible to give a comprehensive account of standardization work, even in the relatively narrow field of data security. Data security standards form a hierarchy based on the application of encipherment; initially the emphasis was almost entirely on the Data Encryption Standard (DES), but the work is now broader based, specifying no particular encryption algorithm; public key algorithms, discussed in chapter 8, are entering the picture. The next higher layer of standards can use any block cipher, such as the DES, and this layer contains the 'Modes of Operation' which were described in chapter 4; definition of modes of operation may also be necessary for public key algorithms. Above the modes of operation are the procedures for employing encipherment (and authentication) in open systems interconnection; the ISO standard which describes the security architecture, ISO 7498/2, was discussed at length in chapter 6. ISO 7498/2 describes the security services which may be required in the OSI context and suggests where these may be located in the hierarchy; however it does not prescribe the detailed implementation of these services. Further standards will be needed for key management and possibly for other purposes not yet defined. Also there are specific applications of data security for which standards are produced, for example the authentication of financial messages, where the ABA and other banking organizations have special interests. In the following sections we shall describe some of the standards work which is taking place in these areas.

11.2. STANDARDIZATION RELATED TO THE DATA ENCRYPTION STANDARD

The early history and the nature of the DES are described in chapter 3. Subsequent work on DES in the national and international bodies has in no way departed from the algorithm that was first published in definitive form in FIPS Publication

46.[1] However, there are a few ways in which the algorithm is implemented which have subsequently been changed.

Publication 46 states 'the algorithm specified in this standard is to be implemented in computer or related data communication devices using hardware (not software) technology'. There is no such requirement in the American national standard entitled Data Encryption Algorithm[2] which is ANSI X3.92-1981.

There are dangers in implementing any cipher by software in a processor which is not adequately protected, particularly one which is shared by many tasks or many users. It is notoriously easy for systems programmers or maintenance engineers to read any part of a store and this could be the part containing the key being used for the cipher (or, equivalently, the 16 DES sub-keys). This was, no doubt, the concern which led to the hardware requirement in FIPS 46. The Federal Standard allows the use of a microprocessor having its program in read-only memory.

The DES has a 56-bit key contained in a 64-bit key variable and in FIPS 46 it is stated that the other eight bits are 'set to make the parity of each 8-bit byte of the key odd, i.e. there is an odd number of ones in each 8-bit byte'. In practice, many users of the DES have departed from this rule, allowing the extra bits to be arbitrary or using them for other purposes. The ANSI Data Encryption Algorithm states in section 3.5: 'one bit in each 8-bit byte of the KEY may be utilized for error detection in key generation, distribution and storge. Bits 8, 16, ... 64 are for use in ensuring that each byte is of odd parity.'

These sentences are exact copies of sentences in FIPS 46, page 16, and in this respect the two standards differ only in the emphasis given to the parity requirement in the FIPS explanation section and in the ANSI foreword, which are not parts of the standards.

Until 1987 an international standard corresponding to the DES was in preparation. Publication of this text has been abandoned by ISO for reasons which we shall discuss shortly. In this text no requirements concerning key parity conditions were specified. The text stated that: 'bits 8, 16, 24, 32, 40, 48, 56 and 64 of the key-block do not take part in the algorithm. Their use is not specified in this standard.' The text noted that: 'in some applications they have been used as odd parity bits'.

The name of the standard was modified a little during its adoption by ANSI. It is not usual to have the word 'standard' in the title of a standard because this is unnecessarily repetitive. Therefore the US national standard is called 'Data Encryption Algorithm' which can be shorted to DEA. The ISO name for the algorithm was to have been 'Data Encipherment Algorithm No. 1', shortened to DEA1; this title recognized the intention of publishing further encipherment standards.

In the event, as we have indicated already, ISO decided not to publish the DEA1 as an international standard. The main reason for this was a feeling that publication as an ISO standard would lead to the algorithm becoming even more popular and widely used. There is no doubt that the algorithm has been adopted as a component in many secure systems, not least by the international banking community. The fact of this widespread adoption has led to unease that the target presented by the large number of systems dependent on DES would be attractive to criminal cryptanalysts; the potential gain accruing from 'breaking' the algorithm could be immense and the possible damage to systems credibility could be catastrophic. Any action further encouraging even more dependency on one algorithm could be undesirable. This belief led ISO to decide against publishing DEA1 and to replace the publication by a register of encipherment algorithms to be operated on behalf of ISO. Publication of the register is intended to encourage a degree of diversification in choosing encipherment algorithms for particular applications. Later in this chapter we shall examine the nature of the register as defined in the draft standard describing its structure and functionality.

Several useful texts have been published by the National Bureau of Standards regarding the DES. 'Guidelines for Implementing and Using the NBS Data Encryption Standard'[3] is a useful introduction to data encryption and it describes security threats, the implementation of the algorithm, key management and transparency. It also describes how to apply the DES with a limited character set — the method which was described on page 90.

'Validating the Correctness of Hardware Implementations of the NBS Data Encryption Standard'[4] describes a test bed for validating an implementation of the DES. The kind of tests suggested by NBS in this publication ensure that any 'normal' implementation conforms entirely, but for a cipher device we must also consider a malicious implementation which is designed, under very special conditions, to betray its key. Such an implementation with a 'Trojan Horse' could be designed to pass the NBS tests with very high probability. Therefore, for reasons of strict accuracy, it is not possible to give a definition of compliance testing. A device complies with the standard if it produces the correct output for all possible inputs and this means that the *mechanism* of the device must be inspected. Since the mechanism is not defined, only the result, there is no way to define the inspection process.

Federal Standard 1027 — General security requirements for equipment using the DES

Federal Standard 1027[5] is mandatory for all DES cryptographic components, equipments, systems and services used by US Government departments and agencies (the mandate was confirmed by a US Treasury Directive dated 16 August 1984). For other users and manufacturers it provides a useful guide to the security features which go beyond the cipher and its modes of operation. It covers physical security, locks, mounting, key entry, alarms, tests, controls and indicators among other things.

Key entry can be carried out *manually* and then the key is entered as 16 hexadecimal digits starting with bits 1 to 4 of the 64-bit key variable. The alternative is entry from a *key loader* and for this purpose a synchronous serial data interface is defined in the specification. The entry of keys is under the control of a lock which requires a physical key to be entered and turned. The same lock controls entry of the initializing variable (IV) when this is entered, along with the key, for cipher block chaining mode. It would seem logical to enter the IV on the same interface as the key variable but this possibility is not mentioned in the specification. Federal Standard 1027 requires odd parity in the octets of the key variable, in strict accordance with the DES of FIPs 46.

The standard FS 1027 specified the automatic testing of its internal functions which the equipment must carry out. One alternative is to duplicate the encryption equipment and compare the results before output. A means must then be provided to test the comparator circuits by forcing an error. An alternative is to carry out a so-called 'S-box test' each time a new key is entered and a 'checkword test' each time an IV is used, i.e. at the beginning of each chain. Alternatively the S-box test can be carried out for each new IV usage. The *S-box text* is a test of DES operation with a set of values which guarantees that all the S-box entry combinations have been applied. Such a set is part of the NBS validation test.[4] The *checkword test* operates by enciphering a known 64-bit input word with the key and storing both the input word and the corresponding cipher text. For the checkword test, the same encipherment is carried out again. The data for this test have to be renewed every time a new key is installed. The comparator used in the checkword test must also be capable of checking by forcing an error. Whichever comparator is used (on all encipherments or on the checkword test only) this comparator itself should be checked no less frequently than each new key installation.

Careful testing is needed to keep an encryption system secure because the failure to encipher correctly could send out plaintext on the line until the intended receiver reported the problem.

A *bypass* mode in an encryption equipment means that encipherment does not take place and plaintext goes out on the channel that would normally carry ciphertext. At the receiving end, the same bypass mode allows the plaintext to reach its destination. Clearly this is a very dangerous procedure; it is used only for diagnostic tests or in dire emergencies where the message is more important than its secrecy. Bypass mode is an optional feature in FS 1027 but it *must* be under the control of a lock and must be announced by a status indicator.

The standard concludes with requirements for electro-magnetic interference and compatibility using criteria from US military standards. There are many other details concerning alarms, indications, key handling, stand-by mode etc.

Using Federal Standard 1027 as a starting point, British Telecom authors have produced a more comprehensive document entitled 'General Security Requirements for Cryptographic Equipment: a Code of Practice'; a second document addresses operational aspects. This material has been further developed and published by the British Institution of Radio and Electronics Engineers.[6] Publication of a British Standard for physical security is being considered by BSI.

A register of cryptographic algorithms

In an earlier section we observed the ISO decision not to proceed with the standardization of cryptographic algorithms. In place of publishing standard algorithms ISO proposes to establish a register of cryptographic algorithms. The procedures under which the register is expected to operate are being developed in a draft international standard[7]; these procedures will specify how the register will be used by those wishing to make entries in the register and by the Registration Authority. The register will serve as a common reference point for the identification of cryptographic algorithms by a unique name. It will also contain the basic parameters associated with a register entry. The main purpose of the register is to enable entities (such as communicating entities in OSI) to identify and negotiate an agreed cryptographic algorithm.

The Registration Authority's role will be to add, modify and delete register entries in accordance with the rules provided, to assign ISO-entry names to registered cryptographic algorithms as needed and to update and to have published the register when required. Note that the Registration Authority does not evaluate or make any judgement of the quality of protection provided by any registered algorithm; thus no guarantee of security quality is implied in the presence of an algorithm on the register.

Submission of algorithms for registration will be via the national member body of ISO in each country (BSI, ANSI, DIN, AFNOR, etc). Likewise any alterations to entries or deletions of entries will be initiated via the ISO member bodies.

The register entry details will include:

(a) a formal ISO-entry name for the algorithm
(b) proprietary name, if any
(c) intended range of use
(d) cryptographic interface parameters (block size, key domain, etc.)
(e) set of functional test words
(f) identity of organization requesting registration
(g) dates of registrations and modifications
(h) whether the subject of a national standard
(i) patent licence restrictions
(j) export restrictions
(k) list of references to associated algorithms
(l) description of the algorithm
(m) statement of modes of operation
(n) other information

It should be noted that items (j) to (n) are optional; some or all may not be present.

At the time of writing nominations for Registration Authority had been called for by ISO and we understand that two member bodies have put forward candidate organizations. The present state of the draft international standard is that agreement on its text has not been reached and further drafting will now take place.

11.3. MODES OF OPERATION

We have described in chapter 4 the four modes of operation which have been the subject of several standards. The original publication was FIPS Publication 81 entitled *DES Modes of Operation.*[8] This was adopted, apparently without change, by ANSI as *Modes of Operation for the Data Encryption Algorithm.*[9]

The modes of operation standard prepared by ISO was published in August 1987.[10] Whereas the FIPS and ANSI documents relate specifically to DES, the ISO standard is worded to apply to any 64-bit block-cipher algorithm.

We need add nothing more to the description in chapter 4 of the electronic code book (ECB) mode of operation and cipher block chaining (CBC). For CBC the standard does not specify how to deal with the incomplete final block (if there

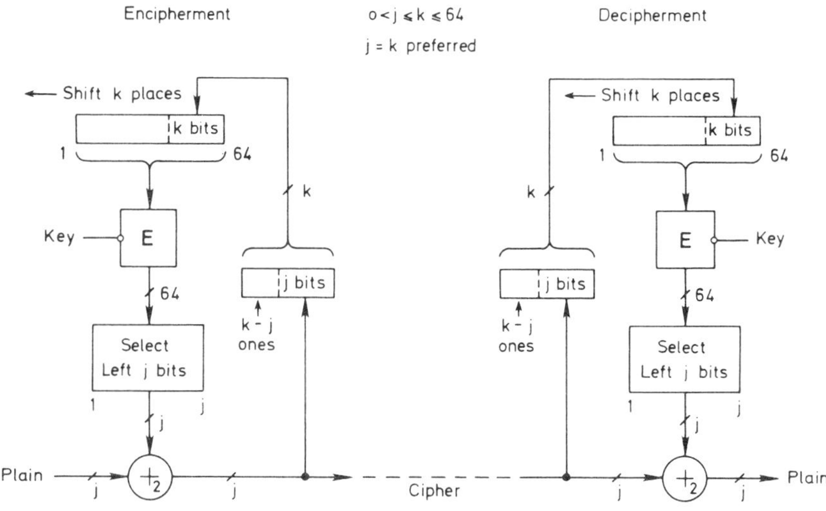

Fig. 11.1 Detail of the cipher feedback standard

is one) and leaves this to be treated in the individual application. For CBC an IV of 64 bits is used.

Cipher feedback (CFB) in these standards is a little more elaborate than the scheme described in chapter 4, which we illustrated only for 8-bit feedback. Figure 11.1 shows the CFB mode as it is described in the standard produced by ISO. In this figure, the feedback path is of width k bits. (The notation k must not be confused with the key.) In the shift registers a left shift by k bits takes place each time k bits are fed back. This creates the new input for the encipherment algorithm. From the output, the left j bits are selected and used for Vernam encipherment of the plaintext character. The difference between the character length j and the feedback length k is made up by $k-j$ ones which are attached to the left of the cipher character in the feedback path. It is recommended in the ISO standard that $k=j$ so that this difference between feedback width and character does not exist. The only case for which it is known to be used is a 7-bit character with an 8-bit feedback width.

For the IV the ANSI standard allows any number of bits to be chosen. This value is placed in the least significant or right-hand end of the shift register and padded on the left with zeros. Federal Standard 1027 requires an IV length of at least 48 bits in CFB operation.

In the ISO standard the 64 bits which fill a register to start any mode of operation are called the *starting variable*. The way these are derived from the given IV, which may be shorter, is not defined.

Output feedback (OFB) mode, sometimes known as 'output block feedback', is defined in FIPS 81 and the ANSI standard with a feedback path of k bits. We demonstrated in chapter 4 that OFB with feedback of less than 64 bits is insecure and that 64-bit feedback should always be used. The ISO standard allows only 64-bit feedback in OFB mode, but does provide that the leftmost j bits can be selected from the output block if this is convenient for the data stream being enciphered. Figure 11.2 shows, on the left, the OFB configuration in FIPS 81 and

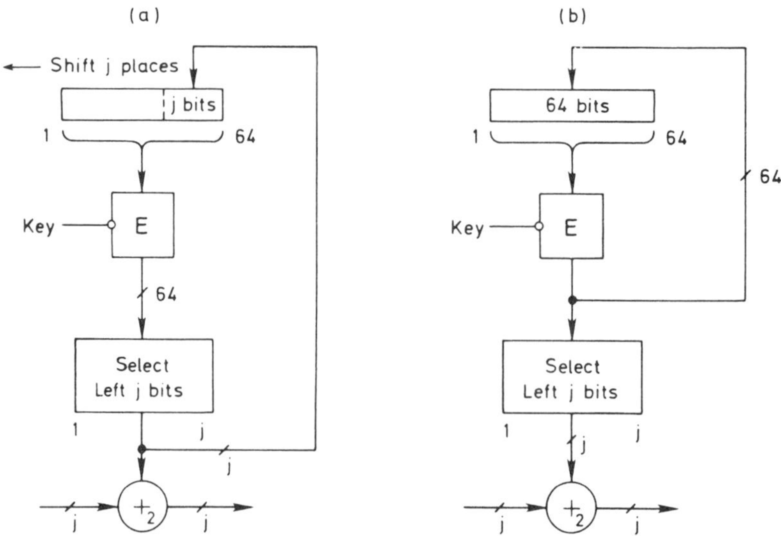

Fig. 11.2 Two versions of output feedback

the ANSI standard. The right side of the figure shows the configuration in the ISO standard. Since OFB mode is frequently used for high-speed synchronous operation it is unlikely that anything less than 64-bit output will be used in practice.

Note that ISO has now a draft standard[11] in preparation which provides definition of modes of operation for an *n*-bit block cipher; this carries the process of generalization one step further.

11.4. ENCIPHERMENT IN THE PHYSICAL LAYER OF DATA COMMUNICATIONS

One of the first international standards to be agreed in the area of secure communications was ISO 9160,[12] entitled 'Information Processing — Data encipherment — Physical layer interoperability requirements'. The physical layer is the lowest layer of OSI at which encipherment can be performed. Usually the encipherment can be continued only over one link of a network, but it has the advantage that all data bits are enciphered, making it a useful tool for traffic flow confidentiality.

The standard covers both synchronous and asynchronous or start/stop operation. In synchronous operation every element transmitted is enciphered. For start/stop operation, the start and stop elements which frame the characters transmitted must be preserved and only the data bits (normally 8 of them) are enciphered. As a consequence the presence of start/stop characters on the line is made obvious. The whole of the data including headers is enciphered but the quantity of characters carried remains obvious. To complete the traffic flow confidentiality in this case, dummy traffic would have to be added.

During the development of ISO 9160 the use of the DES algorithm in one-bit cipher feedback was assumed. In the eventual standard no algorithm is stated, because ISO's policy has changed. Even the mode of operation is left open but the algorithm should be 'applicable on a single bit or single character at a time basis, as appropriate to the physical layer service provided'. In practice this will mean using one-bit cipher feedback or 8-bit cipher feedback.

For synchronous operation one-bit cipher feedback is ideal because all framing problems are solved. Its only problem might be the achievable data rate. For start/stop operation with 8-bit characters there are the alternatives of one-bit or 8-bit cipher feedback. Systems that have been built favour one-bit cipher feedback for its simplicity. They claim 'protocol transparency' since the only part of the protocol that may need adjustment is the setting up of the channel before transmission.

The use of the cipher feedback mode of operation as specified in ISO 8372 entails the prior transmission of an initializing variable or IV. Annex B of ISO 9160 describes the structure of the IV and its transmission for the use of DES (described here as DEA, i.e. ANSI X3.92). Briefly, for synchronous operation 48 bits are sent and then padded out with zeros on the left to form the 64-bit starting value or SV. As an alternative, all 64 bits can be sent. For start/stop operation the necessary number of characters are sent so that at least 48 bits of IV are obtained, padded out as before. There are two options giving, respectively, minimum IV lengths of 60 and 64 bits. Details are in Annex B which is an integral part of the standard.

The physical arrangement is, in principle, that shown in Figure 11.3, which shows a point-to-point connection with and without encipherment. The aim is to make the *data encipherment equipment* (DEE) as far as possible transparent; therefore it is placed between the data terminal equipment (DTE) and the data circuit terminating equipment (DCE) which is typically a modem. Consequently the DTE/DCE interface is 'split' and the DEE must present the appearance of a DCE on one side and a DTE on the other. In practice it is often more economical to incorporate encipherment in one of the other units, so that only a single DCE/DTE interface remains. To cover all versions, the standards describe encipherment as shown in Figure 11.3 and cover the other cases by requiring them to operate functionally in the same way.

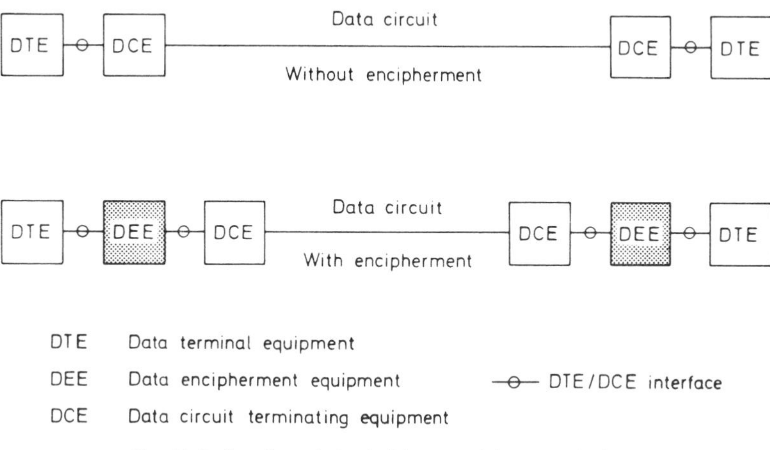

Fig. 11.3 Location of physical layer encipherment devices

Principles for encipherment at the physical layer

Encipherment is applied to the binary stream or stream of characters which is the physical-layer data service unit. Since the aim is to provide bit transparency, all versions of this standard agree in requiring 1-bit CFB. In the case of synchronous transmission every bit is enciphered in this way. In the case of start/stop transmission the start and stop elements must be preserved so that only the data-carrying bits in between them are enciphered. At the end of each character, the encipherment process stops, the stop bit is transmitted and nothing happens until the next start bit arrives; this is transmitted unchanged and encipherment begins again and continues until the next stop bit is reached. The notation employed in our figures is shown in Figure 11.4 with the usual convention that the *mark* condition corresponds to 1 and the *space* condition corresponds to 0. The start elements are represented by space and the stop elements by mark, which is also the condition of the line between characters. The shading indicates the enciphered data, which is framed by the start and stop elements.

Before encipherment and decipherment can begin, the IV must be transmitted over the line and this is sent in clear for CFB operation.

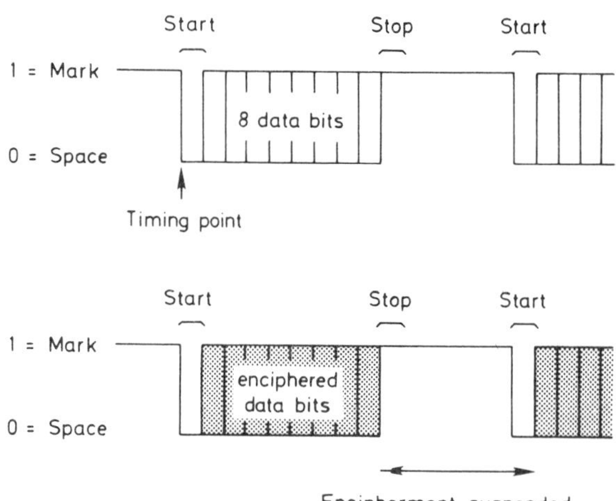

Fig. 11.4 Encipherment of start/stop characters

Signalling the start of transmission

Transmission can begin as soon as a connection has been made across the DTE/DCE interface by the exchange of suitable signals. We will use V24 terminology. New data networks employ CCITT recommendation X20 for asynchronous circuits and X21 for synchronous circuits and there are also recommendations X20 bis and X21 bis which allow the use of an interface similar to V24 in connection with these new networks.

The data encipherment equipment (DEE) is almost transparent to the procedure used to set up the circuit. Most of the signals involved are passed through from one interface to the other. The standard has been written to cover the older

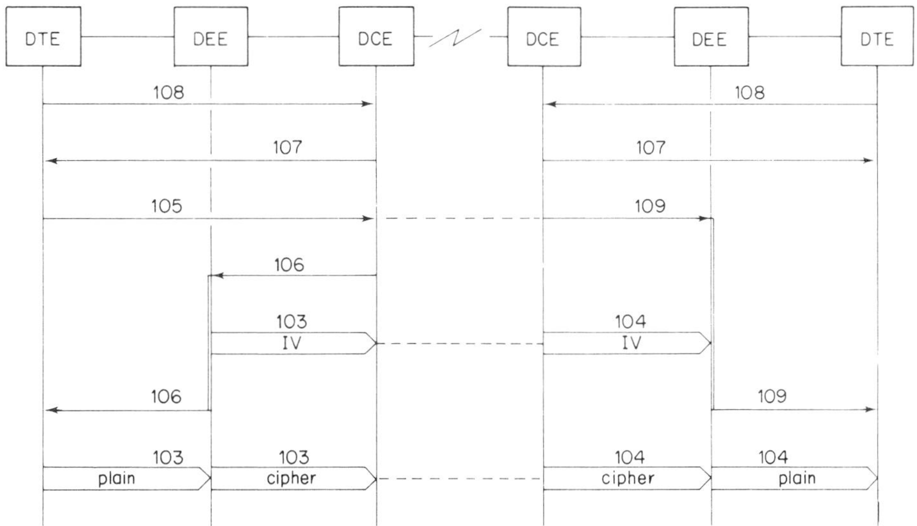

Fig. 11.5 V24 interchange circuits during circuit establishment

DCE/DTE interface V24, which allows both synchronous and start/stop working and the newer X20 (start/stop) and X21 (synchronous) interfaces. The transitional versions X20 bis and X21 bis can be treated as versions of V24.

The main new requirements are to let the DEE know when the circuit is ready and to ensure that the transmission of user data is held up until the IV has been sent and received. Figure 11.5 shows the signals employed in the V24 interface.

107 (data set ready) and 108 (data terminal ready) must be on and, if the DCE requires it, 105 (request to send). At the receiving end 109 (data channel received line signal detector) alerts the DEE but is not passed on yet to the DTE. The DCE at the sending end returns 106 (ready for sending) but this also is held up by the DEE. The DEE then sends the IV over line 103 (transmitted data) and this is received on line 104 (received data) and absorbed by the receiving DEE.

These preliminaries having been completed, the sending DEE forwards the 106 signal and the receiving DEE forwards the 109 signal, after which the two DTEs can communicate normally with the DEEs doing the encipherment and decipherment transparently. Figure 11.6 shows the way that the IV is followed by enciphered data in start/stop operation.

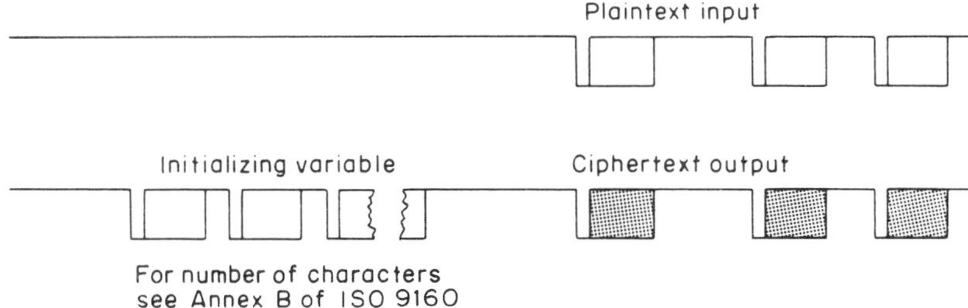

Fig. 11.6 Start of encipherment at the sending data encipherment equipment (start/stop)

For synchronous operation, the IV is sent as one continuous block of data, preceded by a single zero bit which marks its beginning, since the line is earlier in a MARK or '1' condition. Following the transmission of the IV, there are two alternatives. In alternative A the first enciphered data bit can follow immediately. In alternative B, the IV is followed by a '1' state on the line for 10 ms to 50 ms, then a single-zero bit heralds the start of transmission of enciphered data.

The physical connection is broken at the end of transmission by the 107 or the 108 circuits going to the off state. Potentially, the DEE can break the connection via both of the circuits simultaneously, but this is a fault condition.

There are corresponding rules, given in the standard, for the start and finish of transmission using the X20 or X21 interfaces.

We have now described the normal operation of the system. In addition, the standard covers how the break condition for start/stop operation is handled and the use of 'bypass control' so that loopback tests can be carried out.

Treatment of the break signal

In start/stop operation it is possible under certain fault conditions for the line to go into a *space* condition for longer than the duration allowed in any character. This can also be used as an 'out of band' signal though this is less used in present-day systems. There must be some convention on how the DEE treats the break signal and it should be, as far as possible, transparent to it. Two possibilities are provided as options in the ISO document, known as Classes A and B, which are illustrated in Figure 11.7.

In Class A operation, the beginning of the break signal is interpreted quite naturally as the start element followed by a character consisting entirely of zeros. Consequently the DEE output consists of enciphered zeros until the point where the input can detect the absence of a stop element. This is recognized by the DEE as signifying a break condition, encipherment is halted and the zero condition is transmitted on the DEE output. At the receiving end, the enciphered zeros are correctly deciphered, but the absence of a stop element signals that a break condition is present so the deciphered zeros can be followed immediately by a continued zero condition while the break signal lasts.

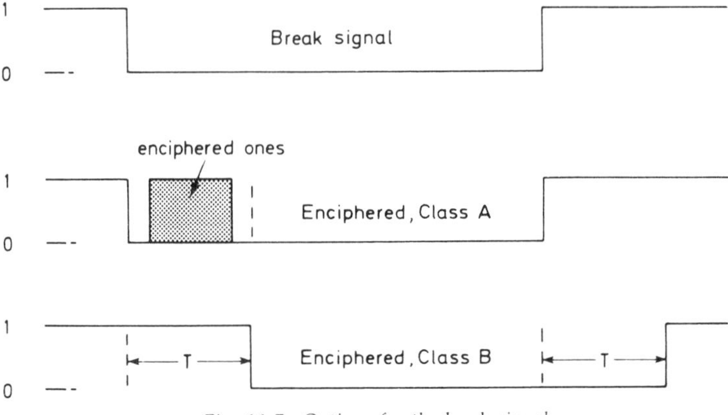

Fig. 11.7 Options for the break signal

In the alternative Class B, the DEE is transparent to the break signal which passes through it unchanged. To achieve this, the absence of the stop element must be detected before any character can be sent. This implies that a DEE operating in this manner must receive the entire character before a character is transmitted. That is, the DEE must have at least a one-character delay both in encipherment and decipherment, shown as T in the figure.

The option of bypass control

In the operation of modems on telephone circuits, the diagnosis of line faults is carried out by inducing a 'loop-back' in one of the modems. This can be a local loop back which tests nothing but the circuit from DTE to DCE and back again or a 'remote loop back' which takes place at the distant modem and tests the entire line without involving the equipment beyond the DCE at the far end. These provisions enable some types of fault to be distinguished without having to call an operator at the far end. The presence of data encipherment equipments would upset this diagnostic method.

To allow this diagnosis in the presence of a DEE when the V24 type of interface is used there is a *bypass* facility which suspends encipherment and transmits the plaintext through the DEE. This bypass is brought into operation by a 'test mode' circuit on the DTE/DCE interface, either 140 or 141 but not both. The occurrence of either signal is responsible for selecting the bypass mode. The appropriate loop-back signal is transmitted through the DEE to the DCE. In this way, the bypass mode should not occur without the loop back taking place. Figure 11.8 illustrates these two bypass–loop-back conditions.

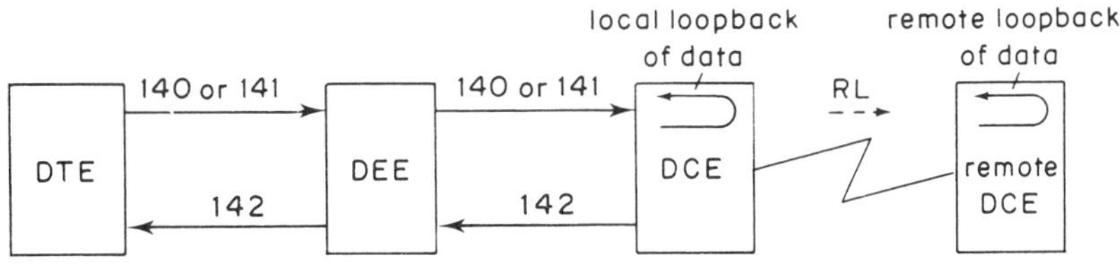

141 = local loopback 140 = remote loopback 142 = test mode indicator

Fig. 11.8 Local and remote loopback configurations

Bypass control is an option which does not have to be provided on data encryption equipments to give compliance with the standard but the existence of this option in the standard will encourage manufacturers to include it. In order to be safe against misuse, bypass must be controlled by a lock with a physical key.

Characteristics of encipherment in the physical layer

The form of encipherment defined in ISO 9160 is well suited to fitting into a data network as an afterthought. It disturbs the existing system to a minimal extent. The delay due to encipherment and decipherment is about one bit time for each, or one character time for each if 8-bit CFB is used with start/stop. Setting up a circuit is delayed by transmission of the IV, but this is significant only for half duplex operation with frequent turn-around. In the case of a leased line, no set-up delay will be apparent.

The extension of transmission errors due to the CFB mode of encipherment may be a problem on analogue circuits. It can be avoided by using an 'additive' stream cipher, but then the need for detecting loss of synchronization and recovering synchronization further complicates the system and prevents full transparency.

The standard does not ensure successful interworking unless the two ends make prior agreement on:

Cipher algorithm
Mode of operation
Size and format of IV
Break option (start/stop)
Immediate/delayed option (synchronous)
Use of bypass facility

Annex B of the standard, section B.4, describes how the choice of the various options should be codified.

ISO 9160 is the basis for a useful cipher facility that is relatively easy to install and should result in encryption products that have a wide range of use. Extension to encompass the ISDN interfaces is an obvious future task for ISO.

11.5. PEER ENTITY AUTHENTICATION

In secure communication two types of authentication are encountered — message authentication (message integrity) and peer entity authentication. The former has been discussed extensively in chapter 5. In the latter the aim is to allow communicating entities (user application processes for example) to obtain mutual assurance that both parties are genuine. Various methods of peer entity authentication have been proposed, but a basic feature of all systems is that the entities satisfy each other that they possess some secret that can be verified. Secret key cryptosystems and public key cryptosystems can be applied to achieve this type of authentication, as can also the recently conceived zero knowledge protocols.

ISO is currently developing several standard texts for peer entity authentication. One of these[13] makes use of an *n*-bit secret key algorithm. Within this draft standard four distinct mechanisms are identified. The first assumes that genuine communicating entities are in possession of the same secret key before authentication is attempted; the protocol consists of challenges that establish to the mutual satisfaction of both parties that each is in possession of the same secret key. The second does not start from the same premise; at the start no common secret key exists, but a third party, an authentication server, is invoked to make up the deficiency. The third protocol improves on the second by assuming that all parties have reliable clocks available; this simplifies the task of preventing fraud by replay of messages. The fourth protocol deliberately sets out to minimize the security relevant information that flows between the authenticating entities.

To illustrate the principle of peer entity authentication we shall explain the first of the above mechanisms. The initiating entity A generates a random number,

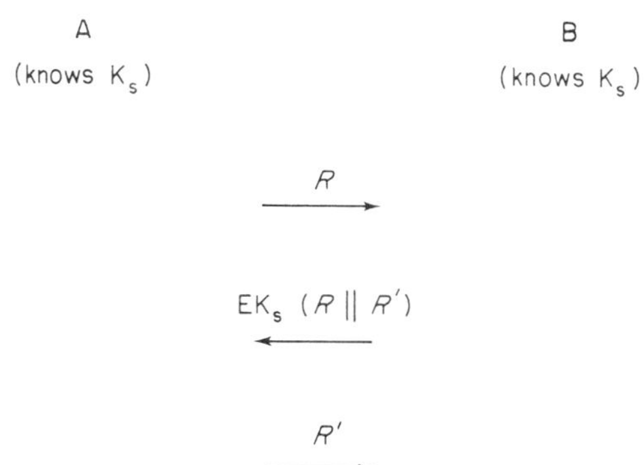

Fig. 11.9 Peer entity authentication with secret key algorithm

R, see Figure 11.9, and sends this to entity B in clear. Entity B prepares a reply in which A's random number R is concatenated with a new random number, R', generated by B. Before sending the reply B enciphers it with key *Ks*, previously agreed by some unspecified method between A and B. When A receives the enciphered reply this can be deciphered with *Ks*, revealing the original random number R and the new one R'. To complete the authentication process A sends back R' in cleartext to B. The result of this process is that entities A and B are each assured that the other entity with which they are communicating is in possession of *Ks*. It is on this knowledge that the authentication is founded. Clearly the process of establishing *Ks* at the two entities in the first place must be secure, with no possibility of a hostile third party learning *Ks*. The remaining three mechanisms are more complicated but are based on the same principle.

Peer entity authentication using public key algorithms is somewhat more elegant. Two draft texts are currently being worked upon. The first[14] is concerned with a two-way handshake; certified copies of entity public keys are provided by an authentication server, the certification being based upon digitial signature. If we call the initiating entity A and the other entity B, the protocol proceeds by A sending a signed 'token' (which includes a time and date stamp, but no random number) to B together with the certified value of A's public key. Using the latter B can check the validity of the former. If satisfied, B responds with B's signed token which A can verify using the certified value of B's public key obtained from the authentication server. Only one message flows in each direction between the parties, hence the name 'two-way handshake'.

The second draft text[15] using public key algorithms makes use of a 'three-way handshake'. The three-way handshake protocol is distinguished from the two-way handshake protocol by the presence in each message of a random number challenge; each entity generates its independent random number. The authentication is based on signature of tokens containing the random numbers and three messages are necessary to complete the process. There is a close logical parallel between this public key method and the secret key method explained above.

Because of the close connection between the three peer entity authentication texts discussed above, it is proposed in ISO that the three texts shall be brought together into a unified text before final publication as an international standard.

In this connection it is noteworthy that CCITT has published a text[16] describing a directory authentication framework. This text is extremely detailed and covers much the same ground as the three ISO texts; unfortunately it is not compatible with the ISO texts. However, it also goes far beyond these in discussion of issues such as key management and authentication servers; these issues are treated in ISO as separate work items and will be the subject of separate standard texts.

11.6. STANDARDS FOR DATA SECURITY IN BANKING

The work on standards development described in the foregoing sections is carried out at ISO within subcommittees of Technical Committee JTC1, notably SCs 6, 20 and 21. There is another area of substantial activity within ISO that is also producing standards for data security. This is within the domain of Technical Committee 68 on Banking Procedures; the subcommittees particularly charged with data security responsibilities are 2 and 6. The work under TC68 has been almost totally independent of that under JTC1.

Before giving a summary of the data security work within TC68 it is well to consider the work of the banking interests within ANSI, because ANSI standards have formed the basis of some of the standards in the data security area published by TC68. The ANSI committee working in the secure banking area is X.9 (which operates in close collaboration with the American Bankers Association). The following is a list of relevant standards produced within X.9:

X9.8 Personal Identification Number (PIN) Management and Security,
X9.9 Financial Institution Message Authentication (Wholesale),
X9.17 Financial Institution Key Management (Wholesale),
X9.19 Financial Institution Message Authentication (Retail),
X9.23 Financial Institution Encryption of Wholesale Financial Messages,
X9.24 Financial Institution Key Management (Retail).

Thus far TC68 has produced several standards in the data security area. The first of these was ISO 8730[17] which describes the logical procedure for message authentication; it is closely based on ANSI X9.9, but does not specify encipherment algorithms for use in the authentication process. The second TC68 standard was ISO 8731[18] which has two parts, each specifying a suitable algorithm for message authentication. The first part of ISO 8731 specifies the DES algorithm, using the ANSI standard X3.92 as a reference. The second part of ISO 8731 specifies the Message Authentication Algorithm which is discussed in chapter 5 of this book. TC68 has also published ISO 8732[19] on wholesale key management; this is based fairly closely on ANSI standard X9.17 and was described in chapter 6. Further standards are in process of development.

11.7. POSTSCRIPT

We close this account of data security standards with a postscript rather than with a set of conclusions because the work is ongoing in a very active sense. Indeed, as we write, the organization of data security standards work within ISO/IEC/JTC1 is undergoing a drastic reorganization. The same subjects will be worked upon that we have described here, with future extensions, but the distribution of responsibilities between the various committees will be different. It is expected that a committee replacing SC20 will take responsibility for 'security techniques'. Committees with responsibility for particular communication protocols will look after security aspects of those protocols.

REFERENCES

1. *Data Encryption Standard,* Federal Information Processing Standards Publication 46, National Bureau of Standards, US Department of Commerce, January 1977.
2. *Data Encryption Algorithm,* X3.92, American National Standards Institute, 1981.
3. *Guidelines for Implementing and Using the Data Encryption Standard,* Federal Information Processing Standards Publication 74, National Bureau of Standards, US Department of Commerce, April 1981.
4. Gait, J. *Validating the Correctness of Hardware Implementations of the NBS Data Encryption Standard,* Special Publication 500/20, National Bureau of Standards, US Department of Commerce, November 1977.
5. *Telecommunications: General Security Requirements for Equipment Using the Data Encryption Standard,* Federal Standard 1027, Federal Telecommunications Standards Committee, April 1981.
6. Serpell, S.C. *Cryptographic Equipment Security: A Code of Practice,* Institution of Electronic and Radio Engineers, 1985.
7. *Data Cryptographic Techniques — Procedures for the Registration of Cryptographic Algorithms,* Draft International Standard 9979, International Standards Organization, 1988.
8. *DES Modes of Operation,* Federal Information Processing Standards Publication 81, National Bureau of Standards, US Department of Commerce, December 1980.
9. *Modes of Operation for the Data Encryption Algorithm,* X3.106-1983, American National Standards Institute, 1983.
10. *Information Processing — Modes of Operation for a 64-bit Block Cipher Algorithm,* International Standard 8372, International Standards Organization, 1987.
11. *Information Processing — Modes of Operation for an n-bit Block Cipher Algorithm,* Draft Proposal 10116, International Standards Organization, 1988.
12. *Information Processing — Data Encipherment — Physical Layer Interoperability Requirements,* International Standard 9160, International Standards Organization, 1988.
13. *Peer Entity Authentication Mechanisms Using an n-bit Secret Key Algorithm,* Draft Proposal 9798, International Standards Organization, 1988.
14. *Peer Entity Authentication Mechanism Using a Public Key Algorithm with a Two-way Handshake,* Draft Proposal 9799, International Standards Organization, 1988.
15. *Peer Entity Authentication Mechanism Using a Public Key Algorithm with a Three-way Handshake,* Draft Proposal 10117, International Standards Organization, 1988.
16. *The Directory — Authentication Framework,* CCITT Draft Recommendation X.509 (also DIS 9594-8), 1988.
17. *Banking — Requirements for Message Authentication (Wholesale),* Draft International Standard 8730, International Standards Organization, 1988.
18. *Banking — Approved Algorithms for Message Authentication, Part 1, DEA,* International Standard 8731/2, 1988, International Standards Organization. *Part 2, Message Authentication Algorithm (MAA),* International Standard 8731/2, 1988, International Standards Organization.
19. *Banking — Key Management (Wholesale),* Draft International Standard 8732, International Standards Organization, 1988.

GLOSSARY

Active card
Thin plastic card of credit-card size containing a store, processor or both, and capable of interaction with a terminal, for example in a point-of-sale transaction. Also called a 'smart card'.

Active line-tap (or wire-tap)
Deliberate interference with a communication system to alter the signals, data or messages it carries. This includes introducing new data or messages, repeating those that were sent earlier, delaying or deleting messages. *See also* passive line-tap.

Algorithm
A precise statement of a method of calculation. It can sometimes be expressed conveniently in a programming language.

Alphabet
The range of values which may be taken by a particular unit of communication, such as a character. When a cipher employs one or more permutations of a character code, each of these is known as an alphabet.

ANSI
American National Standards Institute.

Arbitration
Settling a dispute by referring it to a third party. This arbitrator may take part in a procedure which enables disputes to be settled, as in the production and checking of an 'arbitrated signature'.

ASCII
American Standard Code for Information Interchange. Its full title is USASCII. This code is a national version of the ISO 7-bit coded character set and is mainly compatible with CCITT alphabet No. 5.

ATM
See automatic teller machine.

Authentication
1. Ensuring that a message is genuine, has arrived exactly as it was sent and came from the stated source.

2. Verifying the identity of an individual, such as a person at a remote terminal or the sender of a message.

See also: Data origin authentication and peer-entity authentication.

Authenticator
A number which is sent with a message to enable the receiver to detect alterations made to the message since it left the sender. The number is a function of the whole message and a secret key.

Automatic teller machine (ATM)
A machine at which customers can do business with their bank. Withdrawal of cash is a main part of the service, but the ATM is typically more than a 'cash dispenser' and allows, for example, presenting cheques for payment into an account and enquiry for the account's balance.

Bijection
A mapping from one set of entities $\{x\}$ onto another $\{y\}$ having the same number of elements, such that each x maps onto a unique y and each y is the mapping of a unique x, i.e. it is a one-to-one mapping.

Birthday paradox
How many people must there be in a group to make it likely that at least two of them have the same birthdate? The answer, twenty-three, appears paradoxical. This type of problem is significant for the understanding of security in some systems (*see* page 279).

Block
A block of data is an array or sequence of bits that are usually stored or transmitted together. Many such arrays have specialized names such as 'character', 'field' or 'packet' which suggest their function. Block is a neutral term and often signifies a fixed-length array of bits.

Block cipher
A cipher method which acts on a fixed-length block of data to produce a result which depends only on that block.

BSI
British Standards Institution.

By-pass mode
A mode of operation of data encipherment equipment in which plaintext passes through unchanged. Federal Telecommunications Standard 1027 requires by-pass mode to be controlled with a lock.

CCITT
International Telephone and Telegraph Consultative Committee. The letters come from the title in French. With the CCIR (radio) it forms the International Telecommunications Union (UIT).

Chain
A sequence of message units which are treated together for encipherment, decipherment or authentication. The units may be characters or blocks, for example. Chaining prevents the analysis of ciphertexts being applied to individual message units, or such enciphered units being reassembled in a different sequence to make a meaningful message.

Check field
A field added to a data item which provides redundancy in such a way that the most likely errors will be detected. Check fields are often used to catch keyboarding errors.

CID
Chain Identifier; a sequence number used in the plaintext of a chain which is to be enciphered. Its purpose is to guard against replay and deletion.

Cipher
A method of cryptography by applying an algorithm to the letters or digits of the plaintext. A distinction is made between a cipher and a code, the latter using a table instead of an algorithm. This distinction is not always clearcut.

Ciphertext
The enciphered form of messages or data.

Code
A method of cryptography which employs an arbitrary table (a codebook) to translate from the message to its coded form. Contrasted with a 'cipher'. The word also has non-cryptographic meanings as in 'character code'.

Complexity
The degree of difficulty of a calculation, measured by the number of elementary operations it requires. The nature of the elementary operations should be specified. Complexity is a function of the 'size' of the problem, which is measured by a parameter which should also be stated.

Computationally secure
Secure under the assumption that the computation needed to undermine the security is too extensive to be feasible.

Confidentiality
The property that information is not made available or disclosed to unauthorized individuals, entities or processes.

Cryptanalysis
The study and development of methods by which, without a prior knowledge of the key, plaintext may be deduced from ciphertext, given sufficient working material and computational power. One aim of cryptanalysis would be to discover the key, but that may not always be necessary.

Cryptography
The technique of concealing the content of a message either by a code or a cipher.

Cryptology
The science which includes all aspects of cryptography and cryptanalysis.

Cyclic redundancy check (CRC)
A kind of check field used to guard against errors. This is used in some CCITT recommendations and ISO standards.

DAC
Data Authentication Code. A name used for an authenticator in the proposed Federal Information Processing Standard on Computer Data Integrity. The word 'code' is misused in this term.

Data link layer
The layer of Open Systems Interconnection immediately above the physical layer. At this layer of procedure, error and flow control are introduced.

Data origin authentication
Corroboration that the source of data received is as claimed.

DCE
Data circuit terminating equipment. Data communication local lines typically end at an equipment on the subscriber's premises which belongs to the network and provides the 'customer interface'. This outermost part of the network is the DCE.

DEE
Data encipherment (or encryption) equipment.

DID
Data Identifier; synonymous with 'chain identifier'. A term used in the proposed Federal Information Processing Standard on Computer Data Integrity.

Digital signature
A number depending on all the bits of a message and also on a secret key. Its correctness can be verified by using a public key (unlike an authenticator which needs a secret key for its verification).

DTE
Data terminal equipment. The equipment which is connected to a data communication network in order to communicate. The DTE may be anything that communicates, for example a computer system.

EFT
Electronic Funds Transfer. A term used for various payment methods that employ communication or electronic storage. The most usual context is in point-of-sale transactions.

Electronic code book
The straightforward use of a block cipher to encipher blocks of data. It likens the process to a code book in which the cipher is listed for all possible values of plaintext, but in practice the size of the 'code book' makes this infeasible. It should be used with caution.

Enemy
In this book it is a convention to refer to any adversary of the system as an 'enemy'. In security analysis the possibility of joint action by several enemies must also be considered.

Entity
A person or process taking part in a communications procedure. In 'Open Systems Interconnection' there can be communicating entities at any of the seven levels.

Error extension
See garble extension.

Exhaustive search
Solving a problem (usually that of finding the key that is in use) by searching through all possibilities, assuming that each can rapidly be tested and the true result (key) will show up.

Exponential function
Normally used to mean exp(x) which is e^x, but here we use it also for a^x as a function of x with a as a parameter, called the base.

FIPS
Federal Information Processing Standard. These are issued by the National Bureau of Standards and, though recognized more widely, are intended primarily for US Federal Departments and Agencies.

Flag
1. A part of the format of an item of data which contains a bit (or a few bits) which pertain to the use of the item as a whole. The item is said to be 'flagged'.
2. An easily recognized pattern of bits to delineate an item of data, such as a 'frame' in HDLC.

Garble extension
Noise in communication systems may produce errors in the ciphertext. When this is deciphered, the errors in the plaintext sometimes extend further than those in the ciphertext. This is garble (or error) extension.

Greatest common divisor (gcd)
For two numbers (positive integers), their greatest common divisor is the largest number which divides into both numbers exactly. Example: gcd (75,45) = 15.

Hamiltonian cycle
In a graph it is a cycle which includes all the points of the graph.

Hamming distance
Given fixed-length blocks of binary data, the Hamming distance between two such blocks is the number of bits in which their values are different.

```
Example:  1 0 1 1 0 1 0 1 1 1
          0 1 1 1 0 0 0 1 1 0
          * *       *       *  Hamming distance = 4
```

Hash function
A well-defined function of a message or block of data which appears to generate a random number. Care is needed in selecting hash functions because they are used for several different purposes.

Index of coincidence
Two texts can be placed together and corresponding characters compared for equality (coincidence). The proportion of coincidences among the characters is a very useful statistic. Friedman defined the index of coincidence as 'coincidences minus non-coincidences divided by total number of letters compared'.

Initializing value (or variable), IV
For the encipherment or decipherment of a chain it may be necessary, or desirable, to use a value (e.g. a block of bits or set of characters) which starts off the process. This is the initializing value or IV and is known to both sender and receiver.

Integrity (of data)
The property that data have not been altered or destroyed in an unauthorized manner.

Intelligent token
A small device incorporating a processor and store which can be carried by someone for identification and to make transactions with special terminals. One form of this token is the active card or 'smart card'.

ISO
International Organization for Standardization. The international forum at which representatives from national standards bodies (such as ANSI) meet to agree standards for world-wide use.

Key block
In the context of the Data Encryption Standard it is the 64-bit block of data which contains the 56-bit key. It is also known as the 'key variable'.

Key space
The range of all the values which a cryptographic key may take. A large key space is important for security.

Key variable
See key block.

Knapsack problem
A problem which is important in complexity theory, see page 235. A number of public key ciphers and some signature methods have been based on varieties of the knapsack problem.

Least common multiple (lcm)
For two numbers (positive integers) their least common multiple is the smallest number into which they both divide exactly. Example: lcm (75,45) = 225.

Line level encipherment
On-line encipherment of the data or messages passing over a single line of a communication system. As they leave the line, the data or messages are deciphered. This term does not correspond with OSI terminology, but typically involves either the physical or data link layer.

Longitudinal parity check
A parity check applied throughout a string of data, for example the modulo 2 sum of each of the 8 bits in a string of octets, which together form a 'check octet'.

MAC
Message Authentication Code. A name used for an authenticator, for example in ANSI Standard X9.9 'Financial Institution Message Authentication'. The word 'code' is misused here.

MDC
Manipulation Detection Code or Modification Detection Code. A redundancy check field included in the plaintext of a chain before encipherment, so that changes to the ciphertext (an active attack) will be detected. The word 'code' is misused here.

Modulo, modulus
The expression 'modulo m' is a statement that arithmetic with m as modulus is being used, see page 245.

Multi-drop
A communication line which connects to more than two stations so that (in principle) each may communicate with several others, according to an agreed procedure.

Negative file
In the context of payment systems, a file listing those accounts which are not to be allowed transactions because of some problem, such as a lost or stolen card or a denial of further credit.

Node
An equipment in a communication network at which two or more communication lines converge. Usually it refers to a switching node but, strictly speaking, a multiplexer or concentrator could be called a node.

One-time tape or pad
A tape containing random numbers or bits. Only the original and one copy are made so that one station can use it to encipher and another to decipher messages under the strict rule that each portion of the tape may be used only once. Another version is the one-time pad, where the numbers are written down on two identical pads; used pages are destroyed.

One-way function
A function $y=f(x)$ which is relatively easy to compute but much more difficult to compute the inverse. That is, given x it is easy to find y, but given a value y it is (except in a few cases) difficult to find *any* solution x of $y=f(x)$. The exception allows for the fact that, with luck, the x,y pair may already be known.

Open Systems Interconnection (OSI)
A system of standards which should make useful communication possible between any systems attached to a communications network.

Padding
When a message (or a record of arbitrary length) is divided into fixed-length blocks for any purpose, the last block may not be completely filled. The information used to fill the remaining part of the last block is padding.

Passive line-tap (or wire-tap)
Unauthorized reading of signals, data or messages from a communication system, without changing the signals. *See also* active line-tap.

Peer-entity authentication
Corroboration that a peer entity in an association is the one claimed.

Permutation
Changing the order (or sequence) of a set of data elements. The elements can be bits, characters, bytes, etc. Permutation can be used as one operation in a process of encipherment. *See also* transposition cipher.

Physical layer
The lowest layer of Open Systems Interconnection, in which physical phenomena (voltage, current, light waves) are interpreted as a sequence of bits or other units (such as start/stop characters).

PIC
Personal Identification Code. A sequence of symbols (letters and numbers) used to verify the identity of the holder of a bank card. It is different from a PIN in allowing letters to be used. The word 'code' is misused here.

PIN
Personal Identification Number. A sequence of decimal digits (usually four, five or six) used to verify the identity of the holder of a bank card. A kind of 'password'.

Plaintext
Messages or data which are not in an enciphered state and are therefore in their normal, readable form.

Power function
The function a^n is the nth power of a. This is considered as a function of a with n as parameter. Thus $y=a^4$ is the fourth power of a.

Prime
A prime number is a positive integer n which has no divisors except the trivial ones 1 and n. Sometimes 1 itself is included among the prime numbers, sometimes not. Non-primes are called 'composite numbers'.

Product cipher
A cipher employing both substitution and transposition, in some cases repeated alternately.

Pseudo-random
A sequence of data which are generated by an algorithm but nevertheless appear random and will pass the customary tests for randomness.

Public key
A cryptographic key used for encipherment but not usable for decipherment. It is therefore possible to make this key public.

Relatively prime
Two numbers (positive integers) are relatively prime if they have no common divisor except the trivial divisor 1. That is to say: their greatest common divisor is 1. Example: 65,24.

Replay
An attack on a cipher system (or authenticator) in which a message is stored and re-used later, replacing or repeating the original. There is a similar attack on stored data by restoring an item to an earlier state it once had.

Repudiation
Denial by one of the entities involved in a communication of having participated in all or part of the communication or denial of the content of the communication.

Residue, modulo m
The residue of x, modulo m, is the (positive) remainder r which is left when x is divided by m. It follows that $0 \leq r < m$. Finite arithmetic is the arithmetic of residues.

Round
The Data Encryption Standard algorithm consists of an initial permutation, then sixteen 'rounds', a register interchange, then a final permutation which is the inverse of the initial one. The sixteen rounds are identical operations but each uses its own sub-key *q.v.*

RSA
Rivest, Shamir and Adleman gave their names to a public key cipher which is usually known as the RSA cipher. There is a corresponding 'RSA digital signature'.

S-box
A substitution operation used in the Data Encryption Standard. This algorithm employs eight different 'S-boxes'.

Security policy
The set of rules for the provision of security services.

Seed value
A value (e.g. block of bits or set of characters) which is used to start a pseudo-random sequence generator and provide unpredictability. A seed must be randomly chosen and usually must be kept secret.

Session key
A cryptographic key which is used only for a limited time, such as one communication session, then discarded. Typically it is transported through the network under encipherment with another key.

Settlement
When banks send messages to effect customer payments, customer accounts are updated at once, but these changes do not constitute net payments between the banks because they are balanced in an account held by one bank for the other. The eventual payment between banks is called settlement and usually employs accounts at some other bank.

Smart card
See active card, intelligent token.

Starting variable
The ISO standard for 'modes of operation' distinguishes the initializing variable or IV from the 64-bit block which is loaded into a register before encipherment/decipherment begins. The latter is called the 'starting variable' and is derived from the IV.

Steganography
Concealment that a means of communication exists. Examples are the use of invisible ink or microdots.

Stream cipher
A cipher which is devised in such a way that an indefinitely long stream of data can be dealt with. It contrasts with a 'block cipher', but both kinds of cipher can be adapted to the other purpose.

Sub-key
In the Data Encryption Standard algorithm, the 56-bit key is used in a 'key

generator' to derive (by simple means) sixteen sub-keys, each of 48 bits, which are used in the sixteen 'rounds' of the algorithm.

Substitution, substitution cipher
Replacement of one unit of data (such as an octet or character) by a related one. A cipher in which each character is replaced by one derived from it is a substitution cipher. The substitution boxes (S-boxes) of the Data Encryption Standard each produce a 4-bit function of a 6-bit input.

S.W.I.F.T.
Society for Worldwide Interbank Financial Telecommunication, which provides a specialized communication network for banks, *see* page 289.

Synchronization
In the context of cryptography, synchronization means keeping in step the processes involved in encipherment and decipherment.

Test key
Used in banking to mean the same as authenticator. The secret key used in calculating the authenticator produces a confusion of terms, so the word 'authenticator' is preferable.

Time-memory trade-off
Calculations which require a large number of computational steps can sometimes be carried out with less computation at the cost of having to store a large number of intermediate results. This is trading time of computation for size of memory, called a time-memory trade-off.

Time stamp, time and date stamp
A value inserted in a message to give the time at which the message was originated. If necessary it is extended to include the date, becoming a time and date stamp.

Token
Something possessed by a person to assist in identifying him, etc. Its form could be a card, pen, key, etc. and it can have functions additional to identification, for example, storage, access control, payment.

Traffic analysis
When communication traffic is in cipher form and cannot be understood it may still be possible to get useful information by detecting who is sending messages to whom and in what quantity. This is traffic analysis.

Traffic flow confidentiality
Concealment of the quantity of users' messages or data in a communication system and their sources or destinations, to prevent traffic analysis.

Transparency
The ability of a communication system to pass messages through without any change. In data communication the requirement is to preserve a sequence of bits or characters.

Transposition cipher
A cipher which permutes a set of bits or characters from the plaintext. Mainly of historical interest as a cipher, but it can be used as a component of a more complex cipher.

Trapdoor
A feature in a cipher or a puzzle which, given the necessary information, enables it to be solved easily. The trapdoor is concealed from those who do not have the information.

Type I error
An error in the operation of an identity verification system which causes a genuine user to be rejected — a false alarm.

Type II error
An error in the operation of an identity verification system which allows a false user (someone claiming the wrong identity) to be recognized as valid — an imposter pass.

Unconditionally secure
Secure even though an enemy is assumed to have unlimited computing power available.

Unicity distance
The amount of ciphertext beyond which some knowledge of the statistical properties of the plaintext (its redundancy) allows the key to be deduced, given unlimited computing power. With less than this amount, there are multiple key values that are consistent with the given ciphertext.

Vernam cipher
A cipher which uses a sequence of bits (or characters) called the 'key stream' and enciphers by adding this stream modulo 2 to the plaintext. It deciphers in the same way with an identical stream. In the original Vernam cipher the key stream came from a loop of punched tape.

Weak key
A value of a cipher key which gives that cipher some special properties that might, in certain circumstances, weaken its security. The supposed weakness has to be related to a specific method of application of the cipher. There can also be weak key values for an authenticator or a digital signature.

INDEX